ENERGY SECURITY

ENERGY SECURITY

AN INTERDISCIPLINARY APPROACH

Gawdat Bahgat
National Defense University
Washington, DC, USA

A John Wiley and Sons, Ltd., Publication

Library of Congress Cataloging-in-Publication Data

Bahgat, Gawdat.
    Energy security : an interdisciplinary approach / Gawdat Bahgat.
        p. cm.
    Includes bibliographical references and index.
    ISBN 978-0-470-68904-2 (hardback)
    1. Energy policy.   2. Energy development.   3. Power resources.   I. Bahgat, Gawdat.   II. Title.
    HD9502.A2B335 2011
    333.79–dc22

                                                                    2010046393

A catalogue record for this book is available from the British Library.

Print ISBN: 9780470689042
ePDF ISBN: 9780470980187
oBook ISBN: 9780470980170
ePub ISBN: 9780470980163

Set in 10/12pt Times by Aptara Inc., New Delhi, India
Printed and bound in Singapore by Markono Print Media Pte Ltd

Contents

About the Author

Dr. Gawdat Bahgat is a professor at the Near East South Asia Center for Strategic Studies, National Defense University, in Washington, DC, United States of America. Dr. Bahgat has taught political science and international relations at several universities. His areas of expertise include energy security, counter-terrorism, proliferation of weapons of mass destruction, international political economy, the Middle East, the Caspian Sea and Central Asia, and US foreign policy.

Dr. Bahgat is the author of seven books and about 200 scholarly articles. His work has been translated into several foreign languages. He has been invited to and presented papers at conferences in Australia, Europe, and the Middle East, and is a frequent contributor to several media outlets. He holds a PhD in political science from Florida State University, an MA in Middle Eastern Studies from American University in Cairo, and a BA in political science from Cairo University.

Preface

Energy is the lifeblood of civilization. Both as individuals and nation states we depend heavily on energy. In almost everything we do, we rely on one or several sources of energy. Many people and governments used to take the availability of energy sources for granted. Our deepening reliance on energy and the rise of a combination of geopolitical, geological, and environmental challenges have cast doubt on this assumption that energy will always be there. Little wonder that energy security has become a major concern to almost all countries in the world.

In recent years policy-makers and scholars have examined different aspects of energy security. These include production, consumption, reserves, refining, shipping, and investment among others. Indeed, the last few decades have witnessed a proliferation of political and academic conferences, industry journals, and books on energy security. Each side has sought to promote its interests with little ground for neutrality and objectiveness.

I have been working on energy for more than two decades. The policy of energy, at national and international levels, and the growing literature are immensely stimulating. For a long time, consumers and producers perceived their interests as mutually exclusive. Since the early 1990s, a consensus has emerged that there is common ground. Long-term stability of energy markets and prices is generally seen as more favorable than short-term gains by one side or the other. These shared interests are the main theme of the analysis in this volume. In all the following chapters I argue that interdependence is the underlying characteristic of today's energy markets.

This book reflects what I learned in my teaching, research, and consulting in more than 20 years. The first chapter introduces readers to some of the major themes and concepts used in this study. This is followed by a close examination of energy outlooks in the major producing and consuming regions. In the last part the analysis focuses on the two most important international energy organizations – the Organization of Petroleum Exporting Countries and the International Energy Agency. The concluding chapter summarizes the main findings and discusses the International Energy Forum as an embodiment of the emerging cooperation between producers and consumers. In this volume the concept of energy security is addressed from both consumers' and producers' perspectives.

In my decades-long journey of learning, teaching, researching, and writing about energy I have accumulated a huge debt to many colleagues, friends, and students. In writing this book I had the privilege of working with the most professional editorial team at John Wiley & Sons, Ltd. In particular I would like to thank Clarissa Lim, Neville Hankins, and Shalini Sharma. Nicky Skinner gave me unlimited support at crucial stages while writing the book and Simone Taylor's encouragement inspired me to transform my abstract thoughts into a book proposal.

Writing a book is a huge adventure, with so many ups and downs. Professional and personal support from family and close friends is crucial in this undertaking. I would like to thank Helen Hooker, Sandra Dickson, Beth Sims, Theresa McDevitt, Helen Wedlake, and Patrizia Bassani. Finally, I would like to thank my friends and colleagues at the Near East South Asia Center for Strategic Studies, the National Defense University. Despite all the assistance I have received in the course of writing this book, all errors of facts or judgment are mine alone.

Acknowledgements

To Sandra Dickson, Beth Sims, and Theresa McDevitt: thank you for your love and support all these years.

List of abbreviations

3-D	Three-dimensional
ACG	Azeri, Chirag, and deep-water Guneshli
ACOTA	Africa Contingency Operations Training and Assistance
AEC	Atomic Energy Commission
AFRICOM	US–Africa Command
AIOC	Azerbaijan International Operating Company
ANILCA	Alaska National Interest Lands Conservative Act
ANWR	Arctic National Wildlife Refuge
AOC	Arabian Oil Company
APEC	Asia-Pacific Economic Cooperation
Aramco	Arabian–American Oil Company
Bcf	Billion cubic feet
Bcm	Billion cubic meters
B/d	Barrels per day
BG	British Gas
BI	Baku Initiative
BP	British Petroleum
BPS	Baltic Pipeline System
BTC	Baku–Tbilisi–Ceyhan
BTE	Baku–Tbilisi–Erzurum
CABC	China–Africa Business Council
CADfund	China–Africa Development Fund
CASOC	California–Arabian Standard Oil Company
CCP	Caspian Coastal Pipeline
CCS	Carbon Capture and Storage
CEO	Chief Executive Officer
CERM	Coordinated Emergency Response Measures
CERT	Committee on Energy Research and Technology
CNOOC	China National Offshore Oil Corporation
CNPC	China National Petroleum Corporation
CPC	Caspian Pipeline Consortium
CTL	Coal-to-liquids
DOE	Department of Energy
DST	Daylight Saving Time

EC	European Commission
ECG	Energy Cooperation Group
ECO	Economic Cooperation Organization
ECT	Energy Charter Treaty
EGAS	Egyptian Natural Gas Holding Company
EGPC	Egyptian General Petroleum Corporation
EIA	Energy Information Administration
EISA	Energy Independence and Security Act
ELG	Energy Leading Group
ENEF	European Nuclear Energy Forum
EO	Executive Order
EPA	Environmental Protection Agency
EPCA	Energy Policy and Conservative Act
EPSA	Exploration and Production-Sharing Agreement
ESPO	East Siberia Pacific Ocean
ESSAP	Energy Security and Solidarity Action Plan
ETS	Emission Trading Scheme
EU	European Union
EURATOM	European Atomic Energy Community
FOCAC	Forum on China–Africa Cooperation
FSR	Former Soviet Republics
F–T	Fischer–Tropsch
GCA	Gaffney, Cline and Associates
GCC	Gulf Cooperation Council
GDP	Gross Domestic Product
GECF	Gas Exporting Countries Forum
GIF	Generation IV International Forum
GNEP	Global Nuclear Energy Partnership
GNP	Gross National Product
GoM	Gulf of Mexico
IAEA	International Atomic Energy Agency
ICT	Industrialized and Other High-Income Countries and Territories
IEA	International Energy Agency
IEF	International Energy Forum
IEP	International Energy Program
ILSA	Iran–Libya Sanctions Act
INOC	Iraq National Oil Company
INOGATE	Interstate Oil and Gas Transport to Europe
IOCs	International Oil Companies
IPC	Iraqi Petroleum Company
JODI	Joint Oil Data Initiative
JV	Joint Venture
KMG	KazMunaiGaz
KRG	Kurdistan Regional Government
LNG	Liquefied Natural Gas
LPG	Liquefied Petroleum Gas

Mcf	Million cubic feet
MEND	Movement for the Emancipation of Niger Delta
MEPI	Middle East Partnership Initiative
MGS	Master Gas System
MMS	Minerals Management Service
Mmt	Million metric tons
MOU	Memorandum of Understanding
MPLA	Popular Movement for the Liberation of Angola
NATO	North Atlantic Treaty Organization
NDRC	National Development and Reform Commission
NEA	National Energy Administration
NGO	Non-Governmental Organization
NGVs	Natural Gas Vehicles
NIOC	National Iranian Oil Company
NIORPDC	National Iranian Oil Refining, Production, and Distribution Company
NNPC	Nigerian National Petroleum Corporation
NOC	National Oil Corporation (Libya)
NOCs	National Oil Companies
NPC	National Petroleum Commission
NPT	Non-Proliferation Treaty
NRC	Nuclear Regulatory Commission
NSCSA	National Shipping Company of Saudi Arabia
OCS	Outer Continental Shelf
OECC	Organization for European Economic Cooperation
OECD	Organization for Economic Cooperation and Development
OFID	OPEC Fund for International Development
OLADE	Latin American Organization for Energy Cooperation
OPEC	Organization of Petroleum Exporting Countries
PEZ	Pipeline Exclusion Zone
PSA	Production-Sharing Agreement
PSI	Pan-Sahel Initiative
Sabic	Saudi Basic Industries Corporation
SCO	Shanghai Cooperation Organization
SEQ	Standing Group on Emergency Questions
SGD	Standing Group for Global Energy Dialogue
Sinopec	China Petroleum and Chemical Corporation
SLT	Standing Group on Long-Term Cooperation
SOCAR	State Oil Company of Azerbaijan Republic
SOM	Standing Group on the Oil Market
SPR	Strategic Petroleum Reserve
TAP	Trans-Adriatic Pipeline
Tapline	Trans-Arabian Pipeline
Tcf	Trillion cubic feet
TGI	Turkey–Greece Interconnector
TOE	Ton Oil Equivalent
ToP	Take-or-Pay

TPC	Turkish Petroleum Company
TSB	Technical Service Basis
TSC	Technical Services Contract
TSCTI	Trans-Sahara Counter-Terrorism Initiative
UAE	United Arab Emirates
UK	United Kingdom
ULCCs	Ultra Large Crude Carriers
UN	United Nations
UNCLOS	United Nations Convention on the Law of the Sea
UNITA	National Union for the Total Independence of Angola
UNSC	United Nations Security Council
US	United States
USAID	United States Agency for International Development
USGS	United States Geological Survey
VLCCs	Very Large Crude Carriers
WTI	West Texas Intermediate
WTO	World Trade Organization

Glossary

This glossary explains some of the technical terms that are used in this book or that readers are likely to encounter. It does not purport to be at all comprehensive.

Acquisition (foreign crude oil): All transfers of ownership of foreign crude oil to a firm, irrespective of the terms of that transfer. Acquisitions thus include all purchases and exchange receipts as well as any and all foreign crude acquired under reciprocal buy–sell agreements or acquired as a result of a buy-back or other preferential agreement with a host government.

Alternative-fuel vehicle: A vehicle designed to operate on an alternative fuel (e.g., compressed natural gas, methane blend, electricity). The vehicle could be either a dedicated vehicle designed to operate exclusively on alternative fuel or a non-dedicated vehicle designed to operate on alternative fuel and/or a traditional fuel.

Barrel of oil: Standard oil industry measure of volume: 1 barrel is equivalent to 42 US gallons (159 liters).

Biofuels: Liquid fuels and blending components produced from biomass feedstocks, used primarily for transportation.

Biomass: Organic non-fossil material of biological origin constituting a renewable energy source.

Brent blend: The principal grade of UK North Sea crude oil in international oil trading. Used as the "marker" for other North Sea grades which trade at differentials to it, reflecting quality and location.

British thermal unit: The quantity of heat required to raise the temperature of 1 lb of liquid water by 1°F at the temperature at which water has its greatest density (approximately 39°F).

Buy-back oil: Crude oil acquired from a host government whereby a portion of the government's ownership interest in the crude oil produced in that country may or should be purchased by the producing firm.

Carbon dioxide (CO_2): A colorless, odorless, non-poisonous gas that is a normal part of the earth's atmosphere. It is a product of fossil-fuel combustion as well as other processes. It is considered a greenhouse gas as it traps heat radiated by the earth into the atmosphere and thereby contributes to the potential for global warming.

Climate change: A term used to refer to all forms of climatic inconsistency, but especially to a significant change from one prevailing climatic condition to another.

Coal: A readily combustible black or brownish-black rock whose composition consists of more than 50% by weight and more than 70% by volume of carbonaceous material. It is formed from plant remains that have been compacted, hardened, chemically altered, and metamorphosed by heat and pressure over geological time.

Coal gasification: The process of converting coal into gas. The basic process involves crushing coal to a powder, which is then heated in the presence of steam and oxygen to produce a gas. The gas is then refined to reduce sulfur and other impurities.

Concession: The operating right to explore for and develop petroleum fields in consideration for a share of production in kind (equity oil).

Conventional oil: Crude oil that, at a particular time, can be technically and economically produced through a well, using normal production practice and without altering the natural viscous state of the oil. Non-conventional oil is more expensive to explore and develop, although there have been major cost reductions in the past few years.

Crude oil: A mixture of hydrocarbons that exists in liquid phase in natural underground reservoirs and remains liquid at atmospheric pressure after passing through surface separating facilities.

Deregulation: The elimination of some or all regulations from a previously regulated industry or sector of an industry.

Diesel fuel: A fuel composed of distillates obtained in petroleum refining operation or blends of such distillates with residual oil used in motor vehicles.

Downstream: That part of the petroleum industry that involves refinery, transportation, and marketing operations as contrasted with upstream operations of exploration, development, and production.

Dry hole: An exploratory or development well found to be incapable of producing either oil or gas in sufficient quantities to justify completion as an oil or gas well.

Dubai: A grade of crude oil which has effectively replaced Saudi Light as the "marker" crude oil in the Persian Gulf.

Energy efficiency: Refers to programs that are aimed at reducing the energy used by specific end-use devices and systems, typically without affecting the services provided.

Energy source: A substance, such as oil, natural gas, or coal, that supplies heat or power. Electricity and renewable forms of energy, such as wood, waste, geothermal, wind, and solar, are considered to be energy sources.

Enriched uranium: Uranium in which the U-235 isotope concentration has been increased to greater than the 0.711% of U-235 present in natural uranium.

Ethanol: A clear, colorless, flammable alcohol. Ethanol is typically produced biologically from biomass feedstocks such as agricultural crops and cellulosic residues from agricultural crops or wood. It can also be produced chemically from ethylene.

Flared: Gas disposed of by burning in flares usually at the production sites or at gas processing plants.

Fossil fuel: An energy source formed in the earth's crust from decayed organic material. The common fossil fuels are coal, natural gas, and oil.

Futures market: A trade center for quoting prices on contracts for the delivery of a specified quantity of a commodity at a specified time and place in the future.

Gallon: A volumetric measure equal to 4 quarts (231 cubic inches; 3.79 liters) used to measure fuel oil.

Gas: A non-solid, non-liquid combustible energy source.

Gas-to-liquids: A process that combines the carbon and hydrogen elements in natural gas molecules to make synthetic liquid petroleum products, such as diesel fuel.

Gasification: A method for converting coal, petroleum, biomass, wastes, or other carbon-containing materials into a gas that can be burned to generate power or processed into chemicals and fuels.

Geothermal energy: Hot water or steam extracted from geothermal reservoirs in the earth's crust. This water or steam can be used for geothermal heat pumps, water heating, or electricity generation.

Global warming: An increase in the near-surface temperature of the earth. Global warming has occurred in the distant past as the result of natural influences, but the term is today most often used to refer to the warming that some scientists believe is taking place as a result of increased anthropogenic emissions of greenhouse gases.

Greenhouse gases: Those gases, such as carbon dioxide and methane, that prevent long-wave radiant energy from leaving the earth's atmosphere. The net effect is a trapping of absorbed radiation and a tendency to warm the planet's surface.

Henry hub: A pipeline hub on the Louisiana Gulf coast. It is the delivery point for the natural gas futures contract on the New York Mercantile Exchange.

Hydrocarbon: An organic chemical compound of hydrogen and carbon in the gaseous, liquid, or solid phase.

Kerosene: A light petroleum distillate that is used in space heaters, cooking stoves, and water heaters and is suitable for use as a light source when burned in wick-fed lamps.

Kyoto Protocol: The result of negotiations at the Third Conference of the Parties in Kyoto, Japan, in December 1997. The Kyoto Protocol sets binding greenhouse gas emissions targets for countries that sign and ratify the agreement.

Liquefied natural gas: Natural gas (primarily methane) that has been liquefied by reducing its temperature to $-260°F$ ($-162°C$) at atmospheric pressure.

Liquefied petroleum gas: A light hydrocarbon material which is gaseous at atmospheric temperature and pressure but which can be liquefied by mild pressurization to facilitate transportation, storage, and handling.

Majors: Generally, the vertically integrated, international oil companies.

Manhattan Project: The US government project that produced the first atomic weapons during World War II. The project started in 1942 and formally ended in 1946.

Methane: A colorless, flammable, odorless hydrocarbon gas which is the major component of natural gas. It is also an important source of hydrogen in various industrial processes.

Mineral: Any of the various naturally occurring inorganic substances, such as metals, salt, sand, stone, sulfur, and water, usually obtained from the earth.

Mineral rights: The ownership of the minerals beneath the earth's surface with the right to remove them. Mineral rights may be conveyed separately from surface rights.

Mining: An energy consuming subsector of the industrial sector that consists of all facilities and equipment used to extract energy and mineral resources.

Natural gas: A gaseous mixture of hydrocarbon compounds; methane is the primary one.

Natural gas, associated–dissolved: The combined volume of natural gas which occurs in crude oil reservoirs either as free gas (associated) or as gas in solution with crude oil (dissolved).

Natural gas liquids: Those hydrocarbons in natural gas that are separated from the gas as liquids through the process of absorption, condensation, adsorption, or other methods in gas processing or cycling plants.

Natural gas, non-associated: Natural gas not in contact with significant quantities of crude oil in a reservoir.

Natural reservoir pressure: The energy within an oil or gas reservoir that causes the oil or gas to rise unassisted by other forces to the earth's surface when the reservoir is penetrated by an oil or gas well.

New York Mercantile Exchange: The most successful market for oil futures contracts on which very large volumes of heating oil and crude oil (WTI grade) in particular are traded. It has considerable influence on the physical trade.

Nominal price: The price paid for a product or service at the time of the transaction. Nominal prices are those that have not been adjusted to remove the effect of changes in the purchasing power of the dollar; they reflect buying power in the year in which the transaction occurred.

Non-associated natural gas: Natural gas that is not in contact with significant quantities of crude oil in the reservoir.

Non-renewable fuels: Fuels that cannot be easily made or "renewed," such as coal, natural gas, and oil.

Nuclear fuel: Fissionable materials that have been enriched to such a composition that, when placed in a nuclear reactor, will support a self-sustaining fission chain reaction, producing heat in a controlled manner for process use.

Nuclear reactor: An apparatus in which a nuclear fission chain reaction can be initiated, controlled, and sustained at a specific rate. A reactor includes fuel (fissionable material), moderating material to control the rate of fission, a heavy-walled pressure vessel to house the reactor components, shielding to protect personnel, a system to conduct heat away from the reactor, and instrumentation for monitoring and controlling the reactor's systems.

Offshore reserves and production: Reserves and production that are in either state or federal domains, located seaward of the coastline.

Oil: A mixture of hydrocarbons usually existing in the liquid state in natural underground pools or reservoirs.

Oil reservoir: An underground pool of liquid consisting of hydrocarbons, sulfur, oxygen, and nitrogen trapped within a geological formation and protected from evaporation by the overlying mineral strata.

Oil shale: A sedimentary rock containing kerogen, a solid organic material.

Oil stocks: Stocks that include crude oil, unfinished oils, natural gas plant liquids, and refined petroleum products.

Oil well: A well completed for the production of crude oil from at least one oil zone or reservoir.

OPEC pricing: OPEC collects pricing data on a "basket" of seven crude oils – Algeria's Saharan Blend, Indonesia's Minas, Nigeria's Bonny Light, Saudi Arabia's Arab Light, Dubai's Fateh (or Dubai), Venezuela's Tia Juana Light, and Mexico's Isthmus (a non-OPEC crude oil) – to monitor world oil market conditions.

Outer continental shelf: Offshore federal domain.

Ozone: A molecule made up of three atoms of oxygen. It provides a protective layer shielding the earth from harmful ultraviolet radiation.

Parent company: The parent company of a business entity is an affiliated company which exercises ultimate control over that entity, either directly or indirectly through one or more intermediaries.

Petrochemical feedstocks: Chemical feedstocks derived from petroleum principally for the manufacture of chemicals, synthetic rubber, and a variety of plastics.

Petrochemicals: Organic and non-organic compounds and mixtures that include chemicals, cyclic intermediates, plastics and resins, synthetic fibers, dyes, pigments, detergents, surface active agents, carbon black, and ammonia.

Petroleum: A broadly defined class of liquid hydrocarbon mixtures. The term includes crude oil, lease condensate, unfinished oils, refined products obtained from the processing of crude oil, and natural gas plant liquids.

Petroleum products: Products obtained from the processing of crude oil, natural gas, and other hydrocarbon compounds. They include unfinished oils, liquefied petroleum gases, pentanes plus, aviation gasoline, motor gasoline, naphtha-type jet fuel, kerosene-type jet fuel, petroleum coke, asphalt, road oil, still gas, and miscellaneous products.

Petroleum refinery: An installation that manufactures finished petroleum products from crude oil, unfinished oils, natural gas liquids, other hydrocarbons, and alcohol.

Pipeline: A continuous pipe conduit, complete with such equipment as valves, compressor stations, communication systems, and meters for transporting natural gas and/or crude oil and petroleum products from one point to another, usually from a point in or beyond the producing field or processing plant to another pipeline or to points of use.

Possible reserves: Those unproven reserves which analysis of geological and engineering data suggests are less likely than probable reserves to be commercially recoverable. Most companies assign a certainty value of 10% for possible reserves.

Probable reserves: Those unproven reserves which analysis of geological and engineering data suggests are more likely than not to be commercially recoverable. Most companies assign a certainty value of 50% for probable reserves.

Production capacity: The amount of product that can be produced from processing facilities.

Production costs: Costs incurred in operating and maintaining wells and related equipment and facilities, including depreciation.

Production payments: A contractual arrangement providing a mineral interest that gives the owner a right to receive a fraction of production, or of proceeds from the sale of production, until a specified quantity of minerals (or a definite sum of money) has been received.

Proved energy reserves: Estimated quantities of energy sources that analysis of geological and engineering data demonstrates with reasonable certainty are recoverable under existing economic and operating conditions. Most companies assign a certainty value of 90% for proven reserves.

Radiation: The transfer of heat through matter or space by means of electromagnetic waves.

Radioactive waste: Material left over from making nuclear energy. It can destroy living organisms if it is not stored safely.

Real price: A price that has been adjusted to remove the effect of changes in the purchasing power of the dollar. Real prices, which are expressed in constant currency, usually reflect buying power relative to a base year.

Recoverable proved reserves: The estimated quantities of fuel which geological and engineering data demonstrates with reasonable certainty to be recoverable in the future from known reservoirs under existing economic and operating conditions.

Reference year: The calendar year to which the reported information relates.

Refined petroleum products: Refined petroleum products include gasoline, kerosene, distillates, liquefied petroleum gas, asphalt, lubricating oils, diesel fuels, and residual fuels.

Refiner: A firm that refines liquid hydrocarbons from oil and gas field gases.

Refinery: An installation that manufactures finished petroleum products from crude oil, unfinished oils, natural gas liquids, other hydrocarbons, and oxygenates.

Regulation: The governmental function of controlling or directing economic entities through the process of rule making and adjudication.

Reinjected: The forcing of gas under pressure into an oil reservoir in an attempt to increase recovery.

Renewable energy resources: Energy resources that are naturally replenishing but flow limited. They are virtually inexhaustible in duration but limited in the amount of energy that is available per unit of time. They include biomass, hydro, geothermal, solar, wind, ocean thermal, wave action, and tidal action.

Repressuring: The injection of gas into oil or gas formations to effect greater ultimate recovery.

Reserve: That portion of the demonstrated reserve base that is estimated to be recoverable at the time of determination. The reserve is derived by applying a recovery factor to that component of the identified fuel resource designated as the demonstrated reserve base.

Reserve-to-production ratio: The length of time that the remaining reserves would last if production were to continue at the rate of production of a given year. Such a ratio is obtained by dividing the reserves remaining at the end of any year by the production in that year.

Reservoir: A porous and permeable underground formation containing an individual and separate natural accumulation of producible hydrocarbons (crude oil and/or natural gas) which is confined by impermeable rock or water barriers and is characterized by a single natural pressure system.

Residential sector: An energy consuming sector that consists of living quarters for private households. Common uses of energy associated with this sector include space heating, water heating, air-conditioning, lighting, refrigeration, cooking, and running a variety of other appliances.

Seven Sisters: A phrase denoting the seven major oil companies that controlled most of the cheap Middle East oil between 1945 and 1973: Standard Oil Co. of New Jersey (later Exxon), Standard Oil Co. of New York (originally Socony, later Mobil), Standard Oil Co. of California (Socal, later Chevron), Royal Dutch Shell, Texaco, British Petroleum (BP), and Gulf.

Solar constant: The average amount of solar radiation that reaches the earth's upper atmosphere on a surface perpendicular to the sun's rays.

Solar energy: The radiant energy of the sun, which can be converted into other forms of energy, such as heat or electricity.

Solar power tower: A solar energy conversion system that uses a large field of independently adjustable mirrors to focus solar rays onto a near single point atop a fixed tower (receiver).

Solar thermal collector: A device designed to receive solar radiation and convert it to thermal energy.

Solar thermal panels: A system that actively concentrates thermal energy from the sun by means of solar collector panels.

Solar thermal parabolic dishes: A solar thermal technology that uses a modular mirror system that approximates a parabola and incorporates two-axis tracking to focus the sunlight onto receivers located at the focal point of each dish.

Spent fuel: Irradiated fuel that is permanently discharged from a nuclear reactor. It must eventually be removed from its temporary storage location at the reactor site and placed in a permanent repository.

Spot market: A market in which natural gas is bought and sold for immediate or very near-term delivery. It is more likely to develop at a location with numerous pipeline interconnections, thus allowing for a large number of buyers and sellers.

Spot price: The price for a one-time open market transaction for immediate delivery of the specific quantity of product at a specific location where the commodity is purchased "on the spot" at current market rates.

Stocks: Supplies of fuel or other energy source(s) stored for future use.

Strategic Petroleum Reserve: Petroleum stocks maintained by a federal government for use during periods of major supply interruption.

Subsidiary: An entity directly or indirectly controlled by a parent company which owns 50% or more of its voting stock.

Sulfur: A yellowish non-metallic element that is present at various levels of concentration in many fossil fuels and is considered harmful to the environment.

Tanker and barge: Vessels that transport crude oil or petroleum products.

Tar sands: Naturally occurring bitumen-impregnated sands that yield mixtures of liquid hydrocarbons and that require further processing other than mechanical blending before becoming finished petroleum products.

Transportation sector: An energy consuming sector that consists of all vehicles whose primary purpose is transporting people and/or goods from one physical location to another. The sector includes automobiles, trucks, buses, motorcycles, trains, subways, and other rail vehicles. It also includes aircraft, ships, barges, and other waterborne vehicles.

Unconventional oil and natural gas production: Oil and natural gas that are produced by means that do not meet the criteria for conventional production. These include resource characteristics, exploration and production technologies, economic and environmental conditions. Perceptions of these factors inevitably change over time and accordingly what is considered unconventional today is likely to be conventional tomorrow.

Underground gas storage: The use of subsurface facilities for storing gas that has been transferred from its original location. The facilities are usually hollowed-out salt domes, geological reservoirs (depleted oil or gas fields) or water-bearing sands topped by an impermeable cap rock (aquifer).

Undiscovered recoverable reserves: Those economic resources of crude oil and natural gas, yet undiscovered, that are estimated to exist in favorable geological settings.

Unfinished oils: All oils requiring further processing. They are produced by partial refining of crude oil and include light oils, kerosene and light gas oils, heavy gas oils, and residuum.

Uranium: A heavy, naturally radioactive, metallic element. Its two principally occurring isotopes are uranium-235 and uranium-239.

Vessel: A ship used to transport crude oil, petroleum products, or natural gas products. Vessel categories are: ultra large crude carrier (ULCC) and very large crude carrier (VLCC).

Well: A hole drilled in the earth for the purpose of (a) finding or producing crude oil or natural gas; or (b) producing services related to the production of crude or natural gas.

Wellhead price: The price of oil or natural gas at the mouth of the well.

West Texas Intermediate: The "marker" crude in North America and the contract grade for the New York Mercantile Exchange's crude oil futures contract. It is widely accepted as the basis for pricing most US and Canadian crude oil.

Wet natural gas: A mixture of hydrocarbon compounds and small quantities of various non-hydrocarbons existing in the gaseous phase or in solution with crude oil in porous rock formations at reservoir conditions.

Wind energy: Kinetic energy present in wind motion that can be converted to mechanical energy for driving pumps, mills, and electric power generators.

Wind power plant: A group of wind turbines interconnected to a common utility system through a system of transformers, distribution lines, and substations. Operation, control, and maintenance functions are often centralized through a network of computerized monitoring systems, supplemented by visual inspection.

1

Introduction

For centuries energy has played a major role in the evolution of human civilizations. In the last two centuries fossil fuels (coal, oil, and natural gas) were crucial for the birth and development of the Industrial Revolution and global economic prosperity. Energy products are certain to maintain their character as the "engine" for maintaining and improving our way of life.

A major characteristic of energy is the mismatch between resources and demand. Generally speaking, major consuming regions and nations (the United States, Europe, Japan, China, and India) do not hold adequate indigenous energy resources to meet their large and growing consumption. On the other hand, major producers (i.e., the Middle East, Russia, the Caspian Sea, and Africa) consume a small (albeit growing) proportion of their energy resources. This broad global mismatch between consumption and production has made energy products the world's largest traded commodities. Almost every country in the world imports or exports a significant volume of energy products. This means the wide fluctuation of energy prices plays a key role in the balance of payments almost everywhere.

The heavy reliance on energy in conjunction with the asymmetric global distribution of energy deposits have underscored the importance of energy security. This sense of vulnerability is not new. Despite the abundance of energy resources and a favorable political and economic environment, industrialized countries started expressing their concerns over energy security as early as the first part of the twentieth century. First Lord of the Admiralty Winston Churchill's decision that the Royal Navy needed to convert from coal to oil in order to retain its dominance signaled a growing intensity of global competition over energy resources (mainly oil). This rivalry between global powers was played out in World War II when the Allies enjoyed access to significant oil deposits while Germany's and Japan's strategies to gain access to oil resources failed and led, among other developments, to their eventual defeat.

The availability of cheap energy resources played a major role in the reconstruction and development of Europe and Japan in the aftermath of World War II. This prolonged era of relative confidence in the availability of abundant and secure energy resources came to an abrupt end following the outbreak of the 1973 Arab–Israeli War. Arab oil producing countries cut their production and imposed oil embargos on the United States and a few other countries to force a change in their political support for Israel. This use of oil by major producers to gain political leverage has shattered consumers' sense of energy security. Since then, the fluctuation

Energy Security: An Interdisciplinary Approach, First Edition. Gawdat Bahgat.
© 2011 John Wiley & Sons, Ltd. Published 2011 by John Wiley & Sons, Ltd.

of energy prices (partly due to geopolitical developments and partly in response to supply and demand changes) has reinforced this sense of vulnerability.

In the last few decades there has been a growing understanding of the challenges that climate change poses to life on earth. More people have come to realize that our way of life (i.e., human activities) contributes and accelerates global warming and that something needs to be done to restrain this human-made environmental deterioration. This slowly growing consensus has added a new dimension to energy security. The concept is no longer limited to the availability of energy resources at affordable prices. Environmental considerations restrain the exploration and development of these resources and urge consideration of less polluting alternative sources of energy.

This brief overview of energy history suggests that there is a wide variety of threats to energy security. These include geological, geopolitical, economic, and environmental threats. In the following chapters I thoroughly examine these challenges on both the consumer and producer sides. In the remainder of this chapter I provide a detailed discussion of the concept of energy security followed by an analysis of the different forms of energy (i.e., oil, natural gas, coal, nuclear power, and renewable sources). The discussion highlights the main themes that characterize the global energy markets.

1.1   Energy Security

The 1973–1974 oil embargo served as a turning point in global and domestic energy markets. The availability of energy supplies at affordable prices was no longer taken for granted. The turmoil in the world economy focused on the disruption of supplies to consuming countries. These oil consumers have implemented several measures (individually and collectively) to mitigate the impact of such disruptions and to reduce their energy vulnerability. The measures include the creation of the International Energy Agency (IEA), the storage of oil supplies in strategic petroleum reserves, and encouraging energy conservation, among others.

Not enough attention was given to the other side of the energy equation – producing nations. The concept of "energy security" is not static. Since the mid-1970s a broader definition has emerged that addresses all the energy players' concerns. In the past few decades, while the industrialized countries have successfully diversified their sources of crude oil imports and greatly reduced their relative dependence on energy (albeit at different degrees), the major oil exporters remained dependent on oil revenues. Petroleum revenues have continued to be the principal source of income for almost all major oil exporting countries. As a result, oil exporters have as many reasons to worry about the security of their markets as importers have to worry about the security of supplies [1]. In short, the security of demand is considered as important as the security of supply. Abdullah Salem El-Badri, Secretary General of the Organization of Petroleum Exporting Countries (OPEC), summed up the argument: "Energy security should be reciprocal. It is a two-way street" [2].

Within this context energy analysts have provided different definitions of energy security highlighting different aspects of the concept. Barry Barton, Catherine Redgwell, Anita Ronne, and Donald Zillman define it as a condition in which "a nation and all, or most of its citizens and businesses have access to sufficient energy resources at reasonable prices for the foreseeable future free from serious risk or major disruption of service" [3]. Daniel Yergin underscores a number of "fundamentals of energy security." The list includes diversification; high-quality and timely information; collaboration among consumers and between consumers and producers; investment flows; and research and development technological advance [4]. Yergin argues that the experience since the early 2000s has highlighted the need to expand the concept of energy security in two critical dimensions: globalization of the energy markets and the need to protect the entire energy supply chain and infrastructure [5].

Christian Egenhofer, Kyriakos Gialoglou, and Giacomo Luciani distinguish between short-term and long-term risks. The former are generally associated with supply shortages due to accidents, terrorist attacks, extreme weather conditions or technical failure of the grid. The latter are associated with the long-term adequacy of supply, the infrastructure for delivering this supply to markets, and a framework for creating strategic security against major risks (such as non-delivery for political, economic, force majeure or other reasons) [6].

Finally, a report by the IEA argues that energy insecurity stems from the welfare impact of either the physical unavailability of energy, or prices that are not competitive or overly volatile. Analysts at the Paris-based organization add that the more a country is exposed to high-concentration markets, the lower is its energy security [7].

All these definitions underscore the fact that energy security is a multi-dimensional concept that incorporates cooperation between producers, consumers, and national and international companies. The experience of the last few decades indicates that the availability of clean energy resources at affordable prices cannot be addressed only at a national level. Rather, international cooperation is a necessity. Thus, energy is part of broader international relations between states. A major theme of today's energy markets is interdependence between consumers and producers. Calls for self-sufficiency or energy independence are more for domestic constituencies. Indeed, energy interdependence fosters cooperation between countries in other areas such as economic development and world peace.

Another major theme of the energy security literature is the importance of diversification of energy mix and energy sources. The less dependent a country is on one form of energy (i.e., oil, natural gas, coal, nuclear power, and renewable sources), the more secure it is. Similarly, the more producing regions there are around the world, the better.

1.2 Diversification of Energy Mix

To a great extent coal was the dominant fuel for most of the nineteenth century and was overtaken by oil in the twentieth century. The transition from coal to oil was due to the general superiority of oil. It has a higher energy density (about 1.5 times higher than the best bituminous coals, commonly twice as high as ordinary steam coals), it is a cleaner as well as a more flexible fuel, and it is easier both to store and to transport [8]. In the early years of the twenty-first century many countries took steps to utilize the world's endowment of natural gas, renewable sources, and nuclear power. The IEA projects that fossil fuels will account for 80% of the world's primary energy mix in 2030 [9]. This means that, despite the renaissance in nuclear power and the growing interest in other alternative fuels, oil, natural gas, and coal will continue to dominate the global energy mix. This projection suggests that countries from all over the world should keep investing and developing these alternative sources of energy while pursuing strategies to produce and deliver fossil fuels to end-users in an efficient, timely, sustainable, economic, reliable, and environmentally sound manner.

1.2.1 Oil

Oil is the world's most vital source of energy and is projected to remain so for many years to come, even under the most optimistic assumptions about the pace of development and deployment of alternative fuels.

Crude oil is classified by density and sulfur content. Crude oil with a lower density (referred to as light crude) usually yields a higher proportion of the more valuable final petroleum products, such as gasoline and other light petroleum products, by a simple refining process known as distillation. Light crude oil is contrasted with heavy crude oil, which has a low share of light hydrocarbons and requires much more severe refining processes than distillation, such as coking and cracking, to produce similar proportions of the more valuable petroleum products.

Sulfur, a naturally occurring element in crude oil, is an undesirable property and refiners have to make heavy investments in order to remove it from crude oil. Crude oil with a high sulfur content is referred to as sour crude, while that with a low content of sulfur is referred to as sweet crude. Crude oil that yields a higher proportion of the more valuable final petroleum products and requires a simple refining process (the light/sweet crude variety) is more desirable and considered superior to the one that yields a lower fraction of the more valuable petroleum products and requires a more severe refining process (the heavy/sour crude variety) [10].

The birth of the oil industry is generally attributed to the famous well drilled for oil in 1859 by Colonel Edwin L. Drake at Titusville, Pennsylvania [11]. Also, it is claimed that F.N. Semyenov was the first to drill a well on the Apsheron Peninsula, near Baku in Azerbaijan, in 1848 [12]. In the succeeding years the oil industry grew rapidly in both the United States and on the shores of the Caspian Sea. For most of the following century the United States and its oil companies dominated the industry. This US domination was seriously challenged in the 1960s and 1970s due to at least two significant developments. First, US oil production peaked and the nation ceased to be self-sufficient and started a steady and growing dependence on foreign supplies. Second, major oil producing nations founded OPEC to defend their interests and the opportunity came in the aftermath of the 1973 Arab–Israeli War. In the twenty-first century, the oil industry is no longer dominated by one player or a small number of international oil companies. Rather, multiple producers, consumers, national and international companies compete with one another and also work together to explore, develop, and deliver approximately 85 million barrels of oil a day.

The IEA projects that oil will continue to dominate the global energy mix, so its share will slightly decline from 34% in 2007 to 30% by 2030 [13]. This persistent domination raises a key question – does the world hold enough oil to meet the growing demand? Furthermore, sustainable supplies require adequate investment. The flow of investments needs a supportive geopolitical environment. The following sections address these issues.

Unlike solar, wind, and other renewable energy forms, oil (and other fossil fuels) is a finite resource. This fact suggests that global production will peak one day and eventually the world will run out of oil. This is known in the oil literature as peak oil theory. Its roots go back to Marion King Hubbert, a Shell geologist, who in 1956 correctly predicted that US production would peak between 1965 and 1970 [14]. His model maintains that the production rate of a finite resource follows a largely symmetrical bell-shaped curve. This theory has since ignited an intense debate regarding the availability of enough supplies to meet global demand and the future of oil in general. Peter Odell agrees that production does indeed go up, and then down, and that the downside usually falls off gradually, "following a depletion pattern modeled fairly accurately by production that is a fixed percentage of what remain (i.e., exponential decline)" [15].

Most of the world's oil executives, government ministers, analysts and consultants reject the peak oil theory on both technological and economic grounds. They argue that technological advances and market laws have always expanded the lifespan of the world's endowment of proven oil reserves.

In the oil industry a distinction is made between proven, probable and possible reserves. Proven reserves are those quantities of petroleum which geological and engineering data indicate with reasonable certainty (90% probability that the actual quantities recovered will exceed the estimate) can be recovered in the future from known reservoirs, under existing economic and operating conditions [16]. Probable reserves are those unproven reserves which analysis of geological and engineering data suggests are more likely than not (50% probability) to be commercially recoverable. Possible reserves are those unproven reserves which analysis of geological and engineering data suggests are less likely than probable reserves to be commercially recoverable (10% probability) [17]. It is important to point out that in most oil producing countries data on reserves are considered state secrets and foreign observers are not allowed to verify the accuracy of official figures [18].

Another distinction is made between conventional and non-conventional oil. The former flows at high rates and with a good quality. Much of conventional oil comes from giant fields discovered a long time ago. Most of the oil that the world currently consumes is considered conventional oil. On the other hand, non-conventional oil comes from enhanced recovery achieved by changing the characteristics of the oil in the reservoir through steam injection and other methods. Non-conventional oil exists in hostile environments, usually in small accumulations and with a poor quality. It is difficult and expensive to produce and is environmentally challenging. Examples include heavy oil and tar-sand deposits in western Canada, Venezuela, and Siberia [19].

Oil extraction techniques are advancing all the time. Technological advances have enabled oil companies to extract more oil from existing fields and avoid unsuccessful drilling. The clear successes of the late 1950s, 1960s, and 1970s in finding oil were largely due to the expanding use of seismic surveys, with digital seismic surveys, in particular, being introduced from the mid-1960s. Furthermore, a substantial increase in world oil production in the last few decades has come from offshore fields. Modern technology has enabled oil companies to find and develop oil deep at the bottom of the oceans. Offshore oil production started in the early 1940s and has grown from a modest 1 million barrels per day (b/d) in the 1960s to nearly 25 million b/d in 2005 to represent one-third of world crude oil production [20]. In short, what was considered non-conventional is increasingly regarded as conventional.

Technology is also reducing the cost of exploration and development. When the world comes close to exhausting oil deposits, prices will gradually move higher as the costs of alternative energy decline. In short, it can be argued that the world is running out of easy and cheap oil, but there is still plenty to explore and develop. The IEA projects that the world's total endowment of oil is large enough to support the anticipated rise in consumption in the foreseeable future [21].

1.2.2 Natural Gas

Natural gas is a fossil fuel that contains a mix of hydrocarbon gases, mainly methane, along with varying amounts of ethane, propane, and butane. Carbon dioxide, oxygen, nitrogen, and hydrogen sulfide are also often present. Natural gas is "dry" when it is almost pure methane, absent the longer-chain hydrocarbons. It is considered "wet" when it contains other hydrocarbons in abundance. "Sweet" gas possesses low levels of hydrogen sulfide compared to "sour" gas [22]. Natural gas found in oil reservoirs is called "associated gas." When it occurs alone it is called "non-associated gas."

Natural gas is rapidly gaining importance in global energy markets. Prized for its relatively clean and efficient combustion, gas is becoming the fuel of choice for a wide array of uses, notably the generation of electric power. World natural gas reserves are abundant, estimated at about 185.02 trillion cubic meters (6534.0 trillion cubic feet), or 60.4 times the volume of natural gas used in 2008 [23].

Ancient civilizations used gas on a small scale but it has been used extensively as a fuel source since the nineteenth century. With the discovery of oil in Pennsylvania, associated gas was used for both industrial and domestic purposes. The growing need for energy during and in the aftermath of World War II gave momentum to gas exploration and development. An extensive pipeline network was built in parallel with the expansion of gas production. Thus, by the middle of the twentieth century, natural gas provided about a third of total primary energy in the United States and the nation was by far the main natural gas producer and consumer in the world [24].

In the 1950s and 1960s several natural gas discoveries were made in Europe, particularly in and around the North Sea. The turmoil in oil markets, caused by the 1973–1974 Arab embargo, gave more incentives to consuming countries to diversify their energy mix. Since then natural gas has become an important source of energy worldwide. Still, the problem of transporting natural gas slowed down the full utilization of global deposits. Pipelines, the main method of transporting natural gas, imposed severe limitations on trade in the fuel. By nature, pipelines are economical for trade over relatively small (though growing) distances, and thus markets made through pipes were regional in nature.

The introduction of liquefied natural gas (LNG) in the early 1960s changed the dynamics of the gas industry and trade. LNG is natural gas that is stored and transported in liquid form under atmospheric pressure at a temperature of $-260\,°F$ ($-160\,°C$). Like the natural gas that is delivered by pipeline into homes and businesses, it mainly consists of methane. Liquefying natural gas provides a means of moving it long distances when pipeline transport is not feasible. Natural gas is turned into a liquid using a refrigeration process in a liquefaction plant. The unit where LNG is produced is called a train. Liquefying natural gas reduces its volume by a factor of 610. The reduction in volume makes the gas practical to transport and store. In international trade, LNG is transported in specially built tanks in double-hulled ships to a receiving terminal where it is stored in heavily insulated tanks. The LNG is then sent to regasifiers which turn the liquid back into a gas that enters the pipeline system for distribution to customers as part of their natural gas supply [25].

The development of LNG was slow due to the costly technologies associated with producing, storing, and shipping it. In the late 1950s and early 1960s the technology for shipping LNG was developed and the world's first major LNG export plant opened in Arzew, Algeria, in 1964, exporting gas to buyers in France and the United Kingdom. By 1972, LNG plants were up and running in the United States (Alaska), Brunei, and Libya, with a second plant added in Algeria at Skikda. In the ensuing decades Algeria, Indonesia, Malaysia, Australia, Qatar, Nigeria, Trinidad and Tobago, Oman, and Egypt have emerged as major LNG exporters [26]. The expansion in LNG trade is due mainly to technological advances which substantially reduced the costs. Furthermore, the speedy rise of LNG has the potential to transform the natural gas market from a regional to an international one. In other words, high costs made it more convenient to transport natural gas short distances. Declining costs are making it easier to ship LNG almost anywhere in the world.

Still, compared to oil, gas is more capital intensive; project time horizons are longer, and wariness about uncertain political environments appears to be greater. In addition, natural gas is used mainly in electric power generation, where it has to compete with coal, nuclear power, and hydroelectricity. Oil, by contrast, is still the unrivaled king of energy sources for mobility.

1.2.3 Coal

Coal is a readily combustible black or brownish-black rock whose composition, including inherent moisture, consists of more than 50% by weight and more than 70% by volume of carbonaceous material. It is formed from plant remains that have been compacted, hardened, chemically altered, and metamorphosed by heat and pressure over geological time [27].

Compared to other fuels, coal enjoys several advantages. It is abundant and more evenly distributed around the world than oil or natural gas. It is cheap and costs are continuously being reduced by competition [28]. The many suppliers and the possibility of switching from one to another means security of supply. The global ratio of coal reserves to production is 120 years [29]. Coal is widely used in electricity generation (about 40% of the world's electricity) [30]. On the other hand, coal faces significant environmental challenges in mining, air pollution and emission of carbon dioxide (CO_2). Indeed, coal is the largest contributor to global CO_2 emissions from energy use and its share is projected to increase [31].

CO_2 is a colorless, odorless, non-poisonous gas that is a normal part of the earth's atmosphere. It is a product of fossil-fuel combustion as well as other processes. It is considered a greenhouse gas as it taps heat (infrared energy) radiated by the earth into the atmosphere and thereby contributes to the potential for global warming. The challenge for governments and industry is to find a path that mitigates carbon emissions yet continues to utilize coal to meet urgent energy needs. This will require not only clean coal technologies for new plants, but also rehabilitation and refurbishment of existing inefficient plants. And this must happen not only in industrialized countries, but also in developing countries, which are expected to account for most coal consumption in the foreseeable future.

Faced with the reality that coal will be a major source of energy for a long time, it becomes clear that cleaner, lower-carbon, coal-based energy technologies will play a central role in solving the global climate challenge. Those technologies include coal gasification, which makes clean gas from coal and strips out the CO_2 before burning the gas, and post-combustion capture, which strips CO_2 out of the exhaust gas left after coal is burned. Another rapidly developing method is carbon capture and storage (CCS), a technique that has been around for decades. This approach is designed to mitigate the contribution of fossil-fuel emissions to global warming, based on capturing CO_2 from large point sources such as fossil-fuel power plants. It can also be used to describe the scrubbing of CO_2 from ambient air as a geo-engineering technique. The CO_2 might then be permanently stored away from the atmosphere [32].

The intense fluctuation in oil and natural gas prices has revived interest in the use of Fischer–Tropsch (F–T) technology to produce transportation fuels from coal. The F–T process is a catalyzed chemical reaction in which carbon monoxide and hydrogen are converted into liquid hydrocarbons of various forms. The principal purpose of this process is to produce a synthetic petroleum substitute for use as synthetic lubrication oil or as synthetic fuel [33]. The process was invented in petroleum-poor but coal-rich Germany in the 1920s to produce liquid fuels. It was used by Germany and Japan during World War II to produce ersatz fuels. Later, the process was used in South Africa to meet its energy needs during its isolation under the apartheid regime. This process has received renewed attention in the quest to produce low-sulfur diesel fuel in order to minimize the environmental impact from the use of diesel engines. The F–T process is an established technology and already applied on a large scale, although its popularity is hampered by high capital costs, high operation and maintenance costs, and the uncertain and volatile price of crude oil.

These decades-long efforts to mitigate emissions suggest that coal will continue to be used to meet the world's energy needs in significant quantities. Indeed, the IEA projects that coal's share of global energy demand will climb from 26% in 2006 to 29% in 2030 [34].

1.2.4 Nuclear Power

The fact that nuclear power releases virtually no environmentally damaging emissions of carbon dioxide, sulfur dioxide, and nitrogen oxide could make it an attractive option for many countries seeking technologies leading to reduced greenhouse gas emissions or abatement of local and regional pollution. In the 1960s and 1970s, particularly after the Arab oil embargo, nuclear power promised to be a viable solution for industrialized countries looking for energy security and cheap power. However, most of the promise of nuclear energy has evaporated as a result of loss of investor and public confidence in the technology.

At the beginning of the twenty-first century there were approximately 440 nuclear reactors in use around the world and about 26 under construction. Most of these reactors are concentrated in 31 countries. Just six countries – the United States, France, Japan, Germany, Russia, and South Korea – produce almost three-quarters of the nuclear electricity in the world [35]. Nuclear power is almost exclusively used for electricity generation and globally it produces about 16% of electricity.

Since the early 2000s there has been a global revival of interest in nuclear power. Almost all over the world, governments are taking a second look at nuclear power, particularly in Europe, North America, Asia, and most recently in the Middle East. As a result, several reactors are being built or under consideration. Several developments have contributed to this "nuclear renaissance." First, the surge in oil and natural gas prices in the early 2000s and the great uncertainty surrounding the future of these two fuels have prompted many governments to diversify their energy mix and reduce their over-dependence on hydrocarbon fuels. Second, Russia's politically motivated, frequent use of its oil and gas deposits to punish and/or reward clients has deepened Europe's sense of vulnerability and put more pressure on European leaders to reduce their dependence on Moscow. Third, the emerging and growing consensus on climate change has made many countries more determined to contain pollution and honor their commitments on climate change international agreements, particularly the Kyoto Protocol. However, many world leaders have come to realize that they cannot maintain their non-nuclear energy policy and simultaneously fulfill their commitments to reduce CO_2 emissions. Finally, since the Chernobyl disaster in 1986 there has not been a major nuclear accident. Indeed, the industry safety record has made substantial improvements. Furthermore, several countries have figured out ways to deal with nuclear waste without endangering the health of their populations [36]. These developments have made nuclear power more attractive and contributed to the wave of new construction of nuclear plants.

Despite this renewed global interest in nuclear power, several issues need to be addressed before it reaches its full potential. These include costs, safety, radioactive waste disposal, and proliferation of nuclear weapons.

Cost: Like the other sources of energy, nuclear power will succeed in the long run only if it has a lower cost than competing fuels. Nuclear power plants have relatively high capital costs and very low marginal operating costs. Construction costs reflect a combination of regulatory delays, redesign requirements, and construction management and quality control problems. The specter of high construction costs has been a major factor leading to little credible commercial interest in investments in new nuclear plants. However, a closer look suggests that nuclear power costs might not be very high if the costs of CO_2, produced by fossil fuels, are taken into consideration. Furthermore, as engineering companies acquire more expertise, there will be substantial reductions in construction costs.

Safety: After the 1979 accident at Three Mile Island, Pennsylvania, in the United States and the 1986 disaster at Chernobyl in the Soviet Union, public concern about reactor safety increased substantially. There is also concern about the transportation of nuclear materials and waste management. The September 11, 2001 terrorist attacks on the World Trade Center and the Pentagon have heightened concerns about the vulnerability of nuclear power stations and other facilities, especially spent-fuel storage pools, to terrorist attack. There is concern about the exposure of citizens and workers to radiation from the activities of the industry despite good regulations. There are also significant environmental impacts, ranging from long-term waste disposal to the handling and disposal of toxic chemical wastes associated with the nuclear fuel cycle.

Radioactive waste disposal: Spent nuclear fuel discharged from nuclear reactors will re-main highly radioactive for many thousands of years. The management and disposal of this radioactive waste from the nuclear fuel cycle is one of the most difficult problems facing the nuclear power industry. The primary goal of nuclear waste management is to ensure that the health risks of exposure to radiation from this material are reduced to an acceptably low level for as long as it poses a significant hazard. One strategy involves the separation of individual radionuclides from the spent fuel. Another strategy is to dispose of the waste in repositories constructed in rock formations hundreds of meters below the earth's surface. Each strategy has its own advantages and disadvantages. The lack of consensus on the most appropriate way to deal with radioactive waste stands as one of the primary obstacles to the expansion of nuclear power around the world.

Proliferation of nuclear weapons: A major challenge facing nuclear power is the so-called "dual use" of nuclear material and know-how. In other words, the same material (enriched uranium and plutonium) and applied technology that are used to make peaceful nuclear power can be used to make nuclear weapons. This means that nations wishing to acquire or enhance a nuclear weapons capability can use commercial nuclear power as a source of technological know-how or usable material for nuclear weapons. The possession of a complete nuclear fuel cycle, including enrichment, fuel fabrication, reactor operation, and reprocessing, moves any nation closer to obtaining a nuclear weapons capability [37]. The crisis with North Korea and the international concern over Iran's nuclear program illustrate this dilemma.

Since the dawn of the nuclear age, proliferation concerns have led to an elaborate set of international institutions and agreements, none of which have proven entirely satisfactory. The nuclear Non-proliferation Treaty (NPT) is the major international mechanism to prevent nations from acquiring nuclear weapons capability. The International Atomic Energy Agency (IAEA) is responsible for verifying NPT compliance with respect to fuel cycle facilities through its negotiated safeguards agreements with NPT signatories. In addition, many policy-makers and proliferation experts have proposed the creation of an international fuel bank, under IAEA supervision, that would assure nations' supply of nuclear fuel as long as they observe the NPT's provisions [38].

To sum up, unless these issues (costs, safety, radioactive waste disposal, and proliferation of nuclear weapons) are satisfactorily addressed, nuclear power is unlikely to realize its potential. Indeed, nuclear power's share of the global energy mix is projected to decline slightly from 6% in 2008 to 5% by 2030 and its share of electricity output to drop from 16% to 10% during the same time span [39].

1.2.5 Biofuels

In recent years, biofuels have attracted increasing attention. Their main attraction is that they are made from renewable feedstocks that can be grown by farmers, and substituting them for petroleum products reduces greenhouse gases and dependence on foreign oil. In short, biofuels have been promoted as serious solutions to the twin challenges of climate change and energy security. It is no surprise, then, that global interest in bio-energy has grown rapidly in recent years. In the early 2000s, bio-energy became a multi-billion-dollar business. The United States and Brazil dominate the current liquid biofuels industry, but many other governments, particularly Australia, Canada, and Europe, are now actively considering the appropriate role for biofuels in their future energy portfolios.

Bio-energy is defined as energy produced from organic matter or biomass [40]. A wide range of biologically derived feedstocks can be transformed into liquid fuels. The technologies used to make that transformation are also numerous. At present, the predominant liquid biofuels in use are ethanol and biodiesel. Ethyl alcohol, or ethanol, can be produced from any feedstock that contains relatively dense quantities of sugar or starchy crops. The most common feedstocks are sugar cane, sugar beet, maize (corn), wheat, and other starchy cereals such as barley, sorghum, and rye. Biodiesel is based on vegetable oils such as those obtained from oil palm, rapeseed, sunflower seed, and soybean; some is made from tallow, used cooking oil, and even fish oil [41].

The global interest and impressive development of the biofuel industry have raised serious concerns about its impact on food prices, climate change, and energy security.

Food prices: To the extent that increased demand for biofuel feedstock diverts supplies of food crops (e.g., maize) and diverts land from food crop production, global food prices will increase. The competition over land and water has heightened the so-called "food-versus-fuel debate." This competition favors large producers, as illustrated by the prevailing trend toward concentration of ethanol ownership in Brazil and the United States. The transition to liquid biofuels can be especially harmful to farmers who do not own their own land, and to the rural and urban poor who are net buyers of food, as they could suffer from even greater pressure on already limited financial resources. Though it is true that increased use of biofuels has contributed to a surge in grain and vegetable oil prices, other factors such as droughts and the rise in demand for meat and milk products have probably played a role in the overall high food prices [42].

Climate change: The potential impact of biofuels on the environment is uncertain and needs to be closely scrutinized. Several elements need to be taken into account. These include the type of crop, the amount and type of energy embedded in the fertilizer used to grow the crop, emissions from fertilizer production, the energy used in gathering and transporting the feedstock to the bio-refinery, alternative land uses, and the energy intensity and fuel types used in the conversion process. In addition, water availability will influence feedstock choice and the location of conversion facilities. Finally, it is important to point out that the ability of various bio-energy types to reduce greenhouse gas emissions varies widely.

Energy security: Reducing dependence on fossil fuels has been a major reason for investing in bio-energy. The idea of producing energy at home and becoming self-sufficient instead of being vulnerable to interruption of foreign supplies appeals to many policy-makers. Again, this strategy needs to be scrutinized. At least two dynamics should be taken into consideration. First, fossil fuels are used in the production and transportation of the feedstock. Second, the energy content of a liter of ethanol is typically only two-thirds of the energy content of a liter of gasoline [43].

In order to avoid the potential negative impact of biofuels, there has been a growing interest in developing biofuels produced from non-food biomass. Feedstocks from ligno-cellulosic materials include cereal straw, forest residues, and purpose-grown energy crops such as vegetative grasses and short-rotation forests. These so-called "second-generation biofuels" could avoid many of the concerns facing "first-generation biofuels" and potentially offer greater cost reduction potential in the longer term [44]. Once the "second-generation biofuel" technologies are fully commercialized, it is likely they will be favored over many "first-generation biofuel" alternatives by policies designed to pursue national objectives such as environmental performance and food security.

The future of bio-energy is uncertain. In many countries biofuels cannot compete on their own with other sources of energy. They survive by receiving generous governmental subsidies, which will not last forever. In short, the rapid development of modern bio-energy worldwide clearly presents a broad range of opportunities, but it also entails many tradeoffs and risks.

1.2.6   Other Renewable Sources

The IEA's definition of renewable energy sources includes energy generated from solar, wind, biomass, the renewable fraction of municipal waste, and geothermal sources, hydropower, ocean tidal and wave resources, and biofuels [45]. Recently there has been growing global interest in developing these renewable sources for at least two reasons. First, these sources provide an alternative to dependence on foreign supplies of fossil fuels. Usually, these alternatives are indigenous. They also contribute to the diversification of the energy mix. In short they enhance national and global energy security. Equally important, most renewable sources are environmentally friendly. They produce very little pollution. Indeed, the renewed interest in renewable sources is largely driven by mounting concern over climate change.

Despite these advantages, the share of renewable sources in world total primary energy supplies is currently very small. However, prospects for renewable energy "have never looked better" [46]. From 2008 to 2030, wind, solar, geothermal, tide and wave energy are together projected to grow faster than any other source of energy worldwide, at an average rate of 7.2% per year [47]. The evolution of the economic potential of renewable sources over the coming decades will depend both on their technological development and on cost in relation to competing conventional energy technologies.

Renewable energy systems are diverse – each type has its own unique characteristics. Solar power offers some substantial advantages over other energy sources. Solar generating facilities are most productive in the middle of the day, when demand for electricity typically is at its peak. Unlike fossil-fuel-fired generating capacity they produce no toxic emissions and unlike nuclear plants they leave no radioactive waste. Rooftop solar panels can be installed in homes and businesses, reducing the need for centralized power plants and transmission lines. And, of course, the sun's rays are free and available in infinite quantity. Heat storage and/or fossil-fuel backup may help fully cover the mid-peak demand during a few hours after sunset. While round-the-clock operation is technically possible, industrial heat storage options are currently not economically feasible. Although the costs of converting sunlight into usable power have dropped in recent decades, generating electricity from conventional power sources (i.e., coal or natural gas) is still cheaper. Government tax incentives are closing the gap in many countries [48]. Solar power has usually been thought of as a way of supplying electricity or heating water to a single building. But in several countries (Spain, Portugal, Australia, and the United States among others) solar power plants capable of powering thousands of homes have been built [49].

Wind power has been used for a long time, but in the last few decades several countries have allocated more investments in installing wind turbines. In recent years Germany, Denmark, the United States, China, and India, among others, have increased their reliance on wind power. Wind power is directly dependent on the cube of the wind speed within the operating range. The wind blows more reliably offshore and is often stronger, making turbines sited in the sea attractive (additionally, they can also be hidden from view). But siting turbines offshore is both more difficult and more expensive [50]. Wind power can become unavailable at times of low wind speeds, but also at times of very high wind speeds when wind turbines need to be shut down in order to avoid damage to equipment.

The ocean represents four-fifths of the surface of the earth, and humankind has always been impressed with the kinetic energy contained in the moving water of the waves and tides. Still, the mechanical technology has not been demonstrated to routinely convert this immense available energy source into economic electric power. There has been very limited success with tides, which are cyclic, depending on the relative position of the moon. There has not been commercial success with waves either. Wavers are a reciprocating motion that vary greatly in height and so require considerable mechanical apparatus to convert them to the steady rotary motion needed for electric power production. In short, neither tides nor waves have the economics of very large-scale operations available to them because they are both local and cyclic [51].

Hydroelectricity is considered a renewable source because it depends on rainfall, which is a recurrent phenomenon in different seasons every year. In sites where a waterfall exists or where it can be built with the construction of a barrage, the potential energy of the falling water can be harnessed to generate electricity. The hydroelectricity score of growth is limited by the availability of suitable sites and the serious and complex environmental problems that affect many of them. There is little scope for growth in developed countries so that future expansion is most likely to take place in the developing world [52].

Geothermal energy depends on the availability of permeable hot rock and hot water. It can provide base load capacity since variability is not an issue. Near-surface geothermal heat is only accessible in limited regions worldwide. Geothermal energy is largely untapped in many areas of the world and is available in many developing economies in South and Central America, Africa, and South-East Asia [53].

Most of these renewable sources share a number of characteristics. They are more likely to develop at a local level. Unlike oil, natural gas, and coal, which are shipped all over the world, solar, wind and water power are limited in their potential to expand geographically. Most of these renewable sources supply energy intermittently. In many cases the technology is available to deal with this drawback. Still, this "irregularity" suggests that renewable sources are less reliable than conventional fuels. Further, the costs to develop most of the renewable sources are still high. In recent years technological advances have substantially lowered costs. Still, most of these sources cannot compete with fossil fuels on their own and depend heavily on government subsidies. These characteristics should not discourage the development of renewable sources. Technological, environmental, financial, and political incentives are making renewable energy more attractive. In short, renewable energy is very promising but has some way to go to realize its full potential.

This brief survey of sources of energy highlights the advantages and disadvantages of each fuel. Political, economic, geological, and environmental considerations shape each consumer's choice of energy mix. These choices are also influenced by the availability of adequate investments, by decision-makers' willingness to welcome foreign investments into their energy sectors, and by geopolitical dynamics.

1.2.7 Investment

Energy security depends on sufficient levels of investment in mineral development, generation capacity, and infrastructure to meet demand as it grows. Fossil fuels suffer from natural decline. The rate of natural decline varies from one region to another and from one fuel to another. Energy analysts estimate that the global average rate of natural decline of oil fields is around 10% [54]. This means a need for more investment to combat natural decline and to explore and develop new fields to meet growing demand.

The amount of capital that national and international energy companies and governments are willing to allocate is conditioned, at least, by two factors – namely, economic and political factors. Generally, high energy prices mean more capital accumulation in the producers' coffers. Part of these financial resources is invested to expand production and make more profit. On the other hand, low energy prices mean less money available for investment. For example, systematic under-investment characterized the oil industry for most of the 1990s.

The decision to invest in one country or one sector is usually driven by a number of considerations. One of them is the investment environment. Capital flows will not materialize without a reasonable and stable investment framework, timely decision-making by governments, and open markets. How much a government is willing to partner with a private or foreign company varies from one region to another. Ironically, for a long time most Middle Eastern oil producers (the largest in the world) did not welcome private/foreign investment in their energy sector (particularly oil) on the grounds that it is a strategic sector. Saudi Arabia, the world's largest oil producer and exporter, rejects any foreign investment in upstream projects (exploration and development). In the 1990s, the kingdom reluctantly allowed some foreign companies to explore for natural gas, but not oil. Other major Middle Eastern producers such as Kuwait and Iran impose very strict conditions on foreign participation. Thus, most capital goes to exploring and developing high-cost reserves due to limitations on international oil companies' access to the cheapest resources.

The IEA projections imply a need for a cumulative investment in the upstream oil and gas sector of around \$8.4 trillion (in year 2007 dollars) over 2007–2030, or \$350 billion per year on average [55]. This necessary investment is not likely to materialize without an agreement between energy producers and international companies on the appropriate investment environment.

1.2.8   Resource Nationalism

The re-emergence of a phenomenon known as resource nationalism has further complicated the investment environment and altered the dynamics of the relationship between national oil companies (NOCs) and international oil companies (IOCs) from cooperation to confrontation. The term is assumed to have two components: limiting the operations of private IOCs and asserting greater national control over natural resource development. Another driver is the perception among ordinary people that they have seen little or no benefit from the extraction of "their" oil and minerals. Finally, there is also an important ideological component to the phenomenon, strongly linked to the perceived role of the state in the operation of the national economy [56].

The first NOC was created in Austria in 1908 [57]. Prior to the 1970s there were only two major incidents of successful oil nationalization, the first following the Bolshevik Revolution of 1917 in Russia and the second in 1938 in Mexico. During the 1970s, however, virtually all of the oil resources outside of North America passed from international petroleum companies to the governments of the oil producers. A clear extension of resource nationalism was control over oil prices by the oil exporting countries. Thus, the politics of resource nationalism were integral to the politics of the so-called new international economic order, a Third World movement whose aim was to correct the perceived structural inequities inherent in the global balance of power [58].

In the twenty-first century the relationship between NOCs and IOCs is ambivalent. The former hold nearly 80% of global reserves of oil and dominate the world's oil production [59]. Analysts at the James A. Baker III Institute for Public Policy at Rice University project that NOCs will control a greater proportion of future oil supplies over the next two decades [60]. In addition to making a profit, the NOCs serve the strategic interests of their governments. Thus, the rising role of the NOCs suggests a growing influence of geopolitical considerations at the expense of commercial interests.

1.2.9   Geo-policy

Energy and energy products are considered both commercial and strategic commodities. Almost all human activities depend on different forms of energy, most obviously mobility. Accordingly, decisions on production, prices, trade, and investment are not exclusively subject to supply and demand equilibrium. Rather, political and strategic considerations shape these decisions substantially.

Given this overlap between economic interests and strategic considerations, energy security is challenged by a number of geopolitical threats:

- Internal instability, civil wars, and sectarian or ethnic violence have disturbed production in producing countries. The ongoing ethnic strife in Nigeria and sectarian conflict in Iraq following the demise of Saddam Hussein's regime are cases in point.
- Terrorist attacks on energy infrastructure threaten the free flow of energy shipments and require huge expense to protect energy installations. Attacks on Saudi Arabia's main refineries and Iraq's oil pipelines have been reported.
- Politically motivated suspension of oil or natural gas supplies by a major exporter can threaten energy security in several receiving countries. In the last few years Russia stopped the flow of its natural gas to Ukraine. The reasons are a combination of disagreement over prices and Moscow's displeasure at political developments in Kiev. These interruptions have had a broad impact on several European countries since a substantial proportion of Russian gas to Europe transits Ukraine.
- Economic sanctions against a major producing country can deprive it of badly needed foreign investment and deal a heavy blow to its hydrocarbon production. The severe reduction in Libya's oil production for most of the 1990s can be largely explained by the international sanctions that Tripoli was under. Similarly, US sanctions on Iran since 1979 have deprived the country from fully utilizing its energy potential. Before the 1979 Revolution, Iran's oil production reached 6 million b/d. Despite massive efforts to update and modernize the country's energy infrastructure in the last three decades, Tehran's oil production has never reached the pre-Revolution level.
- War between energy producers can lead to the destruction of their energy infrastructure and installations and to a surge in prices. The Iran–Iraq War (1980–1988) and the First Gulf War (1990–1991) took Iranian, Iraqi, and Kuwaiti production off the market and caused turmoil in the global energy markets.
- Territorial disputes can increase tension between the concerned parties and slow down the full development and utilization of their hydrocarbon deposits. The five countries that share the Caspian Sea (Azerbaijan, Iran, Kazakhstan, Russia, and Turkmenistan) have failed to agree on how to divide the Basin between them. This failure has not stopped the IOCs from investing in the region, but the absence of a legal framework has complicated the speedy utilization of the Caspian oil and gas deposits.

This list is not exclusive, rather these examples illustrate some of the major internal and external challenges that threaten energy security. Another major challenge is the security of shipping lanes. Energy trade depends on the security of the thousands of tankers which carry millions of tons of oil, natural gas, and coal from producing regions to consuming ones. These tankers cross narrow and strategic straits. In 2007, total world oil production amounted to approximately 85 million b/d [61], and around 55 million b/d or 64% of the world's total oil flows through these fixed maritime routes [62]. The international energy market is dependent upon reliable transport. The blockage of a chokepoint, even temporarily, can lead to substantial increases in total energy costs. In addition, chokepoints leave oil tankers vulnerable to theft from pirates, terrorist attack, and political unrest in the form of wars or hostilities as well as shipping accidents which can lead to disastrous oil spills.

The Strait of Hormuz connects the Persian Gulf with the Gulf of Oman and the Arabian Sea and is considered the most important oil chokepoint due to its daily oil flow of 16.5–17 million barrels (2008), which is roughly 40% of all seaborne traded oil (or 20% of oil traded worldwide) [63]. Closure of the Strait of Hormuz would require the use of longer alternate routes at increased transportation costs.

The Strait of Malacca links the Indian Ocean to the South China Sea and Pacific Ocean and is the shortest sea route between Persian Gulf suppliers and the Asian markets – notably China, Japan, South Korea, and the Pacific Rim. Oil shipments through the Strait of Malacca supply China and Indonesia, two of the world's most populous nations. It is the key chokepoint in Asia with an estimated 15 million b/d flow (2008) [63].

Other important transit chokepoints include the Suez Canal, which connects the Red Sea with the Mediterranean Sea, and Bab El-Mandab, a strategic link between the Mediterranean Sea and Indian Ocean. The Turkish Straits (Bosporus and Dardanelles) connect the Black Sea with the Sea of Marmara and the latter with the Aegean and Mediterranean Seas respectively. Finally, the Panama Canal connects the Pacific Ocean with the Caribbean Sea and Atlantic Ocean.

In addition to tankers crossing maritime routes, transit pipelines are used to ship oil and natural gas to consumers. A transit pipeline is defined as an oil or gas pipeline which crosses another sovereign territory to get its throughput to market. Normally there are at least three parties to any transit pipeline agreement, each located in different sovereign entities. These are the producer of the oil or gas, the consumer, and the third party – the transit country (there can be more than one transit country). Any reading of the history of transit oil and gas pipelines suggests a tendency to produce conflict and disagreement. Paul Stevens explains these conflicts as follows: (a) different parties with different interests are involved in the pipeline project; (b) there is no overarching legal jurisdiction to police and regulate activities and contracts; and (c) the projects attract profit and rent to be shared between the various parties [64].

To sum up, energy security does not reside in a realm of its own, but is part of the larger pattern of relations among nations. How those relations go will do much to determine how secure we are when it comes to energy. Furthermore, energy security is no longer the sole purview of any individual state. Increasingly its challenge is met at the level of transborder, regional, and international interactions.

1.3 Conclusion

Energy markets are rapidly evolving to meet growing and changing needs all over the world. Energy security is certain to remain a major concern for policy-makers and analysts for a long time. In closing, several conclusions need to be highlighted. First, in recent years, nuclear power and renewable sources have received increasing attention. These fuels have great potential and are likely to increase their contribution to the global energy mix. However, most energy analysts and organizations project that oil, coal, and natural gas will continue to dominate energy markets.

Second, despite legitimate concerns about the availability of enough energy resources to meet the world's growing demand, it seems that the world's combined fuel deposits are adequate to meet this challenge in the foreseeable future. Stated differently, geology poses less of a challenge to energy security than geo-policy. What happens "above ground" is more likely to shape global energy markets than what is available "underground." These above-ground challenges include relations between producers and consumers, investment policies, and environmental issues, among others. This is why an interdisciplinary approach is needed to address these challenges.

Third, within this context, the concept of energy security should reflect the interests of all concerned parties (consumers and producers, national and international companies, and environmentalists, among others). The concept should also include the whole energy chain (exploration, development, production, transportation, refining, and final consumption). Finally, energy security should not be seen in zero-sum terms where one party's gains are another party's losses. Energy could be, and indeed in many cases is, a win–win situation. Creating greater certainty and stability would benefit all parties. Cooperation, not confrontation, is a key strategy in pursuing energy security.

References

[1] Zanoyan, V., Global energy security. *Middle East Economic Survey*, **41** (15). Available at http://www.mees.com (accessed April 14, 2003).

[2] El-Bardi, A.S., *Energy Security and Supply*. Available at http://www.opec.org (accessed February 14, 2008).

[3] Barton, B. et al. (eds) (2004) *Energy Security: Managing Risk in a Dynamic Legal and Regulatory Environment*, Oxford University Press, London, p. 4.

[4] Yergin, D. (2007) *The Fundamentals of Energy Security*, Testimony before the US House of Representatives Committee on Foreign Affairs, March 22, 2007. Available at http://www.cera.com (accessed March 23, 2007).

[5] Yergin, D. (2006) Ensuring energy security. *Foreign Affairs*, **85** (2), 69–82, 77.

[6] Egenhofer, C., Gialoglou, K., and Luciani, G., *Market-based Options for Security of Energy Supply*, Center for European Policy Studies. Available at http://www.ceps.be (accessed March 21, 2004).

[7] International Energy Agency, *Energy Security and Climate Policy*, Available at http://www.iea.org (accessed May 9, 2007).

[8] Smil, V. (1998) Future of oil: trends and surprises. *OPEC Review*, **22** (4), 253–276, 268.

[9] International Energy Agency (2008) *World Energy Outlook*, International Energy Agency, Paris, p. 38.

[10] Fattouh, B. (2008) *The Dynamics of Crude Oil Price Differentials*, Oxford Institute for Energy Studies, Oxford, p. 8.

[11] Yergin, D. (1990) *The Prize: The Epic Quest for Oil, Money & Power*, Free Press, New York, p. 27.

[12] Campbell, C.J. (1997) *The Coming Oil Crisis*, Multi-Science Publishing, Brentwood, Essex, p. 32.

[13] International Energy Agency (2008) *World Energy Outlook*, International Energy Agency, Paris, p. 38.

[14] Hubbert, M.K. (1962) *Energy Resources*, National Academy of Sciences, Washington, DC, pp. 16–38.

[15] Odell, P. (2000) The global energy market in the long term: the continuing dominance of affordable non-renewable resources. *Energy Exploration & Exploitation*, **18** (2&3), 131–206, 154.

[16] Haider, G. (2000) World oil reserves: problems in definition and estimation. *OPEC Review*, **24** (4), 305–327, 312.

[17] Martinez, A.R. and McMichael, C.L. (1999) Petroleum reserves: new definitions by the society of petroleum engineers and the world petroleum congress. *Journal of Petroleum Geology*, **22** (2), 133–140, 138.

[18] Campbell, C.J. (2005) Just how much oil does the Middle East really have, and does it matter? *Oil and Gas Journal*, **103** (13), 24–26, 24.

[19] Campbell, C.J. (1998) Running out of gas: this time the Wolf is coming. *National Interest*, (**51**), 47–55, 50.

[20] Sandrea, I. and Sandrea, R. (2007) Exploration trends show continued promise in world's offshore basins. *Oil and Gas Journal*, **105** (9), 34–40, 34.

[21] International Energy Agency (2008) *World Energy Outlook*, International Energy Agency, Paris, p. 41.

[22] Energy Information Administration, *Glossary*. Available at http://www.eia.doe.gov (accessed August 15, 2009).

[23] British Petroleum (2009) *BP Statistical Review of World Energy*, British Petroleum, London, p. 22.

[24] Banks, F. (1987) *The Political Economy of Natural Gas*, Croom Helm, London, p. 6.

[25] Energy Information Administration (2003) *The Global Liquefied Natural Gas Market: Status and Outlook*, Energy Information Administration, Washington, DC, p. 3.

[26] Victor, D., Jaffe, A., and Hayes, M. (2006) *Natural Gas and Geopolitics from 1970 to 2040*, Cambridge University Press, Cambridge, p. 11.

[27] Energy Information Administration, *Glossary*. Available at http://www.eia.doe.gov (accessed August 16, 2009).

[28] Editorial (2000) Energy: a new United Nations report examines the issues in oil, gas, coal, and nuclear, *Energy*, **25** (4), 4–10, 9.

[29] British Petroleum (2009) *BP Statistical Review of World Energy*, British Petroleum, London, p. 32.

[30] Land, T. (2004) Can coal claim its crown back? *Energy Economist*, (**278**), 22–24, 22.

[31] Katzer, J. (2007) *The Future of Coal*, Massachusetts Institute of Technology Press, Cambridge, MA, p. 1.

[32] Stein, S. (2008) Energy independence isn't very green. *Policy Review*, (**148**), 3–18, 6.

[33] Robinson, K. and Tatterson, D. (2007) Fischer-Tropsch oil-from-coal promising as transport fuel. *Oil and Gas Journal*, **105** (8), 20–36, 22.

[34] International Energy Agency (2008) *World Energy Outlook*, International Energy Agency, Paris, p. 39.

[35] Froggat, A., *Experts' Comment – Nuclear Power*. Available at http://www.chathamhouse.org.uk (accessed January 24, 2008).

[36] Jun, J., *Is the World Going Nuclear Again to Offset Rising Cost of Fossil Fuels?* Radio Free Europe. Available at http://www.rferl.org (accessed January 10, 2006).

[37] Beckjord, E. (2003) *The Future of Nuclear Power*, Massachusetts Institute of Technology Press, Cambridge, MA, p. 22.

[38] Pascual, C., *The Geopolitics of Energy: From Security to Survival*, Council on Foreign Relations. Available at http://www.cfr.org (accessed February 1, 2008).

[39] International Energy Agency (2008) *World Energy Outlook*, International Energy Agency, Paris, p. 39.

[40] United Nations, *Sustainable Bio-energy: A Framework for Decision Makers*. Available at http://esa.un.org/un-energy/publications.htm (accessed December 24, 2007).

[41] Organization for Economic Cooperation and Development, *Bio-fuels: Is the Cure Worse Than Disease?* Available at http://www.oecd.org/dataoecd/33/41/39276978.pfd (accessed September 29, 2007).

[42] Hunt, S. (2008) Bio-fuels, neither savior not scam: the case for a selective strategy. *World Policy Journal*, **25** (1), 9–17, 11.

[43] Organization for Economic Cooperation and Development (2008) *Bio-fuels: Is the Cure Worse than Disease?* Organization for Economic Cooperation and Development, Paris.

[44] Sims, R. and Taylor, M. (2007) *From 1st – to 2nd – Generation Bio-fuel Technologies: An Overview of Current Industry and R&D Activities*, International Energy Agency, Paris, p. 5.

[45] Olz, S., Sims, R., and Kirchner, N. (2007) *Contribution of Renewables to Energy Security*, International Energy Agency, Paris, p. 7.

[46] Leggett, J. (2009) It's Time for Renewables. Financial Times (May 26).

[47] International Energy Agency (2008) *World Energy Outlook*, International Energy Agency, Paris, p. 39.

[48] Feltus, A. (2008) Solar power's day in the sun. *Petroleum Economist*, **75** (1), 22–25, 23.

[49] Harvey, F. (2008) Supply and Demand Tables Start to Turn. Financial Times (30 June).

[50] Harvey, F. (2008) Winds of Change Blow Across the Global Market. Financial Times (30 June).

[51] Parker, H. (2009) Renewables not adaptable to large-scale installations. *Oil and Gas Journal*, **107** (11), 24–30, 27.

[52] Mabro, R. (2006) Renewable energy, *in Oil in the 21st Century: Issues, Challenges and Opportunities* (ed. R. Mabro), Oxford University Press, Oxford, pp. 327–343, 335.

[53] Olz, S., Sims, R., and Kirchner, N. (2007) *Contribution of Renewables to Energy Security*, International Energy Agency, Paris, p. 25.

[54] International Energy Agency (2008) *World Energy Outlook*, International Energy Agency, Paris, p. 43.

[55] International Energy Agency (2008) *World Energy Outlook*, International Energy Agency, Paris, p. 44.

[56] Stevens, P., *The Coming Oil Supply Crunch*, Chatham House. Available at http://www.chathamhouse.org.uk (accessed August 8, 2008), p. 15.

[57] Stevens, P. (2008) National oil companies and international oil companies in the Middle East: under the shadow of government and the resource nationalism cycle. *Journal of World Energy Law and Business*, **1** (1), 5–30, 14.

[58] Morse, E. (1999) A new political economy of oil? *Journal of International Affairs*, **53** (1), 1–29, 18.

[59] Jaffe, A.M., *Testimony before the Select Committee on Energy Independence and Global Warming*, US House of Representative, Washington, DC. Available at http://www.bakerinstitute.org (accessed June 17, 2008).

[60] Baker, J.A. III, *The Changing Role of National Oil Companies in International Energy Markets*, Institute for Public Policy of Rice University. Available at http://www.bakerinstitute.org (accessed March 21, 2007).

[61] British Petroleum (2009) *BP Statistical Review of World Energy*, British Petroleum, London, p. 8.

[62] Brothers, L., *Global Oil Choke Points*. Available at http://www.lehman.com (accessed August 30, 2008).

[63] Energy Information Administration, *World Oil Transit Chokepoints*. Available at http://www.eia.doe.gov (accessed January 30, 2008).

[64] Stevens, P., *Transit Troubles: Pipelines as a Source of Conflict*. Available at http://www.chathamhouse.org.uk (accessed March 30, 2009).

2

United States

The United States is a major player in the global energy system. In 2010, the nation is the world's largest oil consumer and third producer (after Saudi Arabia and Russia); largest natural gas consumer and second producer (after Russia); second largest coal consumer and producer (after China); largest nuclear power consumer and third hydroelectricity consumer (after Canada and Brazil). Energy is consumed in three broad end-use sectors: the residential and commercial sector, the industrial sector, and the transportation sector. Thus, the US economy and, indeed, way of life depend on the availability of reliable supplies of energy. Furthermore, being the world's only superpower adds more restraints and responsibilities. The US military's energy use (particularly oil) represents a significant proportion of the nation's total consumption. The US Army, Navy, and Air Force play an important role in protecting shipment routes and ensuring security and stability in some of the major energy producing regions.

Despite this dominant role of energy in the US way of life, Washington has never fully articulated a comprehensive energy strategy. For decades different administrations have sought to respond to specific crises, but once these crises had been averted, political will evaporated. Throughout the late 1950s, production and consumption of energy were nearly in balance. Over the following decades, however, consumption started to outpace domestic production and a large gap developed. In the twenty-first century the United States has to address serious energy challenges including over-consumption, declining production, and climate change, among others. The institution in charge of dealing with these challenges is the Department of Energy (DOE).

The origins of the DOE can be traced to the Manhattan Project and the race to develop the atomic bomb during World War II. Following the war, Congress engaged in a vigorous and contentious debate over civilian versus military control of the atom. The Atomic Energy Act of 1946 settled the debate by creating the Atomic Energy Commission (AEC) to maintain civilian government control over the field of atomic research and development.

In response to changing needs in the mid-1970s, the AEC was abolished and the Energy Reorganization Act of 1974 created two new agencies: the Nuclear Regulatory Commission to regulate the nuclear power industry; and the Energy Research and Development Administration to manage the nuclear weapon, naval reactor, and energy development programs. However, the extended energy crisis of the 1970s soon demonstrated the need for unified

Energy Security: An Interdisciplinary Approach, First Edition. Gawdat Bahgat.
© 2011 John Wiley & Sons, Ltd. Published 2011 by John Wiley & Sons, Ltd.

energy organization and planning [1]. On August 4, 1977, President Jimmy Carter signed the US Department of Energy Organization Act, centralizing the responsibilities of the Federal Energy Administration, the Energy Research and Development Administration, the Federal Power Commission, and other energy-related government programs into a single presidential cabinet-level department.

The new department, activated October 1, 1977, was responsible for long-term planning, high-risk research and development of energy technologies, federal power marketing, energy conservation, energy regulatory programs, a central energy data collection and analysis program, and nuclear weapons research, development, and production.

At the outset it was recognized that the formulation of a national energy policy would have to rely on a complete, reliable, and updated database. Much of the existing database had been assembled by industry and, therefore, was considered potentially self-serving and lacking in credibility in policy-making circles. The DOE enabling law met the challenge by establishing the Energy Information Administration (EIA). The EIA's responsibility is to collect, evaluate, assemble, analyze, and disseminate energy data and information. It carries out these responsibilities independently of both the US government and Congress [2].

In the following sections I examine US efforts to explore and develop its fuel deposits and the attempts to articulate a cohesive energy strategy. The analysis suggests that, given the magnitude of the energy challenges, a multi-dimensional strategy is needed. Such strategy comprises diversification of the energy mix, conservation, cooperation with major producers, and an active role in slowing global warming. Furthermore, given the declining indigenous sources, calls for "energy independence" are not realistic. Rather, the United States, like the rest of the world, is likely to remain part of a global system that is based on interdependence between all participants. This interdependence should not be seen as a threat to US energy security.

2.1   Oil

Petroleum was known to native peoples in the north-eastern parts of the colonial United States and was put to various uses by some tribes. As settlement by Europeans proceeded, oil was discovered in many places in north-western Pennsylvania and western New York. In the mid-1800s, expanding uses for oil extracted from coal and shale began to hint at the value of rock oil and encouraged the search for readily accessible supplies. This impetus launched the modern petroleum age, which began when Edwin L. Drake discovered oil in western Pennsylvania. Drake's discovery ignited an oil boom, which was fed by strong demand for lighting fuel and lubricants. Over the next four decades, the boom spread to Texas and California, as well as to foreign countries. With only temporary interruptions, world oil consumption has expanded ever since.

Until the 1950s the United States produced nearly all the petroleum it needed. But by the end of the decade, the gap between production and consumption began to widen and imported oil became a major component of the US supply. Beginning in 1994, the nation imported more petroleum than it produced. Crude oil production in the lower 48 states reached its highest level in 1970 at 9.4 million b/d. A surge in Alaskan oil output at Prudhoe Bay beginning in the late 1970s helped postpone the decline in the overall US production, but Alaska's production peaked in 1988 at 2.0 million b/d and has since fallen [3].

On the other hand, US consumption rose annually until the 1973 oil embargo, which temporally stalled the annual increases. Low oil prices from the mid-1980s to the early 2000s gave a boost to consumption. The surge in oil prices at the end of the 2000s and rising concern about global warming have lowered consumption. The transportation sector accounts for about two-thirds of all petroleum used in the United States. The large and growing gap between indigenous production and consumption has been filled by imports. Traditionally most of the oil imports come from five countries – Canada, Saudi Arabia, Venezuela, Mexico, and Nigeria.

The decades-long decline of oil production has renewed and intensified interest in developing oil deposits in the Arctic National Wildlife Refuge (ANWR) and deep in the Gulf of Mexico.

ANWR consists of 19 million acres (7.7 million hectares) in north-east Alaska. Its 1.5 million acre coastal plain is considered one of the most promising US onshore oil and gas exploration and development prospects. Little wonder it is said to be "America's last great oil frontier" [4]. Early indications of North Slope oil potential led to the establishment in 1923 of the 23 million acre Naval Petroleum Reserve on the western slope to secure a US oil supply for natural security purposes. The US government undertook a program of extensive exploration in the reserve during the 1940s and 1950s, with a few small oil and gas field discoveries resulting.

During World War II, the entire Alaskan North Slope was withdrawn from entry by the public and set aside for US military purposes. That withdrawal was revoked by the Eisenhower administration in 1957 on 20 million acres on the North Slope to make it available for commercial oil and gas leasing. In the 1950s, government scientists concerned about Alaska's wilderness identified the north-eastern corner of Alaska as the best candidate for protection from commercial development and designated 8.9 million acres in that corner as the Arctic National Wildlife Refuge.

The discovery of giant oil reserves at Prudhoe Bay (1968) and the Arab oil embargo (1973) placed North Slope development high on the list of national priorities. Thus, under the Alaska National Interest Lands Conservation Act (ANILCA) of 1980, ANWR was expanded from 9 million acres to 19 million acres.

ANILCA Sec. 1002 directed the Department of Interior to conduct geological and biological studies of the coastal plain and to deliver to Congress the results of those studies with recommendations as to future management of the 1002 area. But the Act's subsequent section banned leasing in the 1002 area unless authorized by Congress [5]. The Act, which set the stage for controversy down to the present, was a type of grand compromise. On the one hand, it more than doubled the total set-aside area to 19.6 million acres, conferred upon it the new title "Refuge," and officially designated 18.1 million acres of it as "wilderness," thereby off-limits to all future development.

On the other hand, ANILCA mandated that the 1.5 million acres of coastal plain be kept off the "wilderness" menu and instead evaluated in terms of both wildlife and petroleum resources. Low oil prices in the late 1980s, peak production from Prudhoe Bay, and the *Exxon Valdez* oil spill of 1989 together weakened the momentum to open the 1002 area. In 1995 Congress moved to allow drilling, but President Clinton vetoed the bill. In the meantime, however, new drilling around the margins of the 1002 area, as well as new technology to reprocess and better interpret vintage seismic data, led to upward revisions of estimated oil recovery. These updated estimates suggested that a significantly greater volume of oil could be economically recovered than previously believed [6].

In 1998, following estimates of technically recoverable oil and natural gas liquids from ANWR, the United States Geological Survey (USGS) indicated that there was a 95% probability that at least 5.7 billion barrels of oil are recoverable and there was a 5% probability that at least 16 billion barrels of oil are recoverable. The USGS calculated that once all the oil has been discovered, more than 80% of the technically recoverable oil is commercially developable at an oil price of $25 per barrel [7].

The opening up of ANWR to oil and gas development can have several significant impacts, including reducing world oil prices, reducing the US dependence on imported foreign oil, improving the US balance of trade, and increasing US jobs. These potential positive impacts should be weighed against a number of uncertainties, particularly regarding both the size and quality of the oil resources that exist in ANWR and potential environmental changes [8]. For example, more than 35 species of land and marine mammals, as well as over 100 species of birds, have been identified in the ANWR area [9]. The impact of oil and gas exploration on this wildlife is uncertain.

The future of ANWR will be shaped by a number of geological, political, and economic forces. These include technological development, variations in oil prices, and changes in legislation. One prediction is almost certain: the debate over ANWR will not cease, no matter what decision is eventually made, or not made. Given this uncertainty, more attention has been given to deep-water drilling in the Gulf of Mexico (GoM).

There is little doubt as to the importance of deep-water sites for the oil industry. Most geologists believe that the world's oceans and seas contain a great many deep-water basins with undiscovered oil reserves. Advanced technologies have greatly reduced the risks and costs of finding and producing oil at deep-water sites, increasing their attractiveness. Many companies see investment in deep-water and ultra-deep-water prospects to be the best chance to replace their depleting reserve bases. The offshore United States is an important source of domestic oil and gas. More than 25% of domestic production of oil comes from the offshore areas of six states: Alabama, Alaska, California, Louisiana, Mississippi, and Texas [10]. The GoM is one of the largest single sources of oil and gas supply to the US market.

Offshore exploration is divided into shallow- and deep-water exploration. The Minerals Management Service (MMS) considers projects in less than 1000 feet (305 meters) water depth to be shallow-water wells and those above are deep-water ones [11]. The first offshore well on the GoM shelf was drilled in 1947 off Louisiana. Since then many fields have been discovered and developed. Most of the largest fields have already been developed; however, the gulf shelf displays significant resilience, continuing to make substantial contributions to the US oil and gas industry. Largely due to technological advances, smaller fields have increasingly become economic to discover and develop [12].

A major milestone was reached early in 2000 when more oil was produced from the deep-water GoM than from the shallow-water GoM [13]. The enacting of the Deep Water Royalty Relief Act in 1995 paved the way for such significant development. The legislation provided economic incentives for operators to develop fields in water depths greater than 656 ft (200 m). Deep-water production began in 1979 with Shell's Cognac field, but it took another five years before the next deep-water field (ExxonMobil's Lena field) came on line. Both developments relied on extending the limits of platform technology used to develop the GoM shallow-water areas [14]. Deep-water exploration and production grew with the tremendous advances in technology since those early days. One of the first impacts was a dramatic increase in the acquisition of three-dimensional (3-D) seismic data. Such 3-D seismic data are huge volumes of digital energy recordings resulting from the transmission and reflection of sound waves through the earth. These large "data cubes" can be interpreted to reveal likely oil and gas accumulations [15].

In addition to technological advances, several governmental initiatives have had an impact on the evolution of offshore exploration. In 1982, the MMS was created as a bureau within the Department of Interior to manage the GoM Outer Continental Shelf (OCS) mineral resources in an environmentally responsible manner. One of the MMS's first steps was to introduce area-wide leasing, which greatly expanded the available OCS areas of interest to industry. However, accelerated leasing in the GoM was offset by legislation that established a leasing moratorium [16]. In December 2006, Congress passed, and President George W. Bush signed, the US Gulf of Mexico Energy Security Act into law. The measure gave access to 8.3 million acres in the Eastern and Central Gulf, while providing a 125 mile (200 km) buffer for the Florida coast.

The deep-water arena has made great strides in the last few years, establishing itself as an expanding frontier. The future of deep-water GoM exploration and production seems very promising. This future is likely to be influenced by advances in technology, costs, and environmental issues. This promise was severely tested on 20 April, 2010 when a drilling rig exploded at the Macondo oil field, operated by BP, off the coast of Louisiana. The explosion produced the biggest oil spill in US history. In response President Obama issued a moratorium halting drilling in more than 500 ft of water in the GoM and off Alaska.

The uncertainty regarding sufficient indigenous oil production in conjunction with real and potential disruption of foreign supplies laid the ground to create the strategic petroleum reserve (SPR). The US SPR is the largest stockpile of government-owned emergency crude oil in the world. It is considered the nation's first line of defense against an interruption in petroleum supplies. It is an emergency supply of crude oil stored in huge underground salt caverns along the coastline of the GoM. This location was chosen as an oil storage site because it is also the location of many US refineries and distribution points for tankers, barges, and pipelines.

The need for a national oil storage reserve has been recognized for several decades. In 1944, Secretary of the Interior Harold Ickes advocated the stockpiling of emergency crude oil. In 1952, the Truman administration proposed a strategic oil supply and President Eisenhower suggested an oil reserve after the 1956 Suez Crisis in the Middle East. The Nixon administration made a similar recommendation. The 1973–1974 oil embargo was the turning point. President Ford set the SPR into motion when he signed the Energy Policy and Conservation Act (EPCA) in December 1975. The legislation declared it to be US policy to establish a reserve of up to 1 billion barrels of petroleum [17]. The Energy Policy Act of 2005 directed the Secretary of Energy to fill the SPR to its authorized 1 billion barrel capacity [18].

Since its establishment the SPR has released crude oil on a few occasions. In 1985 Congress and the administration decided to conduct test sales for up to 5 million barrels that involved the private sector and a competitive bidding process for the first time. Following the Iraqi occupation of Kuwait in 1990, the United States launched Operation Desert Storm and President George H.W. Bush ordered the release of oil from the SPR in early 1991. In the aftermath of the 2005 hurricane Katrina, President George W. Bush authorized the sales of crude oil from the SPR to compensate for the massive destruction of oil production facilities in the GoM. In addition to these emergencies, oil has been released from the SPR several times under exchange arrangements with private companies [19].

In recent years some energy analysts have questioned the role of the SPR in pursuing an energy security strategy. The SPR is popular with politicians because it allows them to tell their constituents that they have done something to protect the nation against a future oil embargo. It is also popular with many economists because it is seen as a hedge against the economic impact of future supply disruptions.

On the other hand, opponents of the SPR raise at least three objections. First, the operation of the SPR has become politically controversial in recent years. The central question is whether the SPR should be used only during a "national emergency" or whether it should be used occasionally as a means to alleviate high domestic oil and gasoline prices [20]. Second, the total cost of the SPR includes the costs of building and operating the storage facilities, facility depreciation costs, the purchase of the oil, the higher oil prices that result from reserve additions, and the opportunity cost of holding the oil until a disruption occurs. Some argue that the SPR is the world's "costliest system of oil caches and it is a tremendous waste of money" [21]. Finally, some analysts suggest that the idea of having a government-owned-and-operated inventory has lost any attraction due to the increasing globalization of oil markets. The thought of using oil as a weapon to achieve political goals is outdated [22]. Oil will always be available to those willing to pay the posted price in global spot markets.

Better management of strategic reserves will not by itself eliminate the excessive dependence of the United States on oil. Solving that problem will require a comprehensive strategy that limits overall demand for oil, develops more sources of supplies, and encourages the use of alternative types of energy.

2.2   Natural Gas

Natural gas was used extensively in North America in the nineteenth century as a lighting fuel, until the rapid development of electricity beginning in the 1890s ended that era. The development of steel pipelines and related equipment, which allowed large volumes of gas to be easily and safely transported over many miles, launched the modern natural gas industry. The first all-welded pipeline over 200 miles (320 km) in length was built in 1925, from Louisiana to Texas. Demand for natural gas grew rapidly thereafter, especially following World War II. Residential demand grew 50-fold between 1906 and 1970 [3].

The United States had large natural gas reserves and was essentially self-sufficient in natural gas until the late 1980s, when consumption began to significantly outpace production. Imports rose to make up the difference, nearly all coming by pipeline from Canada. North American international trade predominantly occurs through major long-haul pipeline systems originating from supply basins in British Columbia, Alberta, and Nova Scotia, Canada. Although ownership of individual pipeline networks often changes at the US–Canadian border, the integrated network transports Canadian gas to markets in nearly every major northern consuming region of the United States. On the other border, Mexican supplies are currently inadequate to meet the country's demand. As a result, Mexico currently relies on the United States for a small volume of imports. Demand for natural gas in both Canada and Mexico is rising, which restrains US imports from North America.

These restraints on the North American market have provided incentives to increase the volume of imported gas in the form of liquefied natural gas (LNG). LNG is not new to the United States. Natural gas was stored as LNG for the first time commercially in West Virginia in 1912. Before the United States began importing LNG from overseas, liquefaction technology was used to condense natural gas so that it could be stored more easily. Natural gas is stored as LNG until it is needed, at which time it is converted back to gas and shipped via pipeline to market.

In 1959, the industry transported the first cargo of LNG by ship. The *Methane Pioneer* carried a cargo of LNG from Lake Charles, Louisiana to Canvey Island in the United Kingdom [23]. The United States exported natural gas for a number of years, but in the 1970s natural gas demand rose, due to a number of factors – primarily consumers' desire for independence from oil shocks. During this period, the United States began using more of its own gas. Between 1971 and 1980, four import terminals were built in the United States. The terminals are located at Lake Charles, Louisiana; Everett, Massachusetts; Elba Island, Georgia; and Cove Point, Maryland. In 2003, the Federal Energy Regulatory Commission, which has jurisdiction over onshore facilities, approved the construction of Sempra's Cameron LNG terminal, in Louisiana, the first regulatory approval for an LNG import terminal in 25 years. The second approval was granted in June 2004 to the Freeport LNG project in Texas.

In the 1980s, deregulation of the federal government's price controls on natural gas resulted in an increase of supplies domestically, which caused the natural gas companies to temporarily shut down the LNG import terminals of Elba Island and Cove Point. The terminals at Lake Charles and Everett suffered from very low utilization. New LNG supplies from Trinidad and Tobago and increasing natural gas demand contributed to the Elba Island and Cove Point LNG import terminals' return to operations in 2001 and 2003 respectively. In 2005, an offshore facility, Gulf Gateway Energy Bridge, was added in the GoM to allow for additional imports.

Trinidad and Tobago is a Caribbean nation that is rich in both oil and gas resources, but has in the past concentrated its efforts on exploiting crude oil deposits. In the 1990s, however, the owner of the Distrigas LNG import facility in Everett, Massachusetts, approached the Trinidad and Tobago government with the idea of developing its gas resources through the construction of an LNG production plant. With producers such as British Gas (BG) and British Petroleum (BP) supportive of the proposal, construction of the first LNG plant began in 1996. Three years later, a first train became operational in May 1999, liquefying gas transported to the facility from offshore fields to the south-east of the island [24]. Since then other trains have become operational.

A long-term supply agreement with the Distrigas facility in Everett, Massachusetts, was successfully negotiated. The backers of the project also anticipated that the proximity of the plant to US markets would provide a commercial opportunity for short-term or "spot" sales, into the largest natural gas market in the world. Thus, Trinidad and Tobago has become the largest source of LNG to the US market.

In the early 2000s, several African countries, including Egypt, Nigeria, and Algeria, also became suppliers of LNG to the United States. Other LNG trading partners include Equatorial Guinea, where Marathon Oil Corporation operates an LNG plant on Bioko Island; Norway, through a contract with StatoilHydro AS; Qatar, the largest LNG exporter in the world; and Yemen [25].

In the late 2000s, natural gas became an important component of the US energy mix representing approximately 22% of total primary energy use. It holds an important place in the US electricity market as the second largest source of fuel after coal and the fastest growing fuel for power generation. Approximately 19% of all electricity generated in the United States is derived from the burning of natural gas. In addition, many industrial users switched to natural gas in the 1980s. Natural gas represents 41% of the fuel used in that sector and has become a popular fuel among residential users for heating and cooking. Approximately 50% of Americans heat their homes with natural gas [26].

In addition to these sectors, natural gas has made some modest advances in the transportation sector. The technology for natural gas vehicles (NGVs) has been available since World War II. Recently, NGVs have increased rapidly, primarily because of environmental concerns and the fluctuation of crude, gasoline and diesel prices. The Department of Energy expects alternative-fueled vehicles to account for 10% of the US automobile market by 2020. To reach that goal, the US government is offering incentives, including fuel excise tax credits for equivalent compressed natural gas and LNG fuels. It also offers tax credits for infrastructure costs. The most established US niche market for gas as a transportation fuel is gas-powered transit bus systems. In 2006, 15% of transit vehicles were powered by natural gas [27].

Natural gas is widely used and is considered the fuel of choice for many reasons. It is considered more secure than oil. The United States holds large gas deposits and most of its gas imports come from Canada and a few other suppliers. In other words, the potential of using natural gas for political intimidation is negligible. Second, gas is environmentally cleaner than both oil and coal. Indeed, gas is one of the earth's cleanest sources of energy. Third, it is competitively priced compared to oil, nuclear power, and renewable energy.

Natural gas production in the United States peaked in the early 1970s, then fell for a decade due to weak prices and declining gas fields in Texas, Louisiana, and elsewhere. Production bounced back in the 1990s with the discovery of new fields in New Mexico and Wyoming, but by 2002 output was falling again. Many analysts thought US natural gas indigenous production would never recover and companies such as ConocoPhillips, El Paso Corp., and Cheniere Energy Inc. spent billions on LNG terminals, pipelines and storage facilities.

The supply fears drove up prices, which spurred innovation. Oil and gas companies had known for decades that there was gas trapped in shale, a non-porous rock common in much of the United States but considered too dense to produce much gas [28]. Energy companies developed two technologies. One is horizontal drilling, in which, instead of merely drilling straight down into the resource, horizontal wells go sideways after a certain depth, opening up a much larger area of the resource-bearing formation. The other technology is known as hydraulic fracturing, or "fraccing." Here, the producer injects a mixture of water and sand at high pressure to create multiple fractures through the rock, liberating the trapped gas to flow into the well [29]. These shale formations were developed in Texas, Louisiana, and elsewhere, and consequently, the decline in production was reversed. The US energy industry says there is enough untapped domestic natural gas to last a century – but getting to that gas requires the injection of millions of gallons of water into the ground to crack open the dense rocks holding the deposits. As a result, US natural gas production has risen dramatically since 2005. Unconventional natural gas is the largest contributor to this growth in production. This rising production drove prices down and since the late 2000s natural gas has gained market share at the expense of oil, coal, and other fuels. Equally dramatic is the effect on US reserves. Proven reserves have risen to 245 trillion cubic feet (tcf) in 2008 from 177 tcf in 2000, despite having produced nearly 165 tcf during those years [31].

2.3 Coal

US coal consumption rose slowly in the nineteenth century. However, the arrival of the Industrial Revolution and the development of the railroads in the mid-nineteenth century inaugurated a period of generally growing production and consumption of coal that continues to the present. From 1885 through 1951, coal was the leading source of energy produced in the United States. Crude oil and natural gas then vied for that role.

Since 1950, the United States has produced more coal than it has consumed. The excess production allowed it to become a significant exporter of coal to other nations. It exports coal to a large number of countries, particularly Canada, Brazil, Japan, Italy, and the United Kingdom.

The consumption of coal in the United States has changed dramatically over the years. Initially, most coal was used in the industrial sector, but many homes were still heated by coal and the transportation sector still consumed significant amounts in steam-driven trains and ships. For the last few decades, coal has gradually lost market share to oil and natural gas in the residential, commercial and transportation sectors. In power generation, however, coal still enjoys a dominant role, contributing about half of the nation's electricity [32]. Coal's share of total US electricity generation (including electricity produced at combined heat and power plants in the industrial and commercial sectors) is projected to slightly decline from 49% in 2006 to 47% in 2030 [33].

Increasing coal use for electricity generation at new and existing plants, combined with the startup of several coal-to-liquids (CTL) plants, will lead to modest growth in US coal consumption, averaging 0.7% per year from 2006 to 2030. Although an assumed risk premium for carbon-intensive technologies dampens investment in new coal-fired power plants, the projected increase in coal-fired electricity generation still is larger than for any other fuel. Increased generation from coal-fired power plants will account for 39% of the growth in total US electricity generation from 2006 to 2030 [34].

The future role of coal in the US energy mix will be shaped by at least three dynamics – geological, economic, and environmental. First, the United States has abundant deposits of coal, almost one-third of the world's total [35]. Second, compared to other sources of energy, coal is the least expensive of fossil fuels. Third, it is one of the most polluting fuels. Coal-fired electricity generating units emit gases that are of environmental concern. The timely development of clean-coal technology is crucial to the industry's future.

2.4 Nuclear Power

Since the early 2000s, concern about greenhouse gas emissions and energy security combined with forecasts of strong growth in electricity demand has awakened dormant interest in nuclear energy in the United States and around the world. Yet, the industry has not yet fully addressed the issues that have kept global nuclear energy capacity roughly the same since the early 1980s. Although nuclear safety has improved significantly, nuclear energy's inherent vulnerabilities regarding waste disposal, economic competitiveness, and proliferation remain. Moreover, nuclear security concerns have increased since the September 11, 2001 terrorist attacks.

The United States represents a special case. The nation has the most nuclear capacity and generation among the 31 countries in the world that have commercial nuclear power. In 2010 there are 104 commercial nuclear reactors at 65 nuclear power plants in 31 states [36]. Since 1990, the share of the nation's total electricity supply provided by nuclear power generation has averaged about 20%, with the level of nuclear generation growing at roughly the same rate as overall electricity use. Nuclear power output is expected to grow, but at a slightly lower rate than total electricity generation.

In 1951 an experimental reactor sponsored by the US AEC generated the first electricity from nuclear power. The British completed the first operable commercial reactor, at Calder Hall, in 1956. The US Shippingport unit, a design based on power plants used in nuclear submarines, followed a year later. In cooperation with the US electric utility industry, reactor manufacturers then built several demonstration plants and made commitments to build additional plants at fixed prices. This commitment helped launch commercial nuclear power in the United States [3].

The success of the demonstration plants and the growing awareness of US dependence on imported oil led to a wave of enthusiasm for nuclear electric power that sent orders for reactor units soaring. In the heyday of nuclear power 41 orders for nuclear power plants were placed in just one year (1973) [37]. All commercial nuclear reactors operating in the United States fall into two broad categories: pressurized-water reactors and boiling-water reactors. Because both types of reactors are cooled and moderated with ordinary "light" water, the two designs are often grouped collectively as light-water reactors [38]. The number of operable units increased in turn, as ordered units were constructed, tested, licensed for full power operation, and connected to the electricity grid. In addition, nuclear generation increased as a result of higher utilization of existing capacity and from technical modifications to increase nuclear plant capacity (expressed in megawatts).

Orders for new units fell off sharply after 1974. No new orders were placed after 1978. Although safety concerns, especially after the accident at Three Mile Island in 1979, reinforced a growing wariness of nuclear power, the chief reason for its declining momentum was economic. The promise of nuclear electric power had been that it would make energy "too cheap to meter" [39]. In reality, nuclear power plants have always been costly to build. Furthermore, many units were forced to undertake costly design changes and equipment retrofits, partially as a result of the Three Mile Island accident. Meanwhile, nuclear power plants have also had to compete with conventional coal or natural gas-fired plants with declining operating costs.

Since then, the US nuclear power industry has steadily improved its safety record and operating capacities and has lowered operating costs. Reactors with 40-year operating lives have been allowed to extend another 20 years. Since 2001 national policy has supported new nuclear reactors, providing tax incentives, streamlined licensing, and funds for advanced research and development.

The Nuclear Regulatory Commission (NRC) is responsible for supervising the construction and operation of nuclear reactors. The roots of the NRC go back to the Atomic Energy Act of 1946, which created the AEC. Eight years later, Congress replaced that law with the Atomic Energy Act of 1954, which for the first time made the development of commercial nuclear power possible. The AEC's regulatory programs sought to ensure public health and safety from the hazards of nuclear power without imposing excessive requirements that would inhibit the growth of the industry. By 1974, the AEC's regulatory programs had come under such strong attack that Congress decided to abolish the AEC. Supporters and critics of nuclear power agreed that the promotional and regulatory duties of the AEC should be assigned to different agencies. The Energy Reorganization Act of 1974 created the NRC. It began operations on

January 19, 1975 [40]. The NRC, like the AEC before it, focuses its attention on several broad issues that are essential to protecting public health and safety. These issues include costs, managing spent fuel, and proliferation concerns.

Costs: Projections of costs for building a single nuclear power plant range from $5 billion to $12 billion [41]. Before 1979, it took an average of seven years for plants to go on line. By 1990, the average lag from groundbreaking to operation had reached 12 years [42]. The delays, in turn, have been widely attributed to a ratcheting up of regulatory requirements for health, safety, and environmental reasons. As construction stretches over several years to a decade, a number of things can unpredictably raise the price tag. For example, prices for necessary commodities – such as steel, cooper, and concrete – have risen significantly in the past few years. Thus, capital cost is one of the most important factors determining the economic competitiveness of nuclear energy. On the other hand, nuclear reactors are not expensive to operate compared to other types of fuel [43]. Consequently, nuclear plants tend to be used to their maximum capacity, whereas natural gas plants tend to be turned on intermittently to serve peak demand. This distinction between capital costs and operating costs is usually taken into consideration in determining the feasibility of nuclear power.

Managing spent fuel: Nuclear power's environmental appeal is that it produces less carbon than coal or natural gas. Nuclear power generation itself does not contribute to airborne emissions of carbon dioxide, although related activities such as the production of nuclear fuel for reactors do result in CO_2 emissions. Nuclear power, however, faces another environmental challenge – how to manage spent fuel. Nuclear waste is a solid waste that must be carefully stored because it is radioactive and can contaminate anything with which it comes into contact.

To manage this waste, two spent fuel strategies are being used or under consideration by the countries that generate nuclear power. First, reprocessing of the spent fuel, with the separated plutonium or enriched uranium either stored indefinitely for possible future use or recycled as mixed oxide fuel. Second, interim storage of the spent fuel with the object either of ultimate direct disposal in geological repositories or of making a later decision on ultimate disposal. The United States pursues the second strategy. Currently, most high-level commercial nuclear wastes are stored on-site at 72 nuclear plants in 31 states [44].

In the 1980s, when the Department of Energy (DOE) searched for places to bury waste from civilian reactors and the nuclear weapons program, the idea was for two repositories, one in the west and one in the east. The DOE listed a dozen sites in seven states, ranging from Maine to Minnesota and Mississippi [45]. Congress eventually ordered the DOE to focus on Yucca Mountain in Nevada. In February 2002 President George W. Bush authorized construction of the nuclear waste storage saying that the long-debated project was "essential to the future of the nuclear power industry and the nation's security" [46].

Several years later, the future of this repository is in doubt. Yucca Mountain is an area of active earthquakes and volcanoes. Equally important, people and politicians in Nevada, a state without a single nuclear plant, have expressed strong opposition to the project. The hope is that some time in the future there will be the technical capability to break down the fuel into its constituent components and treat it separately so some can be reused, while some remains as radioactive waste, but no more toxic than the ore that the fuel was produced from, with a non-toxic by-product.

Proliferation concerns: The Non-Proliferation Treaty (NPT) acknowledges that countries have the right to research and develop peaceful nuclear energy. The concerns over global warming and the fluctuation of oil prices are shared by all energy consumers. These concerns have renewed interests in nuclear power all over the world. The challenge facing the United States and other global powers is how to prevent countries with no nuclear weapons capabilities from converting nuclear materials (enriched uranium or platinum) and know-how from civilian to military uses.

In pursuing such a delicate balance, the United States proposed in 2006 the creation of the Global Nuclear Energy Partnership (GNEP). This initiative envisions a consortium of nations with advanced nuclear technology that would provide fuel services and reactors to countries which agree to refrain from fuel cycle activities such as enrichment and reprocessing. In other words, the GNEP is essentially a fuel leasing approach, wherein the supplier takes responsibility for the final disposition of the spent fuel. It would divide the world into two classes: fuel supplier states and fuel client states. The supplier states would provide fuel and perhaps other services such as spent-fuel reprocessing and waste disposal to the client states so that the latter would not have to undertake any activities that could readily power weapons programs [47].

While the current generation of nuclear power plant designs provides an economically, technically, and publicly acceptable electricity supply in many markets, further advances in nuclear energy system design can broaden the opportunities for the use of nuclear energy. The DOE participates in the Generation IV International Forum (GIF), an association of 13 nations that seek to develop a new generation of commercial nuclear reactor designs before 2030. In February 2005, the United States, Canada, France, Japan, and the United Kingdom signed an agreement for additional collaborative research and development of Generation IV systems. The goal is to develop an ultra-safe economic nuclear system that will be designed to produce electricity and hydrogen with substantially less waste and without emitting air pollutants or greenhouse gases [48].

2.5   Ethanol

Biofuels are liquid fuels produced from biomass materials and are used primarily for transportation. The term biofuel most commonly refers to ethanol and biodiesel. Biofuels are made by converting various forms of biomass such as corn or animal fat into liquid fuels and can be used as replacements or additives for gasoline or diesel.

Ethanol has been used intermittently as an octane booster over the years, but received renewed interest with the Energy Tax Act of 1978. Since then ethanol has enjoyed various generous tax breaks, without which it would not be commercial. In the 1980s, Congress passed several inducements to ethanol plant construction, imposed a protectionist import fee, and increased the gasoline excise exemption and income tax credit for blenders [49]. In recent years, several new federal laws designed to increase the production and consumption of domestic biofuels have been enacted. The Energy Policy Act of 2005 established the Renewable Fuel Standard, which mandated that transportation fuels sold in the United States contain a minimum volume of renewable fuels, the level of which increases yearly until 2022. In December 2007, the Energy Independence and Security Act of 2007 increased the mandatory levels of renewable fuel blending credits to a total of 36 billion gallons (136 billion liters) by 2022.

Proponents of ethanol claim that biofuels are superior to fossil fuels – they are renewable, non-toxic, and biodegradable. They are available from a variety of sources – corn, sugar cane, and recently from cellulose. The latter, in the form of existing agricultural and wood waste, is abundant, inexpensive and requires no additional land. It has no food or feed value and therefore no effect on food availability and prices. A number of technologies are pursued for production of cellulosic ethanol and other biofuels. Proponents of ethanol also assert that the US corn ethanol industry is investing in technology improvements to reduce land demand through higher productivity and to minimize its carbon footprint. Cellulosic ethanol will come from existing waste materials, not additional land.

Opponents, on the other hand, argue that biofuels receive heavy governmental subsidies, without which they cannot survive in a competitive market. Ethanol also has contributed to higher food prices because millions of acres of farmland have been diverted to ethanol from food production. Rising demand for ethanol has increased demand for corn, the price of which has dramatically increased. Furthermore, corn acreage has increased at the expense of other crops. The prices of these other crops have risen because of smaller plantings [50]. Finally, opponents claim that the reduction in CO_2 emissions from burning ethanol is minimal and may be negative [51]. After accounting for the energy used to grow the corn and turn it into ethanol, corn ethanol lowers emissions of greenhouse gases by only a negligible rate [52].

These arguments for and against biofuels aside, gasoline replacement by ethanol is constrained by three factors. First, the gasoline ethanol distribution infrastructure does not deliver ethanol for gasoline blending everywhere in the country. Second, there are physical limitations on existing vehicles as to how much ethanol they can use in combination with gasoline. Third, ethanol contains approximately 67% of the energy content of gasoline per gallon, therefore usage of ethanol blends results in decreased gas mileage. This brief review of the US energy deposits suggests that the nation has a variety of options in formulating an energy strategy to meet its large and growing needs. The efforts to articulate and implement a cohesive energy policy have expanded for several decades with a mixed success.

2.6   The Quest for an Energy Strategy

Since the 1930s, policy-makers in Washington have sought to formulate an energy policy. Under Franklin D. Roosevelt's administration, there was a strong belief that the government could not solve the economic problems facing the nation without playing a role in energy policy, which was considered a vital factor in the economic recovery [53]. The intention was not to nationalize or make the industry public, but to coordinate its activities. With US involvement in World War II, the struggle over the formulation of a governmental energy policy (particularly oil) intensified. Despite the heavy drain on its oil supplies during the war, the United States still occupied a strong position with respect to petroleum. This position, however, was gradually and steadily weakened.

President Dwight D. Eisenhower was convinced that the growing share of imported oil in US energy consumption represented a challenge to the country's national security and its prominent role in world affairs. His energy policy had two objectives. It aimed at reducing the share of foreign oil in total consumption and relied more on oil supplies from Canada and Mexico than from faraway producers. Thus, after two years of requesting voluntary import quotas, which oil companies did not comply with, the president made it mandatory in 1959. Under this program the exporting countries were divided into separate groups, depending on preferential treatment. Western hemisphere exporters were favored [54].

The impact of this mandatory import quotas program on US oil policy was mixed. The United States became relatively independent of foreign oil reserves during most of the 1960s. Accordingly, most of its cheap oil reserves were utilized and thereby exhausted. The program stimulated production levels that eroded domestic reserves rather than creating stockpiles and spare capacity. In the late 1960s and early 1970s, oil companies found that it was more profitable to pay additional import fees than to use domestic oil, since domestic production costs were higher than the total cost of imported oil plus the import fee.

The Nixon, Ford, and Carter administrations had to deal with some of the most serious energy crises. In the early 1970s, US domestic oil production began its steady decline and the country's dependence on imported oil increased. In a symbolic move, Richard Nixon announced that, because of the energy crisis, the lights on the national Christmas tree would not be turned on. In addition, he signed the Emergency Highway Conservation Act, setting a speed limit of 55 miles per hour (89 km/h). Most importantly, Nixon announced a plan called "Project Independence," the aim of which was to develop domestic resources to meet the nation's energy needs without depending on foreign suppliers. He wanted to achieve a state of self-sufficiency within a decade. This initiative proposed measures to stimulate investments in domestic oil production, including decontrol of domestic energy prices and subsidizing domestic oil by imposing fees on imported oil. This attempt to achieve self-sufficiency in energy supply was never achieved.

Gerald Ford signed the EPCA, which authorized the establishment of the SPR. Ford also came out in favor of a windfall profits tax on domestic petroleum, the decentralization of oil prices, and an increased reliance on coal, electricity, and nuclear power.

Coming to office in January 1977, Jimmy Carter judged the energy crisis to be a national emergency and offered a program to deal with it – a program that he asked the nation to accept as the "moral equivalent of war." Probably more than his predecessors, Carter focused on the demand side than the supply side of the energy equation. His program called for reduced overall energy consumption, significantly reduced imports, increased reliance on coal and renewable sources of energy like solar, wind, and wood, higher gasoline taxes and various tax credits, and incentives to encourage more efficient automobiles and home insulation. Also, at the president's request, Congress created a new cabinet post, the Department of Energy, in 1977.

The collapse of oil prices that followed the global oil glut in the mid-1980s undermined the sense of urgency to take drastic action to control and restrain the US appetite for more energy. Throughout the 1980s and 1990s, the centerpiece of US energy policy was to foster at home and abroad deregulated markets that efficiently allocated capital, provided a maximum of consumer choice, and promoted low prices through competition [55]. Few significant, new, federal energy policy initiatives emerged during Ronald Reagan's administration or the first years of George H.W. Bush's administration. Both presidents completed the process of deregulating oil and natural gas commodity prices.

The debate about energy policy continued in the early 1990s, although public concern about high oil prices, potential shortages, and dependence on imported oil faded quickly with the end of the Gulf War (1991). Soon after his inauguration, President Clinton proposed the implementation of a large, broad-based tax on energy. The proposal sought to raise revenue to reduce the federal budget deficit, to promote energy conservation, and indirectly to reduce pollution associated with the combustion of fossil fuels. The proposal was widely criticized in

Congress, was unpopular with industry and consumers, and eventually failed. No new major energy policy legislation was passed by Congress during the rest of the decade [56].

The rise of oil prices in the late 1990s and President George W. Bush's and Vice President Dick Cheney's involvement in the oil industry prior to taking office have put energy at the top of the administration's policy. In his second week in office, the president established the National Energy Policy Development Group, headed by the vice president, directing it to develop a national energy policy. After four years of long negotiations between policy-makers in Washington, both houses of Congress approved an energy bill and the president signed it into law in August 2005. The Energy Policy Act of 2005 (Public Law 109-58), more than 1700 pages long, includes the following provisions:

- The Act does not open ANWR to oil and gas leasing. This highly controversial issue is still subject to debate and bargaining between policy-makers and environmentalists.
- The Act requires that amounts of renewable fuel be blended into the nation's gasoline supply. These renewable fuels include ethanol and fuel derived from wood, plants, grasses, agricultural residues, fibers, animal waste, and municipal solid waste.
- The Act does not impose any limits on greenhouse gases, new inventory, or credit trading schemes. It creates a new cabinet-level advisory committee to develop a national policy to address climate change and to promote technologies to reduce greenhouse gas emissions.
- The Act expands daylight saving time (DST) by about a month. Effective in 2007, DST begins the second Sunday in March (instead of the first Sunday in April) continuing through the first Sunday in November (instead of the last Sunday in October).
- The Act contains $14.5 billion in tax incentives. These tax provisions aim at making capital investments in new technology, plant, and equipment cheaper. They also include a two-year extension of the wind energy tax credit and a 30% solar energy tax credit. In addition, the Act significantly expands the federal role in the process of government review and permitting of liquefied natural gas terminals.
- The Act provides incentives to generate electricity from advanced nuclear power plants. It includes several provisions aimed at promoting new construction of nuclear power plants.
- The Act creates new investment tax credits for advanced clean-coal facilities. It authorizes $200 million per year for fiscal years 2006–2014 for distribution by the Secretary of Energy to projects that use or develop clean-coal technology [57].

In short, the Energy Policy Act of 2005 provides incentives to encourage investments in fuel efficiency, renewable sources, clean-coal technology, and nuclear plants. In December 2007 President Bush signed into law the Energy Independence and Security Act (EISA). The EISA is meant to help reduce US dependence on oil by reducing demand for oil by setting a national fuel economy standard of 35 miles per gallon (14.9 km/l) by 2020 – which will increase fuel economy standards by 40% and save billions of gallons of fuel. The bill includes provisions to improve energy efficiency in lighting and appliances, as well as requirements for federal agency efficiency and renewable energy use that will help reduce greenhouse gas emissions [58].

This interest in containing pollution highlights the evolution of US policy on climate change. President Bill Clinton signed the Kyoto Protocol but did not submit it to the Senate for ratification because of strong opposition to the deal, which did not impose greenhouse gas limits on China and other developing economies. President George W. Bush did not submit the Kyoto Treaty for ratification, and largely resisted calls for stronger action on climate change.

That approach began to crumble in 2007, when the Supreme Court found that CO_2 is a pollutant under the Clean Air Act. The Court held that the Environmental Protection Agency (EPA) must determine whether or not emissions of greenhouse gases from new motor vehicles cause or contribute to air pollution which may reasonably be anticipated to endanger public health or welfare, or whether the science is too uncertain to make a reasoned decision [59]. After a thorough scientific review, the EPA issued a proposed finding that greenhouse gases contribute to air pollution that may endanger public health or welfare [60]. In May 2009 the Obama administration set a new national standard that would raise the average fuel efficiency of a new car by 30% by 2016 [61]. A few months later (December 2009), the EPA announced that six gases including carbon dioxide, methane, and nitrous oxide pose a danger to the environment and the health of Americans and that the EPA would draw up regulations to reduce those emissions [62].

2.7   Conclusion: The Way Forward

The energy challenges facing the United States are complex. They range from geological (availability or lack of indigenous supplies) to economic (providing incentives for investment), environmental (addressing climate change issues), and geopolitical (forging relations with foreign suppliers) challenges, among others. Washington's response to these multi-dimensional challenges is a broad combination of measures that addresses both the demand and supply sides of the energy equation. These include improved efficiency, cleaner technology, stimulated investment, and a diversified energy mix and sources, among others.

Given its size, large economy, status as the world's only superpower, the United States is the largest participant in the global energy system. Despite these economic and geopolitical dynamics, US policy-makers have frequently called for energy independence for several decades. The phrase has been a recurrent cry since it was first articulated by President Nixon in 1973. The allure of energy independence is easy to understand. It reinforces the belief that Americans can control their own economic destiny and appeals to a deep-seated cultural feeling that the United States will not be intimidated by hostile foreign powers [63].

These calls for energy independence are unrealistic and, indeed, misleading and counter-productive. It is an empty slogan, not a reasonable ambition for the United States or any other country in the world [64]. A distinction between dependence and vulnerability is needed. Automobile ownership per capita is the highest in the United States. Its mobility runs on oil. The transportation sector uses about two-thirds of the petroleum consumed in the United States. Oil dependence does not necessarily mean that it is vulnerable to an oil disruption. The distinction between dependence and vulnerability suggests that concentration is a key factor in the security of energy supplies.

The United States imports energy from a variety of sources. As energy (particularly oil) became a strategic commodity, the United States considered its flow a matter of national security. Within this context, promoting political and economic stability in major energy producing regions would enhance US energy security. The globalization of energy markets means that a disruption anywhere is a disruption everywhere. There can be no US energy security without a global one.

References

[1] Department of Energy, *Origins & Evolution of the Department of Energy*. Available at http://www.energy.gov/about/print/origins.htm (accessed August 22, 2009).

[2] Merklein, H. and Caruso, G. (2004) The U.S. energy information administration. *Middle East Economic Survey*, **42** (14). Available at http://www.mees.com.

[3] Energy Information Administration, *Energy in the United States: 1653-2000*. Available at http://www.eia.doe.gov (accessed October 8, 2001).

[4] Montgomery, S.L. (2003) ANWR 1002 area and development: one question, many issues. *Oil & Gas Journal*, **101** (15), 38–43, 38.

[5] Williams, B. (2001) New era dawning for Alaskan North Slope. *Oil and Gas Journal*, **99** (32), 58–66, 60–61.

[6] Montgomery, S.L. (2003) Geologic assessment and production forecasts for the ANWR 1002 area. *Oil & Gas Journal*, **101** (16), 35–40, 37.

[7] Energy Information Administration (2000) *Potential Oil Production from the Coastal Plain of the Arctic National Wildlife Refuge: Updated Assessment*, US Department of Energy, Washington, DC, p. 13.

[8] Energy Information Administration (2008) *Analysis of Crude Oil Production in the Arctic National Wildlife Refuge*, US Department of Energy, Washington, DC, pp. 9–12.

[9] Montgomery, S.L. (2003) Wildlife resources and vulnerabilities summarized for 1002 Area of ANWR. *Oil & Gas Journal*, **101** (17), 34–39, 34.

[10] Kallaur, C. (2002) U.S. offshore oil and gas policy: A fresh look is needed. *Oil & Gas Journal*, **100** (32), 28–29, 28.

[11] Minerals Management Service (2004) *Gulf of Mexico Oil and Gas Production Forecast: 2004-2013*, Department of Interior, Washington, DC, p. 2.

[12] Kuuskraa, V., Godec, M., and Kuck, B. (2002) Shallow water gulf oil, gas supply: a glass half full or half empty? *Oil & Gas Journal*, **100** (26), 34–42, 38.

[13] Minerals Management Service (2004) *Deepwater Gulf of Mexico 2004: America's Expanding Frontier*, Department of Interior, Washington, DC, p. 1.

[14] Godec, M., Kuuskraa, V., and Kuck, B. (2002) How U.S. Gulf of Mexico development, finding, cost trends have evolved. *Oil & Gas Journal*, **100** (18), 52–60, 54.

[15] Minerals Management Service (2004) *Deepwater Gulf of Mexico 2004: America's Expanding Frontier*, Department of Interior, Washington, DC, p. 6.

[16] Godec, M., Kuuskraa, V., and Bank, G. (2002) Future gulf supplies: role of the federal government. *Oil & Gas Journal*, **100** (36), 32–38, 35.

[17] Department of Energy, *Strategic Petroleum Reserve*. Available at http://www.fe.doe.gov/spr/spr.html (accessed August 11, 2000).

[18] Department of Energy, *US Strategic Reserve*. Available at http://www.fe.doe.gov/programs/reserves. (accessed February 5, 2006).

[19] Department of Energy, *Releasing Crude Oil from the Strategic Petroleum Reserve*. Available at http://www.fe.doe.gov/programs/reserves/spr/spr-drawdown.html (accessed February 5, 2006).

[20] Taylor, J. and Van Doren, P. (2005) *The Case against the Strategic Petroleum Reserve*, Cato Institute, Washington, DC, p. 2.

[21] Victor, D. and Eskreis-Winkler, S. (2008) In the tank: making the most of strategic oil reserves. *Foreign Affairs*, **87** (4), 70–83, 71.

[22] Emerson, S. (2001) SPR draw downs trigger law of unintended consequences. *Oil & Gas Journal*, **99** (50), 24–30, 24.

[23] Center for Liquefied Natural Gas Essential Energy, *The History of LNG in the U.S.* Available at http://www.ingfacts.org/About-LNG/History.asp (accessed April 8, 2007).

[24] Gaul, D., Energy Information Administration, *U.S. Natural Gas Imports and Exports: Issues and Trends 2003*. Available at http://www.eia.doe.gov (accessed August 24, 2009).

[25] Energy Information Administration, *What Is Liquefied Natural Gas and How Is it Becoming an Energy Source for the United States?* Available at http://www.eia.doe.gov (accessed July 10, 2008).

[26] Baker, J. III (2008) *Natural Gas in North America: Markets and Security*, Institute for Public Policy of Rice University, Houston, TX, p. 1.

[27] Fletcher, S. (2009) Natural gas vehicles gain in global markets. *Oil & Gas Journal*, **107** (7), 20–24, 21.

[28] Casselman, B. (2009) U.S. Gas Fields Go From Bust to Boom. Wall Street Journal (April 30).

[29] Casselman, B. and Gold, R. (2010) Drilling Tactic Unleashes a Trove of Natural Gas and a Backlash. Wall Street Journal (January 21).

[30] Energy Information Administration (2009) *Annual Energy Outlook 2009*, United States Government Printing Office, Washington, DC, p. 77.

[31] Yergin, D. and Ineson, R. (2009) America's Natural Gas Revolution. Wall Street Journal (November 2).

[32] International Energy Agency, *Energy Policies of IEA Countries – the United States*. Available at http://www.iea.org (accessed March 14, 2008).

[33] Energy Information Administration (2009) *Annual Energy Outlook 2009*, United States Government Printing Office, Washington, DC, p. 83.

[34] Energy Information Administration (2009) *International Energy Outlook 2009*, United States Government Printing Office, Washington, DC, p. 50.

[35] British Petroleum (2009) *BP Statistical Review of World Energy*, British Petroleum, London, p. 32.

[36] Energy Information Administration, *What Is the Status of the U.S. Nuclear Industry?* Available at http://www.eia.doe.gov (accessed August 10, 2009).

[37] Squassoni, S. (2007) Risks and realities: the "new nuclear energy revival." *Arms Control Today*, **35** (5), 1–13, 8.

[38] Energy Information Administration, *New Reactor Designs*. Available at http://www.eia.doe.gov (accessed April 17, 2006).

[39] Wald, M. (2006) Slow Start for Revival of Nuclear Reactors. New York Times (August 22).

[40] Nuclear Regulatory Commission, *Our History*. Available at http://www.nrc.gov/about-nrc/history.html (accessed August 4, 2007).

[41] Johnson, T., Council on Foreign Relations, *Challenges for Nuclear Power Expansion*. Available at http://www.cfr.org/publication/16886/nuclear_bottlenecks.html (accessed August 11, 2008).

[42] Nivola, P. (2004) *The Political Economy of Nuclear Energy in the United States*, Brookings Institution, Washington, DC, p. 3.

[43] Lask, E. (2004) U.S. nuclear power: a nuclear future inevitable. *Petroleum Economist*, **71** (11), 4–7, 4.

[44] Lask, E. (2006) Nuclear makes U.S. comeback: the need for energy independence and fuel diversity, environmental concerns and growing public support are behind a resurgence of nuclear power in the U.S. *Petroleum Economist*, **73** (1), 16–17, 17.

[45] Wald, M. (2008) U.S. Decides One Nuclear Dump Is Enough. New York Times (November 7).

[46] Pianin, E. (2002) Nevada Nuclear Site Is Affirmed. Washington Post (February 16).

[47] Ferguson, C.D. (2007) *Nuclear Energy: Balancing Benefits and Risks*, Council on Foreign Relations, New York, p. 23.

[48] Office of Nuclear Energy, Science and Technology, *Bush Administration Moves Forward to Develop Next Generation Nuclear Energy Systems*. Available at http://nuclear.energy.gov/genIV/neGenIV1.html (accessed March 16, 2005).

[49] Editorial (2009) Ethanol's new rescue cry. *Oil & Gas Journal*, **107** (12), p. 16.

[50] Kimins, L. (2007) Energy system limits future ethanol growth. *Oil & Gas Journal*, **105** (44), 18–22, 19.

[51] Wall Street Journal (2009) Editorial: Ethanol's Grocery Bill. Wall Street Journal (June 2).

[52] New York Times (2007) Editorial: The High Costs of Ethanol. New York Times (September 19).

[53] Karlsson, S. (1986) *Oil and the World Order: American Foreign Oil Policy*, Barnes & Noble, Totowa, NJ, p. 36.

[54] Chester, E. (1983) *United States Oil Policy and Diplomacy: A Twentieth-Century Overview*, Greenwood Press, Westport, CT, p. 35.

[55] Morse, E. and Jaffe, A.M., James A. Baker III Institute for Public Policy of Rice University, *Strategic Energy Policy Challenges for the 21st Century*. Available at http://www.bakerinstitue.com (accessed April 16, 2001).

[56] Joskow, P. (2002) United States energy policy during the 1990s. *Current History*, **101** (653), 105–125, 115.

[57] Government Printing Office, *Energy Policy Act of 2005*. Available at http://frwebgate.access.gpo.gov (accessed August 26, 2009).

[58] Council on Foreign Relations, *Energy Independence and Security Act of 2007*. Available at http://www.cfr.org (accessed December 19, 2007).

[59] Environmental Protection Agency, *Proposed Endangerment and Cause or Contribute Findings for Greenhouse Gases under the Clean Air Act*. Available at http://epa.gov/climatechange/endangerment.html (accessed April 19, 2009).

[60] Environmental Protection Agency, *EPA Finds Greenhouse Gases Pose Threat to Public Health, Welfare*. Available at http://yosemite.epa.gov (accessed April 18, 2009).

[61] Broder, J. (2009) Obama Sets New Auto Emissions and Mileage Rules. New York Times (May 20).

[62] Mufson, S. and Fahrenthold, D.A. (2009) EPA Is Preparing to Regulate Emissions in Congress's Stead. Washington Post (December 8).

[63] Fialka, J. (2006) Energy Independence: A Dry Hole? Wall Street Journal (July 5).

[64] Editorial (2006) Democrats and energy. *Oil & Gas Journal*, **104** (44), 19.

3

European Union

The citizens of the 27 nations that constitute the European Union (EU) enjoy one of the highest standards of living in the world. Their economies, and indeed way of life, run on sustainable energy supplies, particularly fossil fuels. The EU suffers from a severe shortage of indigenous energy deposits. This combination of high energy demand and limited production means that the EU is very dependent on foreign supplies.

This deep European dependence on foreign supplies was interrupted by the 1973–1974 Arab oil embargo, when a few European countries, along with the United States, were punished for supporting Israel. The price spike in the mid-1970s and early 1980s (following the Iranian Revolution and the Iran–Iraq War) dealt a heavy blow to European economies and energy security. Furthermore, since the early 2000s, the EU's giant neighbor – Russia – has frequently used its hydrocarbon resources as a political weapon to punish or reward the former Soviet republics, particularly Ukraine. Russia's policy caused severe interruptions of hydrocarbon supplies to several Member States of the EU and, once again, highlighted Europe's energy vulnerability.

The European response to the use of energy as a political weapon by some of the major suppliers has essentially followed three directions: (a) diversification of energy mix; (b) diversification of energy sources; and (c) development of indigenous sources [1]. In order to achieve these goals, the European Commission (EC) has issued a number of comprehensive long-term strategies, most notably the 2008 Energy Security and Solidarity Action Plan (ESSAP). The ESSAP underscores five major themes:

- promoting infrastructure essential to the EU's energy needs;
- a greater focus on energy in the EU's international relations;
- improved oil and gas stocks and crisis response mechanisms;
- a new impetus on energy efficiency; and
- making better use of the EU's indigenous energy reserves [2].

If fully implemented, the package is expected to fundamentally alter the EU's energy outlook. Among its objectives, the ESSAP aims at cutting EU consumption by 15%, reducing imports by 26%, and increasing the contribution of renewable sources in the bloc's energy mix, which will reduce pollution [3].

Energy Security: An Interdisciplinary Approach, First Edition. Gawdat Bahgat.
© 2011 John Wiley & Sons, Ltd. Published 2011 by John Wiley & Sons, Ltd.

In the following sections I examine the EU energy outlook and highlight the large and growing gap between the EU's energy demand and indigenous production. In addressing this gap, the EU has actively pursued a strategy aimed at forging a partnership with major energy producers. These partnerships seek not only to ensure the non-interruption of energy supplies, but also to promote economic and political stability. Such partnerships have been established with Russia, Central Asia/the Caspian Sea, the Mediterranean Sea/North Africa, Arab states on the Persian Gulf, and Turkey.

3.1   The EU Energy Outlook

Most of Europe's indigenous supplies of oil and gas come from the North Sea. The interest in the basin started in the mid-1970s in reaction to the first oil shock. However, the North Sea did not emerge as a key, non-OPEC oil producing area until the 1980s and 1990s, when major projects began coming on-stream. A major impediment to exploration and development operations in the region is the inhospitable climate. In addition, most of the hydrocarbon deposits are located at great depths that require expensive and sophisticated technology. On the positive side, unlike some other major producing regions, Europe enjoys political, financial and legal stability. Such stability and predictability have provided the appropriate investment environment to develop the region's oil and gas deposits.

Five countries operate crude oil and natural gas production facilities in the North Sea: Denmark, Germany, the Netherlands, Norway, and the United Kingdom (UK) [4]. However, it is Norway and the UK that claim most of the action. Offshore oil production in the North Sea as a whole peaked at 6.4 million b/d in 2000, but natural gas production continues to rise [5]. The UK's oil and natural gas production peaked in 1999 and 2000 respectively and has since entered a period of steady decline. With a smaller population and larger reserves, Norway has adopted a policy of carefully managing reserve development. Its oil production peaked in 2001 but natural gas production is growing [6]. Currently, Norway has maintained its status as the largest gas producer and exporter in Western Europe [7].

The role of the North Sea as a major oil and gas producing region and a significant supplier to Europe is far from ending. True, production has declined in the last several years and no major discoveries have been made, but small fields have been found and enhanced techniques have been applied to extract more oil and gas from old fields. The future of the North Sea depends on at least three interrelated dynamics: availability of investments; advances in technology; and oil prices (low prices make operations in the North Sea unprofitable) [8].

At the end of the first decade of the twenty-first century, the EU's energy production mix is dominated by nuclear power (30%), followed by coal (22%), gas (20%), oil (14%), and renewable sources (14%) [9], although the contribution of the latter is expected to increase significantly in the future in line with the ambitious EU policy targets. This energy is largely used in the transportation, industrial, and residential sectors.

3.1.1   Oil

Europe's energy security will continue to depend strongly on the availability of primary energy sources. In the current EU energy mix, oil, gas, coal, and uranium are the major primary energy sources. At the end of the 2000s, the EU's energy consumption is dominated by fossil fuels, which represent about 80% of the energy demand. This figure is likely to remain relatively stable up to 2020. Following a steady increase due to growing consumption for transportation purposes, oil provides the largest contribution to the EU's energy mix (approximately 37%). Oil is likely to remain the most important fuel in 2020 due to limited substitution in the transportation sector.

The EU holds a negligible share of the world's total proven reserves (approximately 5.5 billion barrels or 1.1%) [10]. These reserves are mainly located in the North Sea area (Norway, UK, and Denmark) and in South-East Europe (Romania). The reserve-to-production ratio (R/P), or the current production rate these proven reserves secure, is less than eight years. Given these very modest proven reserves, the EU produces less than one-fifth of its total oil consumption. Most oil imports come from OPEC, Russia, Norway, and Kazakhstan.

Management of emergency stocks is a key element of the EU emergency response system. The EC is responsible for coordinating efforts for ensuring maintenance of emergency stocks and adequacy of response to both internal and external supply disruptions. This includes gathering and publishing regular oil stocks data from Member States [11].

3.1.2   Natural Gas

Concerns about the interruption of oil supplies, restrictions on CO_2 emissions, uncertainty about nuclear power safety, high emissions from coal-based generation, and barriers to rapid development of renewable generation are some of the main reasons for Europe's high dependency on natural gas [12]. The combination of all these factors has made natural gas the fuel of choice for many Europeans. However, the EU holds very limited proven reserves: approximately 2.77 billion cubic meters, or 1.6% of the world's total [13]. These reserves are mainly located in Norway, the Netherlands, the UK, and Romania. At current production rates, the EU's proven reserves secure about 14.4 years of domestic production.

Natural gas accounts for almost one-quarter of the EU's energy mix. Member States which import gas to cover all, or very nearly all, gas consumption are: Belgium, Bulgaria, Czech Republic, Estonia, Lithuania, Latvia, Greece, Spain, France, Luxembourg, Portugal, Slovenia, Slovak Republic, Finland, and Sweden. Gas is not used in Malta and Cyprus. The bloc's natural gas production satisfies about two-fifths of consumption. The only Member States which produce more gas than they consume are the Netherlands and Denmark [14]. Gas is mainly imported from four big suppliers: Russia, Norway, Algeria, and Nigeria.

Natural gas has penetrated all sectors, with the exception of transportation. A lot of the new investments in power generation have gone into combined cycle gas turbine technology. Low natural gas prices, relatively low capital cost, and favorable technology characteristics, especially in terms of environmental emissions, are the drivers of this development.

The EU's small natural gas production and heavy dependence on foreign supplies have highlighted the bloc's vulnerability to interruption. The EC has taken several initiatives to reduce this vulnerability. For example, in 2004 the EU adopted a Directive on security of natural gas supply. This Directive defines general, transparent and non-discriminatory security of supply policies [15]. In order to monitor the security of supply situation and provide a coordination mechanism in case of supply crisis, the Directive establishes a Gas Coordination Group. This is a forum for Member States, the gas industry, and gas customers to exchange information and debate policy developments.

In order to further improve the security of gas supplies and deal with potential crises, the EC adopted new regulations in 2009. These regulations call on Member States to be fully prepared in case of supply disruption, through clear and effective emergency plans involving all stakeholders and incorporating fully the EU dimension of any significant disruption. The regulations also provide a common indicator to define a serious gas supply disruption such as the shutdown of a major supply infrastructure or equivalent (e.g., import pipeline or production facility). The regulations require all Member States to have a competent authority that would be responsible for monitoring gas supply developments, assessing risks to supplies, establishing preventive action plans, and setting up emergency plans. It also obliges Member States to collaborate closely in a crisis, including through shared access to reliable supply information and data [16].

3.1.3   Coal

Proven reserves of coal are much more abundant than those of oil or natural gas, although they represent only a limited share of world reserves. The EU's proven reserves represent 3.5% of the world's total, which would secure 50 years of today's production. Most of these reserves are concentrated in Poland, Czech Republic, Spain, Hungary, and Germany. Most of the imports come from Russia, South Africa, Australia, Colombia, Indonesia, and the United States.

Coal is used mainly in the power generation sector and in some specific industrial applications. The future of coal in the EU will be largely shaped by the growing popular and governmental concern about pollution and technological advances in creating "clean coal." In October 2009 the EC announced that up to six clean-coal power stations will be built across the EU. The EC allocated €1 billion of funding for these carbon capture projects. The technology involves burying greenhouse gas emissions deep underground. The six power stations will be built in the UK, Germany, the Netherlands, Poland, Spain, and Italy [17].

3.1.4   Nuclear Power

Nuclear power plays an important role in the EU energy mix. With 152 reactors spread over the 27 Member States, the EU is the largest nuclear electricity generator in the world, having a mature nuclear industry, spanning the entire fuel cycle with its own technological base and highly skilled workforce. Nuclear energy is currently the main low-carbon source in many Member States, providing more than a third of the EU electricity, and it has proven to be a stable, reliable source, relatively shielded from price fluctuations when compared to the oil and gas markets [18]. Continued use of nuclear energy therefore would contribute to EU energy security as well as to the limitation of CO_2 emissions, but it is also still confronted by a number of outstanding issues that need to be resolved.

The ground for nuclear energy in Europe was laid in 1957 by the European Atomic Energy Community (EURATOM) [19]. Its main functions consisted of furthering cooperation in the field of research, protecting the public by establishing common safety standards, ensuring an adequate and equitable supply of ores and nuclear fuel, monitoring the peaceful use of nuclear material, and cooperating with other countries and international organizations [20].

One of the most crucial factors affecting the prospect of growth of nuclear power is its underlying economics as a nuclear plant involves an upfront investment ranging from €2 to €3 billion. Nuclear energy generation incurs higher construction costs in comparison to fossil fuels, yet operating costs are significantly lower following the initial investments. Furthermore, nuclear power generation is largely immune to changes in the cost of raw material supplies, as a modest amount of uranium can keep a reactor running for decades. Identified sources of uranium in the EU are modest, only about 1.9% of the world's total. Denmark, France, Spain, and Sweden hold the largest deposits within the EU. Its indigenous sources are substantially complemented by reprocessed and re-enriched sources. Therefore, in most industrialized countries new nuclear power plants offer an economic way to generate baseload electricity.

The EURATOM Treaty does not address the particular aspects of nuclear installation safety. As a result, regulatory activities in these areas have developed along national lines under the responsibilities of national authorities. However, with the development of the European nuclear industry, convergence at EC level became necessary in order to support the Member States in their efforts to harmonize safety practices. In July 1975, the European Council issued a resolution on the technological problems of nuclear safety, recognizing that it was the EC's responsibility to act as a catalyst in initiatives taken at international level in the field of nuclear safety. The resolution also called for the harmonization of safety requirement at the EC level. In June 1992 another resolution called on Member States to continue and intensify concerted efforts towards harmonization of safety issues [21].

Another significant development was the adoption of new proposals of Directives dealing with the safety of nuclear facilities and the management of spent fuel and radioactive waste, leading to the creation of the Council Working Party on Nuclear Safety in June 2004 [22]. The EC also undertook consultations with international organizations, such as the International Atomic Energy Agency (IAEA) and the Organization for Economic Cooperation and Development (OECD) Nuclear Energy Agency. Such consultations and efforts led to the establishment of the European Nuclear Energy Forum (ENEF) in 2007. The ENEF involves key decision-makers and organizations at national and EU levels. Its main responsibility is to enhance a better understanding of common approaches that are required in the further development of the safety of nuclear installations [23].

In June 2009, the European Council adopted the Nuclear Safety Directive. This Directive brings legal certainty by clarifying the responsibilities between national authorities and the EC. It sets binding principles for enhancing nuclear safety to protect workers and the general public, as well as the environment. The Directive enhances the role and independence of national regulatory authorities, confirming that license holders bear the prime responsibility for nuclear safety. Member States are required to encourage a high level of transparency of regulatory actions and to guarantee regular independent safety assessments [24].

Given these efforts to improve nuclear safety and the management of spent fuel, as well as the fact that nuclear power is basically free of carbon emissions, several European countries have recently reconsidered their nuclear power policies. Decisions to expand nuclear energy were recently taken in Finland and in France. Similarly, the UK government launched one of the world's most ambitious expansions of nuclear power capacity. The government of Prime Minister Gordon Brown identified 10 sites in England and Wales for new nuclear plants, with the first expected to come online by 2018. Many of the plants are envisioned to replace aging plants that are set to be decommissioned in coming years and are a vestige of a period of accelerated nuclear construction from the 1950s to 1980s [25]. Other EU countries, including the Netherlands, Italy, Poland, Sweden, Czech Republic, Lithuania, Estonia, Latvia, Slovakia, Bulgaria, and Romania have relaunched a debate on their nuclear energy policy. It is important to point out that many restrictions on nuclear power in Europe (and elsewhere) were imposed after the accidents at Three Mile Island in the United States in 1979 and Chernobyl in the Soviet Union in 1986.

To conclude, the EU, like most of the world, is witnessing what can be called a "nuclear renaissance." This renewed interest in nuclear power is fueled by efforts to improve the bloc's energy security, carbon emissions, and nuclear safety. Nuclear power has the potential to make a great contribution to the EU's pursuit of energy security and its future seems promising.

3.1.5   Renewable Sources

Renewable sources of energy – wind power, solar power (thermal and photovoltaic), hydro-electric power, tidal power, geothermal energy and biomass – are an essential alternative to fossil fuels. Using these sources helps not only to reduce greenhouse gas emissions from energy generation and consumption, but also to reduce the EU's dependence on imports of fossil fuels (in particular oil and gas). The share of these renewable sources in the overall EU energy mix is modest, but renewable energy has witnessed the highest growth rate since the 1990s and is projected to maintain this high rate in the foreseeable future.

Europe's leadership in climate change policy has been nearly uncontested over the past few years. In 1997 the EC published a White Paper on renewable energy, "Energy for the future: renewable sources of energy." By issuing this document, the EC aimed to stimulate consultations and discussions with all parties interested in more widespread use of renewable energy sources [26]. The creation of this renewable energy policy was founded on the need to address sustainability concerns surrounding climate change and air pollution, improve the security of Europe's energy supply, and develop Europe's competitiveness and industrial and technological innovation. The White Paper also announced a renewable energy strategy and action plan, highlighting the need to develop all renewable energy resources, create stable policy frameworks, and improve planning regimes and electricity grid access for renewable energy.

Another important step highlighting the EU's concern about climate change was taken in 2003 when the European Parliament and Council adopted the Emission Trading Scheme (ETS). Under the ETS, CO_2 emissions from large sources like power plants and factories are capped, and facilities must have emissions allowances equal to their annual emissions. If they emit more than they have been allocated by the government, they must buy allowances on the carbon market; if they emit less, they can sell their allowances. They may also use project-offset credits from other countries for a portion of their needs [27].

In 2007 the European Council endorsed the so-called "20–20–20 Initiative." This policy refers to cutting greenhouse gases by at least 20% to 1990 levels; increasing use of renewable sources to 20% of total energy production; and cutting energy consumption by 20% by improving energy efficiency, all by 2020. The EC estimates that fulfilling these goals will save between 600 and 900 million tons of CO_2 a year and reduce European annual fossil-fuel demand by over 250 million tons of oil equivalent [28]. Furthermore, at the Copenhagen Climate Conference, held in December 2009, the EU reiterated its conditional offer to move to a 30% reduction by 2020 compared to 1990 levels, provided that other developed countries committed themselves to comparable emission reductions and that developing countries contributed adequately according to their responsibilities and respective capabilities [29].

In 2009 the European Council adopted the climate–energy legislative package. The Directive for the first time sets for each Member State a mandatory national target for the overall share of energy from renewable sources in gross final consumption of energy, taking account of countries' different starting points. The main purpose of mandatory national targets is to provide certainty for investors and to encourage technological developments allowing for energy production from all types of renewable sources. The Directive also adopted a revised ETS for greenhouse gases in order to achieve greater emissions reductions in energy-intensive sectors [30].

All these initiatives and laws have stimulated and encouraged the utilization of renewable sources. For example, several Member States, particularly Germany and Hungary, have made good progress in so-called green electricity (generating electricity from wind and biomass). Similarly, biofuels production has expanded and its share in the transportation sector has increased several fold. For the heating and cooling sector, biomass, solar, thermal and geothermal energy substantially increased their share [31].

This growing European interest in renewable energy is driven by deep concerns about climate change and a strong desire to improve the security and reliability of energy supplies. In addition, investments in renewable sources are helping to create the so-called "green jobs" – jobs in clean energy.

Three conclusions can be drawn from this brief survey of the EU's energy policy. First, EU energy consumption continues to be dominated by fossil fuels – 37% oil, 24% natural gas, 18% coal, 14% nuclear power, and 7% renewable sources [32]. Second, given the EU's very limited proven hydrocarbon reserves, the bloc is projected to remain deeply dependent on foreign supplies to meet its large and growing energy needs. In 2008, it imported 54% of its energy. These imports represent an estimated €350 billion, or around €700 per year for every EU citizen [33]. Oil comprises the bulk of total energy imports (60%) followed by imports of gas (26%) and coal (13%). The EU's energy dependence is projected to stabilize at 56% by 2020, but its reliance on imported fossil fuels will grow deeper (oil 92%, natural gas 74%, and coal 57%) [34]. Third, energy dependence varies widely from one Member State to another, both by fuel and by source. Stated differently, few states produce all their energy needs while others import almost all their fuels. Furthermore, most East European countries and former Soviet republics are more dependent on Russia than other Member States.

This heavy dependence on a few suppliers to meet the EU's energy needs has underscored the bloc's potential vulnerability. Little wonder that energy security has become a major drive of European policy. As José Manuel Durão Barroso, President of the EC, asserts, the issue of energy security is "at the top of the EU's political agenda. It has moved from being a rather theoretical concern of experts to an issue of practical concern to EU citizens" [35].

In the last few years, energy has become a prominent issue in nearly all the EU's external political relations. Generally, EU policy seeks to avoid backing authoritarian regimes and leaders. Instead, its approach is to forge a partnership with major energy producers based on mutual interest. In addition to securing Europe's energy needs, these partnerships aim at promoting political and economic reforms. In recent years the EU has established strategic dialogues with Russia, Central Asia, the Mediterranean Sea, the Arab states on the Persian Gulf, and Turkey.

3.2 Russia

Compared to other oil and gas producers, Russia enjoys two important advantages. First, it holds massive hydrocarbon deposits. Accordingly Russia is the world's largest natural gas producer and second largest oil producer (after Saudi Arabia). Second, geographical proximity to European markets adds another attraction. Thus, in the last few decades Russia has emerged as the major energy partner to several Member States and to the EU as a whole. This EU–Russia energy partnership is further reinforced by close economic and political ties. The EU has more high-level political dialogue with Russia than with any other third country except the United States [36]. On the commercial level, Russia is the EU's third largest trade partner (after the United States and China [37]) and the EU is Russia's most important trading partner and source of foreign investment.

The Cold War of the 1940s and 1950s created a hostile environment for any cooperation between East and West. Thus, in the aftermath of World War II, the Middle East emerged as the main energy supplier to Europe. Several geopolitical developments altered Europe's strategic landscape in the late 1960s and early 1970s. These include the beginning of the détente between the Soviet Union on one side and both the United States and Europe on the other. This reduced security and heightened political tension was driven by and reinforced economic cooperation, including the energy sector. The 1973–1974 Arab oil embargo highlighted Europe's energy vulnerability and strengthened the momentum for diversifying oil and gas sources.

Within this strategic context, several pipelines were built to ship Russia's oil and gas to Europe. This growing energy interdependence between the two sides was further institutionalized with the signing of the Energy Charter Treaty (ECT) in December 1994, which entered into force in April 1998. The ECT is a legally binding multilateral agreement covering the whole energy value chain, from exploration to end-use, and all energy products and energy-related equipment. The ECT is designed to promote energy security through the operation of more open and competitive energy markets. Its provisions focus on four broad areas:

- The protection of foreign investments, based on the extension of national treatment, or most favored-nation treatment (whichever is more favorable), and protection against key non-commercial risks.
- Non-discriminatory conditions for trade in energy materials, products, and energy-related equipment based on World Trade Organization (WTO) rules, and provisions to ensure reliable cross-border energy transit flows through pipelines, grids, and other means of transportation.
- The resolution of disputes between participating states, and – in the case of investments – between investors and host states.
- The promotion of energy efficiency and attempts to minimize the environmental impact of energy production and use [38].

Russia signed the ECT but has never ratified it, partly because it would have obliged Russia to give up its near monopolistic position in energy transit within the former Soviet territory where Central Asian energy producers rely on Russian-controlled pipelines. The EU has spent years trying to get Russia to abide by the provisions of the ECT, which aspires to provide transparent and market-based rules for international energy cooperation. As such, it would oblige Russia to open up the development of its hydrocarbon reserves and the running of its pipelines to foreign commercial involvement. In recent years, the Kremlin has argued that the ECT is outdated and should be replaced by an agreement that would balance the interests of producers, consumers, and transit states. Finally, in August 2009 Prime Minister Vladimir Putin signed an order rejecting Russia's participation in the ECT [39].

In October 2000, Brussels and Moscow launched an energy dialogue to discuss issues of common interest, including the introduction of cooperation on energy saving, rationalization of production and transportation infrastructures, European investment possibilities, and relations between producer and consumer countries [40]. The dialogue involved regular meetings of experts, as well as high-level political discussions during the annual EU–Russia summits. Since then, the two sides have achieved some progress. For example, a technology center in Moscow was established and several pilot projects for energy saving were launched. However, Brussels and Moscow have not reached an agreement on such important issues as pipelines and electricity sector restructuring.

Most notably, the two sides have failed to adequately address the interruption of Russia's natural gas supplies. As early as 1990, Moscow cut energy supplies to the Baltic states in a futile attempt to stifle their independence movement. A similar episode took place in 1992 in retaliation for Baltic demands that Russia remove its remaining military forces from the region. In 1993 and 1994, Russia punished Ukraine, the conduit for about 80% of Russia's gas exports to Europe, by reducing gas supplies to force it to pay for previous supplies and to pressure Kiev into ceding more control to Russia over the Black Sea Fleet and over Ukraine's energy infrastructure. A similar technique was applied to Belarus in 2004. In December 2005 and December 2006, Russia again cut or threatened to cut gas supplies to Ukraine and Belarus respectively to demand higher prices. In January 2009, Russia again cut off gas deliveries to Ukraine.

This frequent cutting off of Russia's gas to Ukraine underscores Kiev's strategic location and significant role as a transit state between Russia and the EU. As mentioned above, about 80% of Russia's natural gas exports to Europe cross Ukrainian territory. Similarly, Kiev's oil pipeline network can transmit roughly 1 million b/d (nearly 7% of total EU demand) to Central and East European destinations [41]. These well-developed pipelines were built during the Soviet era and reflect the strong economic and political ties that connected Moscow and Kiev. However, these traditional ties have been challenged by significant changes in Ukraine's domestic and foreign policy orientation as had been demonstrated by the so-called Orange Revolution in 2005.

Some Ukrainian leaders have sought to distance their country from Russia and forge closer relationships with the EU and the United States. Kiev has sought to join the EU and North Atlantic Treaty Organization (NATO). Such sentiments have triggered Moscow's suspicion. Indeed, Russia's cutting off of gas supplies to Ukraine in the last few years can be seen as an attempt to punish Kiev for its growing relations with the West. Furthermore, Russia has sought to undermine Ukraine's negotiating leverage by building and negotiating a number of pipelines that would bypass it. These pipelines include Blue Stream (already operating at capacity), and Nord Stream and South Stream (under negotiation and construction).

Ukraine has also sought to weaken Russia's bargaining leverage by proposing a pipeline called White Stream, an alternative project for supplies of Central Asian gas to Europe. The scheme would run from Turkmenistan through Azerbaijan to the Georgian port of Supsa and then along the seabed of the Black Sea and through Crimea to the European countries, bypassing Russia [42].

On the other hand, the EU vowed to help upgrade Ukraine's network of gas pipelines in exchange for changes in the country's energy management to avoid a repeat of the disputes that resulted in the cutting off of Russian gas. In March 2009, Brussels and Kiev signed an agreement to improve both the management and capacity of Ukraine's 40-year-old grid of gas pipelines [43]. Russia condemned this agreement as an "unfriendly act" against Moscow and demanded that its interests not be ignored [44].

In a similar development, the EU has established an Eastern Partnership with the former Soviet republics of Armenia, Azerbaijan, Georgia, Moldova, Ukraine, and Belarus – countries that Russia regards as its area of influence. The goal is to increase cooperation in four key areas: democracy and the rule of law, the harmonization of economic systems and rules, energy security, and people-to-people contacts, including visa liberalization [45]. The Eastern Partnership also includes comprehensive free trade agreements and gradual integration into the EU's economy. The Eastern Partnership would promote democracy and good governance; strengthen energy security; promote sector reform and environmental protection; support economic and social development; and offer additional funding for projects to reduce socio-economic imbalances and increase stability [46].

These disagreements between the EU and Russia over Ukraine, the Eastern Partnership, and other issues aside, the main characteristic of this uneasy partnership is interdependence. Russia needs European technology and investment while the EU needs secure and reliable energy supplies [47]. Russia's exports to the EU (largely hydrocarbon products) constitute a major source of government revenues – cutting off oil or gas exports to the EU would cut off a major source of income.

For the future of the EU–Russia energy partnership, several trends can be identified. First, EU overall oil and gas dependence on Russia is unlikely to grow much higher. Russia's oil production has been stagnant in the last few years and its oil fields are being rapidly depleted due to a large production rate. At the same time, gas fields and production are growing but at a slow rate and more gas has been consumed domestically. Meanwhile, EU consumption is rising. Second, given Russia's strategic location between the EU and Asia (two large and growing energy consuming markets), Europe is likely to face more intense competition from China, Japan, and South Korea over Russia's hydrocarbon exports. Third, Europe's efforts to reduce its heavy dependence on Russia are likely to succeed to a certain extent. Central Asian and Middle Eastern exports are likely to reduce this dependence and contribute to EU energy security. However, they are likely to complement, not substitute, Russian exports. Russia's massive resources and geographical proximity make it harder for other producers to compete in the European market. Fourth, due to some disagreements between the EU Member States on how to deal with Russia and also due to a lack of consensus in Moscow on the relationship with Europe, the partnership between the two sides is likely to keep witnessing cycles of cooperation and conflict. However, leaders in both Brussels and Moscow understand how much they need each other. This mutual understanding is likely to prevent any serious damage to their partnership.

3.3   Central Asia/Caspian Sea Region

The Central Asia/Caspian Sea region is one of the oldest hydrocarbon producing areas in the world. Given its massive deposits of oil and natural gas, the region has attracted European attention and investment since the early 1990s. These former Soviet republics lacked the necessary economic infrastructure, civil society, and democratic institutions. The EU partnership with Central Asia/Caspian Sea states is meant to address these issues, as well as energy cooperation.

The Baku Initiative (BI) is a policy dialogue aimed at enhancing energy cooperation between the EU and the countries of the Black Sea, the Caspian Basin, and their neighbors [48]. The BI began as a result of the conclusions reached at the Energy Ministerial Conference which took place in Baku in November 2004 [49]. The BI comprises the EU and Armenia, Azerbaijan, Belarus, Georgia, Kazakhstan, Kyrgyzstan, Moldova, Russia (observer), Ukraine, Uzbekistan, Tajikistan, Turkey, and Turkmenistan. The goals are to consolidate cooperation between the two sides, create predictable and transparent energy markets, stimulate investment and economic growth, and enhance energy security for all partner states. The participants in the BI have agreed on the following steps to achieve these goals:

- Harmonizing legal and technical standards with the aim of creating a functioning integrated energy market in accordance with EU and international legal and regulatory frameworks.
- Enhancing the safety and security of energy supplies and extending and modernizing the existing infrastructure.
- Improving energy supply and demand management through the integration of efficient and sustainable energy systems.
- Promoting the financing of commercially and environmentally viable energy projects of common interest [50].

Four working groups were created to pursue these goals and representatives from all parties have met several times since the BI was launched. It is too early to make any assessment of the outcome, but it is clear that the BI has contributed to more cooperation between the EU and the countries of the Black Sea and the Caspian Basin.

In October 2007 the Council of the EU issued a Strategy for a New Partnership between the EU and Central Asian states (Kazakhstan, Kyrgyzstan, Tajikistan, Turkmenistan, and Uzbekistan). Good governance, the rule of law, human rights, democratization, education, and training are key areas where the EU has expressed its desire to share experience and expertise. The EU pledged to support the exploration of new oil, gas, and hydropower resources, the upgrading of the existing energy infrastructure, and the construction of new pipeline routes and energy transportation networks [51].

In addition to these strategic initiatives, several pipeline schemes are either being negotiated or already under construction to ship oil and natural gas from Central Asia/the Caspian Basin to the EU. Most of these projects seek to achieve a two-fold objective. These large and growing imports are necessary to meet the EU's hydrocarbon energy consumption. Further, these pipeline routes are designed to bypass Russia and, accordingly, reduce the EU's vulnerability to supply interruption.

Norway's oil and gas giant StatoilHydro ASA and Swiss energy trading company Elektrizitats-Gesellschaft Laufenburg AG, or EGL, have formed a consortium to develop the 520 km trans-Adriatic pipeline, known as TAP [52]. The TAP will transport gas from the Caspian Sea and the Middle East via Greece and Albania and across the Adriatic Sea to Italy's Southern Puglia region and further to Western Europe [53].

The Nabucco pipeline is at the heart of European strategy to import gas from Central Asia/the Caspian Basin and the Middle East while simultaneously bypassing Russia. It will run from the Georgian/Turkish or Iranian/Turkish borders to Baumgarten in Austria [54]. In May 2009, the EU held the Southern Corridor–New Silk Road Summit with Azerbaijan, Georgia, Turkey, Turkmenistan, Uzbekistan, Kazakhstan, and Egypt – all either key suppliers of natural gas or crucial transit countries, or both. Azerbaijan, Egypt, Georgia, and Turkey signed an agreement committing themselves to the project. Two months later (July 2009), the five transit countries (Austria, Bulgaria, Hungary, Romania, and Turkey) signed a deal allowing work on the pipeline to start. The EU also gave its support to plans to build a trans-Caspian gas pipeline from Turkmenistan to Azerbaijan. This would allow Turkmenistan, Kazakhstan, and Uzbekistan to feed their gas directly into the Nabucco gas pipeline [55]. Other pipelines are discussed in the following sections.

3.4   Mediterranean Sea

The South Mediterranean shore nations of Algeria, Egypt and Libya hold significant oil and natural gas deposits and their geographical proximity, just across the Mediterranean, makes them an attractive energy partner to the EU. The three countries are major oil and gas suppliers to Europe. This energy partnership is at the heart of a broader strategic relationship between the two sides. In addition to bilateral ties, the EU Member States and their Mediterranean neighbors have engaged in several multilateral forums, a prominent one being the Barcelona Process.

The Barcelona Process was launched in November 1995 by the ministers of foreign affairs of the 15 EU members and 14 Mediterranean partners, as the framework to manage both bilateral and regional relations. It formed the basis of the Euro-Mediterranean Partnership, which has expanded and evolved into the Union for the Mediterranean. It seeks to create a Mediterranean region of peace, security, and shared prosperity. The Partnership is organized into three main dimensions. First, a political and security dialogue aimed at creating a common area of peace and stability underpinned by sustainable development, rule of law, democracy, and human rights. Second, economic and financial partnership including the gradual establishment of a free trade area aimed at promoting shared economic opportunities through sustainable and balanced socio-economic development. Third, social, cultural, and human partnership aimed at promoting and understanding an intercultural dialogue between cultures, religions, and people, and facilitating exchange between civil society and ordinary citizens, particularly women and young people [56].

The Barcelona Process evolved into the Euro-Mediterranean Partnership and was relaunched in 2008 as the Union for the Mediterranean at the Paris Summit for the Mediterranean in July. The Partnership now includes all 27 Member States of the EU, along with 16 partners across the Southern Mediterranean and the Middle East. Some of the most important innovations of the Union for the Mediterranean include the rotating co-presidency with one EU president and one president representing the Mediterranean partners, and a secretariat based in Barcelona that is responsible for identifying and promoting projects of regional, subregional and transnational value across different sectors.

The Union for the Mediterranean has launched a number of initiatives including the de-pollution of the Mediterranean Sea, the establishment of maritime and land highways, civil protection initiatives to combat natural and human-made disasters, a Mediterranean solar energy plan, the inauguration of the Euro-Mediterranean University in Slovenia, and the Mediterranean Business Development Initiative focusing on micro-, small-, and medium-sized enterprises [57].

3.5   The Gulf Cooperation Council

The Gulf Cooperation Council (GCC) is a regional organization consisting of the six Arab states on the Persian Gulf: Bahrain, Kuwait, Oman, Qatar, Saudi Arabia, and the United Arab Emirates (UAE). It was created in May 1981. Its main objectives are to enhance coordination, integration, and interconnection among its Member States [58]. The region holds massive hydrocarbon reserves and is the world's largest oil producer and exporter. The EU and the GCC signed a Cooperation Agreement in 1988. The Agreement identified three objectives: to strengthen relations by placing them in an institutional and contractual framework; to broaden and consolidate their technical cooperation in energy, trade, services, agriculture, fisheries, investment, science, technology, and the environment; and to help strengthen the process of economic development and diversification of the GCC countries and reinforce the role of the GCC in contributing to peace and stability in the region [59].

Two areas of cooperation between the EU and the GCC can be identified. First, since 1990, negotiations to sign a free trade agreement have been initiated [60]. Second, the Financing Instrument for Cooperation with Industrialized and Other High-Income Countries and Territories (ICT) entered into force on January 1, 2007. The ICT is intended to foster long-lasting political and commercial ties between all concerned parties. Among others, it covers the following areas: promote cooperation between academic and scientific communities; stimulate bilateral trade and the flow of investment; initiate dialogue between people and between non-governmental organizations (NGOs); and contribute to job creation [61].

3.6   Turkey

Turkey's strategic location between the EU, one of the world's largest energy consumers, and the Central Asia/Caspian Basin and Middle Eastern regions, with their massive oil and gas deposits, makes it an important energy bridge. Indeed, since the early 2000s Turkey has transformed itself from a transit country to a major energy hub. Oil and gas pipelines from Russia, the Caspian Basin, Iraq, and Iran reach Turkish territories and ports and from there traverse to Europe. Thus, Turkey plays a significant role in EU energy security.

Turkey holds very modest oil and natural gas reserves. Its oil production peaked in 1991 and has since continued its downward trend. There may be significant reserves under the Aegean Sea, but exploration has been limited by a long-running dispute with neighboring Greece over the sovereignty of territorial waters in the Aegean [62]. Meanwhile, most domestic gas production is used for reinjection into Turkey's oil fields as part of enhanced oil recovery projects. In short, Turkey depends on energy imports to an even bigger extent than the EU.

EU–Turkey energy relations are multi-dimensional. The EC supported the project "Encouraged" (i.e., optimization of future energy corridors between the EU and neighboring countries). The project has the following objectives: to assess the optimal energy interconnections and network infrastructure for electricity, gas and hydrogen with and through neighboring regions; to identify, quantify, and evaluate the barriers and potential benefits of a large European energy connected area; and to recommend the necessary measures to ensure and implement these energy corridors and realize a high level of network security [63].

In 2006, in cooperation with the Council of Europe Development Bank and the European Investment Bank, the EC launched the Energy Efficiency Finance Facility Initiative. The goal is to provide financial assistance to the acceding and candidate countries in increasing investments in energy efficiency. Finally, in the framework of its pre-accession assistance for Turkey, the EC provides considerable direct support to the Turkish energy sector, in particular in the areas of legislative alignment and institution building [64].

Several oil and gas pipeline schemes connecting major producers to Europe via Turkey are already in operation, under construction, or being planned. The Baku–Tbilisi–Ceyhan (BTC) pipeline is a major scheme, at a cost of almost $4 billion, and carries Caspian oil to Europe via Georgia and Turkey bypassing the crowded Bosporus and Dardanelles Straits [65]. The South Caucasus pipeline, also known as the Baku–Tbilisi–Erzurum pipeline, is a parallel natural gas pipeline to the BTC one. It carries Azeri natural gas to Turkey. Part of the gas is consumed domestically and the rest is exported to Europe. The Tabriz–Erzurum pipeline carries Iranian natural gas to Turkey. Again, part of the volume is used domestically and the rest is exported to Europe. The pipeline came on line in 2001.

An oil pipeline connecting the Turkish port of Samsun on the Black Sea with the southeastern Mediterranean port of Ceyhan is under construction. It is designed to carry Russian and Kazakh oil between the Turkish ports. It will bypass Turkey's narrow Bosporus Strait, which has seen increasingly heavy tanker traffic in the past few years and has become a major safety concern. The pipeline will also ease traffic in the Dardanelles Strait [66].

The Blue Stream natural gas pipeline connects the Russian system to Turkey via the Black Sea. The pipeline was inaugurated in December 2002. South Stream, a major natural gas pipeline that would carry Russian gas to Europe, is designed to route through Turkey's territorial waters [67].

The Turkey–Greece Interconnector (TGI) was officially inaugurated in November 2007 connecting Karacabey in Turkey with Komitini in Greece via the Sea of Marmara [68]. This pipeline is planned to extend to Italy through the Adriatic Sea and will transport natural gas from the Shah Deniz field off Azerbaijan to Europe [69].

This large and growing network of oil and natural gas pipelines connecting Russia, Central Asia/the Caspian Sea, and the Middle East with the EU underscores the crucial role that Turkey is playing in Europe's energy security. As more of these schemes become operational, Turkey's role will become even more significant.

3.7   Conclusion: The Way Ahead

This brief discussion of EU energy policy highlights some of the dilemmas facing individual Member States and the EU as a bloc. Basically, Europe lacks the necessary indigenous energy sources to sustain the high standard of living for its population. The very limited known deposits mean that the already deep dependence on foreign supplies is certain to grow even deeper in the future. European leaders, however, seem to understand that in a global economy mutual dependence, or interdependence, is inevitable.

In response, they have pursued a multilateral strategy to reduce energy vulnerability and the chances for political intimidation. This strategy emphasizes energy efficiency and diversification of the fuel mix. Most EU Member States, in cooperation with the EC and other institutions, have taken several initiatives to increase their share of renewable sources. Such a policy has a two-fold objective – improving Europe's energy security and addressing European and global climate change issues. Indeed, the EU has been a leading force in international efforts to contain pollution.

Another part of the EU energy strategy is the leading role energy plays in the EU's, and individual Member States' foreign policy. In the last few years Europe has actively sought to establish close cooperation with major energy producing nations and regions. This cooperation is based not only on short-term energy deals, but on promoting economic and political stability. The EU has provided significant financial resources to support civil society, education, and free market mechanisms in Central Asia/the Caspian Sea, the Middle East, Africa, Eastern Europe, and the former Soviet republics. These policies are likely to enhance the prospects for stability and contribute to long-term EU energy security.

References

[1] Kovacovska, L., Association of International Affairs, *European Union's Energy (In)Security – Dependency on Russia*. Available at http://www.amo.cz (accessed June 20, 2007).

[2] European Commission, *EU Energy Security and Solidarity Action Plan: 2nd Strategic Energy Review*. Available at http://ec.europa.eu/energy (accessed November 13, 2008).

[3] European Commission, *Communication from the Commission to the European Parliament: An EU Energy Security and Solidarity Action Plan*. Available at http://ec.europa.eu/energy (accessed November 13, 2008).

[4] Energy Information Administration, *Country Analysis Briefs: North Sea*. Available at http://www.eia.doe.gov/cabs (accessed January 3, 2007).

[5] Westwood, J. (2007) Basin off NW Europe offer opportunities in troubled seas. *Oil & Gas Journal*, **105** (31), 36–42, 36.

[6] British Petroleum (2009) *BP Statistical Review of World Energy*, British Petroleum, London, p. 24.

[7] Westwood, J. (2008) Discoveries, undeveloped opportunities persist as offshore Europe oil output falls. *Oil & Gas Journal*, **106** (29), 35–40, 39.

[8] Nakhle, C. (2008) Can the North Sea still save Europe? *OPEC Energy Review*, **42** (2), 123–138, 124.

[9] European Commission, *Commission Staff Working Document*. Available at http://ec.europa.eu/energy (accessed November 13, 2008).

[10] British Petroleum (2009) *BP Statistical Review of World Energy*, British Petroleum, London, p. 6.

[11] European Commission, *Oil*. Available at http://ec.europa.eu/energy/oil/index_en.htm (accessed October 11, 2008).

[12] Rademaekers, K., Slingenberg, A., and Morsy, S., *Review and analysis of EU wholesale energy markets*. Available at http://www.ecorys.com (accessed September 9, 2008).

[13] British Petroleum (2009) *BP Statistical Review of World Energy*, British Petroleum, London, p. 22.

[14] European Commission, *Q&A: Greater EU coordination of measures to improve the security of gas supplies*. Available at http://ec.europa.eu (accessed June 16, 2009).

[15] European Commission, *Security of Gas Supply*. Available at http://ec.europa.eu/energy/gas/sos/index_en.htm (accessed October 11, 2008).

[16] European Commission, *The Commission adopts new rules to prevent and deal with gas supply crises*. Available at http://ec.europa.eu (accessed July 16, 2009).

[17] Crooks, E. and Pignal, S. (2009) Brussels Backs Clean Coal. Financial Times (October 17).

[18] European Commission, *Nuclear Issues*. Available at http://ec.europa.eu/energy/nuclear/safety/index_en.htm (accessed July 16, 2006).

[19] European Commission, *Nuclear Energy*. Available at http://europa.eu/scadplus/leg/en/s14005.htm (accessed April 8, 2009).

[20] European Parliament, *Euratom Treaty*. Available at http://www.europarl.europa.eu/parliament/archive (accessed August 30, 2009).

[21] European Commission, *Laying down basic obligations and general principles on the safety of nuclear installations*. Available at http://ec.europa.eu (accessed September 8, 2004).

[22] European Commission, *Proposal for a Council Directive (Euratom) Setting up a Community framework for nuclear safety*. Available at http://ec.europa.eu (accessed October 10, 2008).

[23] European Nuclear Energy Forum, *Facts and figures*. Available at http://www.foratom.org (accessed August 30, 2009).

[24] European Commission, *The EU establishes a common binding framework on nuclear safety*. Available at http://ec.europa.eu (accessed June 25, 2009).

[25] Faiola, A. (2009) Britain Seeks to Expand Nuclear Energy. Washington Post (November 10).

[26] European Commission, *Energy for the future: renewable sources of energy*. Available at http://www.europeanenergyforum.eu (accessed August 30, 2009).

[27] Anderson, J. (2009) Can Europe catalyze climate action? *Current History*, **108** (716), 131–137, 133.

[28] Piebalgs, A., *Renewable Energy: Developing a Green Energy Market*. Available at http://ec.europa.eu/external_relations/energy/events/renewable_energy_conference_2009/index_en.htm (accessed October 9, 2009).

[29] Council of the European Union, *Presidency Conclusions on Copenhagen Climate Conference*. Available at http://www.consilium.europa.eu/Newsroon (accessed December 22, 2009).

[30] Council of the European Union, *Council adopts climate-energy legislative package*. Available at http://ec.europa.eu (accessed April 6, 2009).

[31] European Commission, *The Renewable Energy Progress Report*. Available at http://ec.europa.eu (accessed April 24, 2009).

[32] European Commission, *Commission Staff Working Document*. Available at http://ec.europa.eu (accessed November 13, 2008).

[33] European Commission, *Communication from the Commission to the European Parliament*. Available at http://ec.europa.eu (accessed November 13, 2008).

[34] European Commission, *Commission Staff Working Document*. Available at http://ec.europa.eu (accessed November 13, 2008).

[35] Barroso, J.M.D., *Towards a new energy future: showing solidarity and embracing our inter-dependence*. Available at http://ec.europa.eu (accessed April 24, 2009).

[36] Youngs, R., Center for European Policy Studies, *Europe's External Energy Policy: Between Geopolitics and the Market*. Available at http://www.ceps.eu (accessed September 14, 2008).

[37] Gomart, T. (2008) *EU–Russia Relations: Toward a Way Out of Depression*, Center for Strategic and International Studies, Washington, DC, p. 4.

[38] Energy Charter Treaty, *1994 Treaty*. Available at http://www.encharter.org (accessed September 1, 2009).

[39] Moscow Times (2009) Putin Rejects Energy Charter Treaty. Moscow Times (August 7).

[40] Grant, C. and Barysch, K., Center for European Reform, *The EU-Russia Energy Dialogue*. Available at http://www.cer.org (accessed April 17, 2009).

[41] Chow, E. and Elkind, J. (2009) Where east meets west: European gas and Ukrainian reality. *Washington Quarterly*, **32** (1), 77–92, 78.

[42] New Europe, *Tymoshenko puts new White Stream pipeline on EU table*. Available at http://www.neurope.eu (accessed February 5, 2008).

[43] Wall Street Journal (2009) EU, Ukraine in Gas-Pipeline Deal. Wall Street Journal (March 24).

[44] Moscow Times (2009) EU Pledge to Ukraine Unfriendly. Moscow Times (March 27).

[45] Radio Free Europe, *EU Announces Eastern Partnership with Former Soviet Neighbors*. Available at http://www.rferl.org (accessed December 3, 2008).

[46] European Commission, *Eastern Relations: The EU launches program to forge closer ties with six countries in Eastern Europe and the Southern Caucasus*. Available at http://ec.europa.eu/news/external_relations/090508_en.htm (accessed May 8, 2009).

[47] Smith, K.C., Center for European Policy Studies, *Security Implications of Russian Energy Policies*. Available at http://www.ceps.eu (accessed September 14, 2008).

[48] Embassy of the Republic of Azerbaijan to the Kingdom of Belgium and Grand Duchy of Luxembourg, *Baku Initiative*. Available at http://www.azembassy.be/RELATIONS-WITH-THE-EU/Baku-Initiative/24.html (accessed April 15, 2009).

[49] European Commission, *What is the Baku Process?* Available at http://delkaz.ec.europa.eu (accessed April 15, 2009).

[50] European Commission, *Baku Initiative.* Available at http://ec.europa.eu/external_relations/energy/baku_initiative/index.htm (accessed January 6, 2008).

[51] Council of the European Union, *European Union and Central Asia: Strategy for a New Partnership.* Available at www.consilium.europa.eu/uedocs/cms_data/librairie/PDF/EU_CtrlAsia_EN-RU.pdf (accessed November 11, 2008).

[52] Torello, A. (2009) Two Solutions to Same Gas Problem: Both the Planned TAP and ITGI Pipelines Face Real Obstacles in Getting Around Ukraine. Wall Street Journal (June 23).

[53] Trans-Adriatic Pipeline, *TAP Project.* Available at http://www.trans-adriatic-pipeline.com (accessed September 3, 2009).

[54] Nabucco Pipeline, *Project description.* Available at http://www.nabucco-pipeline.com (accessed September 3, 2009).

[55] Blua, A., Radio Free Europe, *EU Agrees Deal on Energy Links with Southern Corridor.* Available at http://www.rferl.org/articleprintview/1624233.html (accessed November 20, 2009).

[56] European Commission, *The Barcelona Process.* Available at http://ec.europa.eu/external_relations/euromed/barcelona_en.htm (accessed September 2, 2009).

[57] European Commission, *The Euro-Mediterranean Partnership.* Available at http://ec.europa.eu/external_ relations/euromed/index_en.htm (accessed September 2, 2009).

[58] Cooperation Council for the Arab States of the Gulf, *The Charter.* Available at http://www.gccsg.org (accessed September 3, 2009).

[59] European Council, *Cooperation Agreement between the EU and the GCC.* Available at http://eur-lex.europa.eu/lexUriServ/LexUriServ.do?uri=CELEX:21989A0225(01) (accessed September 2, 2009).

[60] European Commission, *The EU and the Gulf Cooperation Council (GCC).* Available at http://ec.europa.eu/external_relations/gulf_cooperation/index_en.htm (accessed September 2, 2009).

[61] Independent European Development Portal, *Financing Instrument for Cooperation with Industrialized and Other High-Income Countries and Territories.* Available at http://www.developmentportal.eu/wcm/subsite/snv1v2/content/view/63/89 (accessed September 3, 2009).

[62] Energy Information Administration, *Country Analysis Briefs: Turkey.* Available at http://www.eia.doe.gov (accessed April 3, 2009).

[63] European Commission, *Energy Corridors: European Union and Neighboring Countries.* Available at http://ec.europea.eu (accessed September 4, 2009).

[64] European Commission, *EU Energy Policy and Turkey.* Available at http://ec.europa.eu (accessed June 1, 2007).

[65] Energy Information Administration, *Country Analysis Briefs: Azerbaijan.* Available at http://www.eia.doe.gov/emeu/cabs/Azerbaijan/pdf.pdf (accessed November 30, 2007).

[66] Red Orbit News, *Tapco Begins Construction of Samsun Ceyhan Oil Pipeline.* Available at http://www.redorbit.com/news/display/?id=914154 (accessed April 20, 2009).

[67] Recknagel, C., Radio Free Europe, *Putin Wins Turkey's Approval of South Stream Route.* Available at http://www.rferl.org/articleprintview/1793851.html (accessed August 6, 2009).

[68] . Winrow, G. (2009) *Problems and Prospects for the Fourth Corridor: The Positions and Role of Turkey in Gas Transit to Europe*, Oxford Institute for Energy Studies, Oxford, p. 29.

[69] Editorial (2007) Greece-Turkey gas pipeline link inaugurated. *Oil & Gas Journal*, **105** (44), 10.

4

China

The People's Republic of China (henceforth China) is a major energy consumer. The International Energy Agency highlights some of the main themes of the country's energy outlook. China's primary energy demand is projected to more than double from 1.742 million ton oil equivalent (toe) in 2005 to 3.819 toe in 2030 – an average annual rate of growth of 3.2%. According to the IEA, since 2010 China has become the world's largest energy consumer (overtaking the United States). Oil imports are projected to jump from 3.5 million b/d in 2006 to 13.1 million b/d in 2030, while the share of imports in demand rises from 50% to 80%. Natural gas imports will also increase quickly, as production growth lags demand. Similarly, China became a net coal importer in 2007. By 2030 coal imports will reach 3% of its demand [1].

Figures from the *BP Statistical Review of World Energy* draw a similar picture. In 2009 proven reserves were 15.5 billion barrels (1.2% of the world's total), the production volume was 3795 million b/d (4.8% of the world's total), and the consumption level was 7999 million b/d (9.6% of the world's total). The numbers for natural gas were: proven reserves, 2.46 trillion cubic meters (1.3% of the world's total); production, 76.1 billion cubic meters (2.5% of the world's total); and consumption, 80.7 billion cubic meters (2.7% of the world's total). The figures for coal were 114 500 million tons (13.9% of the world's total), 1414.5 million toe (42.5% of the world's total), and 1406.3 million toe (42.6% of the world's total) [2].

Finally, the Energy Information Administration sums up China's main energy characteristics: China has emerged from being a net oil exporter in the early 1990s to become the world's third largest net importer of oil in 2006. Natural gas usage has also increased rapidly in recent years, and China has looked to raise natural gas imports via pipeline and liquefied natural gas (LNG). China is also the world's largest producer and consumer of coal [3].

China's huge energy consumption volume can be explained by two major factors: population and economic growth. With a population of approximately 1.4 billion people, China is the most populous country in the world. This population is not only the largest in the world, but equally important, economically the fastest growing. Since the late 1970s China's economy has changed from a centrally planned system that was largely closed to international trade to a more market-oriented economy. Reforms started with the phasing out of collectivized agriculture and expanded to include the gradual liberalization of prices, fiscal decentralization, increased

Energy Security: An Interdisciplinary Approach, First Edition. Gawdat Bahgat.
© 2011 John Wiley & Sons, Ltd. Published 2011 by John Wiley & Sons, Ltd.

autonomy for state enterprises, the foundation of a diversified banking system, the development of stock markets, the rapid growth of the non-state sector, and opening up to foreign trade and investment. As a result, in the last three decades China's economy has been the fastest-growing economy in the world and the nation has become the chief economic driver of Asia.

This large and more affluent population consumes more electricity, lives in bigger homes, buys more appliances, and drives bigger cars and longer distances. Despite this steady rise in energy consumption, it is important to point out that Chinese energy consumption per capita is considerably lower than that of the United States or Europe. The nation's large population, however, makes up for this low per capita use [4]. In short, growing population and wealth mean higher energy consumption.

China's population is about 20% of the world's population, but the country holds only 1.2% of the world's proven oil reserves, 1.3% of natural gas reserves, and 13.9% of coal reserves. This disparity between energy consumption and fossil-fuel reserves leaves China with few options. Calls for energy self-sufficiency are long gone. Indeed, the inadequacy of indigenous resources to meet the large and growing demand has heightened the nation's sense of energy vulnerability and made the securing of sufficient energy supplies a matter of national security. Simply stated, China cannot maintain its high economic growth without adequate energy supplies. Most of these supplies are increasingly coming from overseas. China is already a large importer of a number of fuels. In order to secure energy supplies, the Chinese government and companies have pursued an aggressive strategy in equity acquisition and in establishing partnerships with major energy producers in the Middle East, Africa, Central Asia, and Russia, among others.

China's large and growing volume of energy consumption and imports has significant environmental and geopolitical consequences. It has already overtaken the United States as the world's largest polluter. Beijing's participation in the global efforts to address climate change issues is essential. Chinese oil and gas companies compete with their American, European, Russian, and other Asian counterparts. In order to protect the interests of these state-owned companies, the Chinese government has established and strengthened diplomatic ties with producing countries. Beijing is also involved in several schemes to build and secure pipelines and shipping lanes from Russia, Central Asia and other regions.

In the following sections I examine these issues in some detail. First, China's energy mix (oil, natural gas, coal, nuclear power) is analyzed. This is followed by a discussion of the nation's regulatory authority and the efforts to acquire energy assets in foreign countries. Finally, Beijing's energy diplomacy, particularly in Africa and the Middle East, will be examined. As in other major consuming countries, China believes its energy security lies in diversifying its energy mix and suppliers. Compared to the United States, and to a lesser extent the European Union, China's energy diplomacy is less driven by geopolitical considerations and instead is focused on securing commercial interests.

4.1 Regulatory Authority

Despite the significant role played by energy in China's economic development and the great official concern about securing adequate supplies, Chinese officials have experimented with different bureaucratic agencies to control and guide the energy sector. Unlike many other countries, including free market economies such as the United States, China lacks the tradition of a central ministry or department coordinating its energy policy [5]. Shortly after the People's Republic of China was created, the Ministry of Fuel Industries was abolished in 1955 and separate ministries for coal, electricity and oil were established. Fifteen years later (1970), a new Ministry of Fuel and Chemical Industries replaced the former three ministries. This ministry, in turn, was dissolved five years later. In 1988, a Ministry of Energy was launched to oversee coal, oil, nuclear, and hydroelectric development, but similarly, it was dissolved in 1993 [6].

In order to address this institutional confusion, in 2005 China established the Energy Leading Group (ELG), a cabinet minister-level organization chaired by the then Prime Minister Wen Jiabao. The objectives of the ELG include the strengthening of energy strategies by the establishment and improvement of a comprehensive long-term energy development plan; promoting healthy and orderly development of the coal, electricity, oil, and natural gas industries; promoting the use of new and renewable energy sources; encouraging energy conservation; and initiating reform in energy industries and corporations [7].

Within this framework, the National Development and Reform Commission (NDRC) was created as an integrated central authority responsible for studying energy development both at home and abroad, articulating long-term energy strategies, making recommendations to the State Council on energy policies, and administering oil, natural gas, coal, electricity, and other parts of the energy sector [8]. A new organization, the National Energy Administration (NEA), was launched in 2008 to merge all departments working on energy issues. The NEA assumed NDRC's responsibilities in key areas including charting energy strategy and policies as well as managing separate sectors for oil, gas, coal, electricity, nuclear, and renewable energy. The NEA is also responsible for managing China's strategic oil reserves, including building and releasing such reserves and supervising the management of commercial reserves. The NEA also oversees investments in energy resources and negotiates and signs energy contracts with foreign governments and institutions [9].

These efforts to create a centralized and coherent national authority in charge of energy both at home and abroad have some way to go. The coal industry is largely decentralized with the majority of the production capacity controlled by small-town/village mines. The oil and gas sector is dominated by three state-owned corporations: China National Petroleum Corporation (CNPC), China Petroleum and Chemical Corporation (Sinopec), and China National Offshore Oil Corporation (CNOOC). These three companies and their subsidies wield a significant amount of influence in the nation's oil and gas sectors. Between 1994 and 1998, the Chinese government reorganized most state-owned oil and gas assets into two vertically integrated corporations: CNPC and Sinopec.

CNPC is China's largest oil and gas producer and supplier, as well as one of the world's major oil field service providers. It covers a variety of businesses including petroleum exploration and production, natural gas and pipelines, refining and marketing, oil field services, engineering construction, petroleum equipment manufacturing, and new energy development, as well as capital management, finance, and insurance services [10]. It has assets and interests that are widely distributed both at home in China and all over the world, particularly in Africa, the Middle East, Central Asia, Russia, and South America [11]. CNPC was established on September 17, 1988. In November 1999, PetroChina, largely owned by CNPC, was founded. PetroChina operates most of the nation's pipeline networks [12].

Sinopec is the second largest producer of crude oil and natural gas and the largest petrochemical producer in China. In addition to exploration and production, Sinopec focuses on refining, marketing and distribution, and chemicals. It is the largest oil refiner in China, producing gasoline, diesel, kerosene including jet fuel, lubricants, fuel oil, chemical feedstocks, and other petroleum products. In 2008 its primary refining capacity ranked the third largest in the world. Sinopec also ranked third in the world in marketing and distributing oil products. Finally it is the largest producer and distributor of chemical products in China. Its major chemical products include ethylene, synthetic resins, monomers and polymers, synthetic fibers, and synthetic rubbers. In 2008 its ethylene capacity ranked fifth in the world [13].

CNOOC is responsible for offshore oil exploration and production and has recently received more attention due to the growing interest in offshore zones. Like CNPC and Sinopec, CNOOC has expanded its operations to include both upstream and downstream activities. It is also involved in LNG projects and natural gas pipelines. Its international assets are largely in the Pacific Rim and involve foreign partners [14].

4.2   Oil

Following coal, oil is China's most important fuel. In the last few decades there have been major changes in the nation's oil consumption, production, and the overall policy to secure oil needs. Since the mid-1990s China has grown more dependent on foreign oil supplies. In 2008 it was the world's second largest oil consumer (after the United States) and third oil importer (after the United States and Japan). This was not the case just a few decades ago. While major industrial countries in Europe, Japan, and the United States were suffering from strong fluctuations in oil prices in most of the 1970s and early 1980s, China was insulated from these oil shocks. Then, the nation's oil production exceeded its consumption and China was able to export oil to some of its Asian neighbors [15].

A combination of rising consumption and declining production has ended the state of self-sufficiency and made the country a major oil importer. As in other consuming countries, the transportation sector is the driving force in China's oil consumption, accounting for about one-third of total oil use [16]. The impressive economic growth of the last few decades has created a large and growing middle class in most big Chinese cities. This, in turn, has led to a higher rate of motorization (i.e., ratio of automobile ownership per 1000 residents). Private automobile ownership has substantially increased from the early years of economic reform. In 1985 the ratio was 0.3 cars per 1000 people and in 2003 it was 1 car per 1000 people [17].

The history of oil discovery in China can be divided into five periods:

- 1907–1949, oil discoveries in the Ordos Basin and a few other areas.
- 1950–1964, large reserves found in the Songliao Basin.
- 1965–1975, Bohai Bay Basin development.
- 1976–1990, exploration and development in the east, offshore, and west.
- After 1990, development began in the Tarim, Junggar, and Ordos Basins, with production still increasing in these basins [18].

Currently most oil production comes from the following: the Daqing oil fields in Helongjiang Province, Manchuria, north-eastern China, which began producing in 1963; the Shengli field complex, centered on the Yellow River in Shandong Province, adjacent to Bohai Bay; three basins in the Xinjiang Uygur Autonomous Region in far north-western China (Tarim, Turpan,

Karamay); the Changqing oil field in central China's Shaanxi Province; the Sulige oil and gas field in the Ordos Basin, Inner Mongolia Autonomous Region [19].

Although production in eastern China continues to decrease, China's total production has remained constant because of increases in the middle, west, and offshore regions. Indeed, most of the new production comes from offshore fields. These fields have been largely developed in cooperation with IOCs, particularly American and European ones. As in some other countries, foreign investment in the upstream oil and gas sector has been more significant in China's offshore areas than onshore.

Historically, China has been reluctant to open its public utilities and energy sector to foreign investment. But the need for highly sophisticated Western technology and the necessity to comply with the World Trade Organization's (WTO's) regulations prompted Chinese leaders to change their stance. Thus, Beijing has amended regulations governing foreign investment in both its onshore and offshore oil and gas sectors. The requirement of Chinese majority ownership in some sectors has been removed. Foreign parties are no longer required to sell their share of production to their Chinese partners and the preference for using Chinese personnel, goods, and supplies has been diluted. Beijing has also allowed foreign ownership in retail and wholesale oil and gas markets. Tariffs on imported crude oil, gasoline, and lube oil have been either entirely eliminated or substantially reduced. Import quotas and license requirements for refined or processed oil were relaxed [20].

Despite significant efforts to increase domestic production, new discoveries have been proven insufficient to meet the growing demand. Accordingly, China has grown increasingly dependent on foreign supplies. This heavy dependency is likely to remain a major characteristic of China's energy policy in the foreseeable future [21]. Before 1992, China's oil imports primarily came from the Asia-Pacific region. Since 1993 the volume of crude imports from the Middle East has exceeded imports from the Asia-Pacific region [22]. As recently as 1996, China imported 70% of its oil needs from just three countries: Indonesia, Oman, and Yemen [23]. Since then, it has established close energy ties with a variety of producing regions, particularly the Middle East and Africa.

4.2.1   Imports from the Middle East

Together, Middle Eastern countries hold the world's largest oil proven reserves and are the world's largest producing region. China's rising consumption and deepening dependence on foreign supplies mean that the two sides are bound to share mutual interests. The Gulf states need to secure markets for their major product and China wants to enhance its energy security. Little wonder that, since the early 2000s, China has imported most of its oil needs from Middle Eastern producers.

Compared to other major consumers such as Europe and the United States, China does not have a legacy of colonial rule or major political disputes [24]. Instead, for the first three decades after its founding in 1949, China was considered a developing country with a strong resentment against foreign powers (both Western and Soviet) [25]. Stated differently, for most of the post-war period, China's relations with the Middle East were shaped primarily by strategic considerations (i.e., supporting resistance to US and Soviet imperialism). There were fundamental changes in this strategic/ideological framework in the late 1970s and early 1980s for at least two reasons. First, the reduction in tension between major global powers meant that international relations, including in the Middle East, became less about confrontation and more about cooperation and accommodation. Second, the inauguration of domestic modernization programs in China in 1978 suggested that the country's foreign policy had become increasingly driven more by economic and commercial interests and less by ideological and strategic considerations [26]. As one analyst put it, China appeared to be behaving "more geo-economically than geo-strategically" [27].

Against this background, and in order to promote its commercial interests, China sought to establish and strengthen diplomatic relations with all Middle Eastern states. For example, Egypt was the first regional power to establish diplomatic ties with Beijing in 1956, then Iraq in 1958, Iran in 1970, Kuwait in 1971, Oman in 1979, the United Arab Emirates in 1984, Qatar in 1988, Bahrain in 1989, and Saudi Arabia in 1990. In addition to these bilateral ties, China established relations with regional organizations such as the Gulf Cooperation Council (GCC) to encourage trade and investment between the two sides. Similarly, in 2004, China and Arab countries set up a China–Arab Cooperation Forum to accelerate and strengthen cooperation between them [28]. China also established diplomatic relations with Israel in 1992 and the two countries have since cooperated on a variety of issues, particularly on arms sales [29]. This broad diplomatic representation means that China is involved in almost all regional issues including the Arab–Israeli peace process, security, economic development, and trade.

In pursuing its commercial interests in the Middle East, Beijing has particularly focused on two important issues: arms sales and oil. For the last few decades China has become an important weapons supplier to a number of Middle Eastern countries, including Iran, Iraq, and Saudi Arabia. Some Middle Eastern countries are banned from or face restrictions on buying weapons from the United States and Europe. China, along with Russia and North Korea (among others), has become a major supplier to these Middle Eastern countries.

Sino-Iraqi relations were established in 1958. Since then Beijing has been in competition with Soviet and US influence in Baghdad. During the Iran–Iraq War, China supplied arms and ammunition to Iraq. Strategic and commercial relations with Saddam Hussein remained until the Iraqi invasion of Kuwait in 1990. Throughout the Gulf crisis, Beijing advocated a peaceful resolution of the conflict. China did not welcome a strong US military presence in the region and instead called for a peaceful resolution within the scope of the Arab countries. China supported economic sanctions to force Saddam Hussein out of Kuwait but abstained on United Nations Security Council Resolution 678, which permitted the Coalition forces to use force to evict Iraq from Kuwait [30].

China participated in the comprehensive UN economic sanctions imposed on Iraq in the aftermath of the Gulf War. As a result, trade between Beijing and Baghdad came to a halt. However, by the end of that decade, trade relations rose dramatically as a result of the UN oil-for-food program, which allowed Iraq to sell a small volume of its oil production to buy necessary food and medical supplies. In June 1997, a consortium of Chinese energy companies signed a 22-year production sharing agreement with Saddam Hussein's regime to develop al-Ahdab, Iraq's second largest oil field, after the lifting of UN sanctions [31].

In the aftermath of Saddam Hussein's toppled regime, China has sought to restore and strengthen the commercial ties it had with Baghdad. China canceled Iraq's debt and provided reconstruction assistance in order to secure cooperation with the post-Hussein regime in Baghdad. This policy seems to have paid off. In August 2008 the first major oil deal was signed between the Iraqi government and CNPC, worth $3.5 billion. The 22-year contract is a renegotiated version of the 1997 agreement. The latter included production sharing rights, but under the former China will be paid for its services, though it will not share in the profits [32].

The Chinese policy in the Kurdistan region of northern Iraq underscores the rise of pragmatism and waning of ideology as the main driver of Beijing's foreign policy. On one hand, opposition to separatism has been one of the basic components of China's policy. This policy reflects China's uncompromising adherence to the maintenance of territorial integrity – primarily with regard to Taiwan, but also to Tibet, Xinjiang, and Inner Mongolia. Similarly, the Chinese are fundamentally and officially opposed to separatist movements elsewhere, suggesting that self-determination should not necessarily involve national independence and that stateless nations should not necessarily form, or be given, states [33].

On the other hand, Beijing has looked for commercial and energy opportunities in post-Hussein Iraq. Against this background, Beijing hosted Jalal Talabani, Chairman of the Patriotic Union of Kurdistan, who was later to become the first president of post-Hussein Iraq. Similarly, Massoud Barzani, Chairman of the Kurdistan Democratic Party, held several meetings with Chinese officials. As a result, Chinese oil companies have negotiated oil agreements with the Kurdish authority.

China has established the strongest ties in the Persian Gulf region with Iran. The foundations of this strong relationship between Beijing and Tehran are numerous. First, with more than 70 million people, Iran is by far the largest market in the Gulf. Second, for geopolitical reasons, Western powers, particularly the United States, have imposed a variety of sanctions and restrictions on Iran. This means that Chinese products, both civilian and military, face less competition in Iran than in other countries. Third, Iran's proven oil and natural gas reserves are largely underdeveloped and underutilized due to the lack of the necessary investments. China, with its substantial financial assets and its growing need for energy, seems the appropriate partner to the Islamic Republic. Fourth, in addition to these shared commercial and energy mutual interests between Beijing and Tehran, the two sides resent the prominent US role in the Persian Gulf and Central Asia. Finally, strategically, Iran and China need each other. The United States is likely to remain the dominant power and the main partner to post-Hussein Iraq and to Saudi Arabia. Since 1980, diplomatic relations between Tehran and Washington have been severed. For the last few decades, Iran has sought to establish close ties with other global powers to counter the United States; China has proven a reliable partner.

Within this context, the close alliance between Beijing and Tehran has been a major development in the regional and international landscape for recent decades. Diplomatic relations were established in 1970, but real improvements came following the 1979 Iranian Revolution and outbreak of the Iran–Iraq War. China supplied arms to both Baghdad and Tehran. Arms sales to Iran, particularly missiles, have continued and, indeed, expanded. Since the late 1980s China has sold several hundred Silkworm anti-ship cruise missiles to Iran. Unlike several European powers and the United States, China is reluctant to support economic sanctions against Iran regarding its nuclear program. Beijing calls strongly for peaceful negotiations and a resolution within the International Atomic Energy Agency framework. China, along with Russia, strongly opposes using military strikes against Iran's nuclear facilities.

The energy sector is another area where Chinese–Iranian cooperation has made tremendous progress. Chinese companies have recently been successful in concluding a number of high-profile deals in Iran. Chinese activities in Iran include refinery upgrades, pipeline construction, and engineering services, among others. Chinese companies are actively involved in developing Iran's huge South Pars natural gas field and Yadavaran oil field. Two overlapping reasons explain this close cooperation: (1) Iran is one of a few Persian Gulf states where Chinese companies can participate in upstream exploration and development operations; (2) unlike European and Japanese companies, Chinese corporations have proven less likely to be deterred from investing in Iran's energy sector. As a result, since the mid-2000s Iran has become a major oil supplier to the Chinese market. This partnership is likely to endure for many years to come.

It was almost inevitable that the world's largest oil producer and exporter should have relations with the world's fastest-growing economy and a top oil consumer and importer. Saudi Arabia and China established diplomatic relations in 1990 in the middle of the Gulf crisis (e.g., Iraq's invasion of Kuwait). This coincided with China becoming a net oil importer. Naturally, energy has served as the foundation for the growing Sino-Saudi partnership. Visit exchanges by top officials from the two sides have further strengthened cooperation between Beijing and Riyadh.

Chinese President Jiang Zemin's visit to Saudi Arabia in 1999, the first ever by a Chinese head of state to the kingdom, represented a significant turning point. The two nations agreed to pursue close cooperation in several areas, most notably energy. They invested in each other's oil and natural gas sectors, including refineries, petrochemical plants, and gas marketing and distribution. In 2004 Sinopec won a concession to develop the kingdom's natural gas in the Rub al-Khali (Empty Quarter) Desert. Another milestone came in early 2006 when King Abdullah visited China. It was the first visit of a Saudi king to China and the king's first trip outside the Middle East since assuming the throne in 2005. A few months later President Hu Jintao reciprocated and visited the kingdom.

This close cooperation has expanded the trade volume between Beijing and Riyadh. Saudi Arabia buys significant amounts of cheap Chinese products while China imports a substantial proportion of Saudi oil. Indeed, the kingdom has emerged as a top oil exporter to China since the early 2000s.

Several conclusions can be drawn from this brief survey of China's relations with Iraq, Iran, and Saudi Arabia. First, the Persian Gulf has been and is likely to continue to be China's largest source of oil and natural gas. The two parties meet each other's energy security needs: gulf producers seek a large and growing consumer market for their hydrocarbon products while China needs to secure reliable supplies. Second, like other major consumers (i.e., Europe, Japan, and the United States), China is interested in promoting political stability and economic prosperity in the Persian Gulf. Conflicts and turmoil are likely to threaten the steady flow of oil and gas supplies. Third, global powers' commercial interests in the Gulf should not be seen in zero-sum terms. They are not mutually exclusive. For example, China's participation in developing oil and gas in Iran means more supplies in the global market. Fourth, generally, Middle Eastern countries have a positive perception of China. Indeed, the "Chinese model" (i.e., promoting economic modernization while maintaining an authoritarian form of government) is popular among political elites in the Middle East and elsewhere. Finally, though oil trade is the core of the China–Persian Gulf partnership, strategic considerations play a significant role. China is a global power with a permanent seat on the United Nations Security Council. China is also a major arms supplier to the Persian Gulf and the broader Middle East.

4.2.2   Africa

In recent years China has become a major trade and energy partner to several African countries. Chinese state-owned oil companies are aggressively pursuing oil exploration and development operations in Angola, Equatorial Guinea, Nigeria, Republic of Congo, and Sudan among others. Trade volumes between African nations and Beijing have grown several fold since the early 2000s. Africa's crude oil and petroleum products and Chinese cheap products comprise the bulk of this trade.

Two distinctive stages can be identified in Beijing's heavy involvement in Africa. First, for most of the 1960s and 1970s, Chinese policy was driven by ideological considerations. Beijing's interest centered on building ideological solidarity with other developing nations to advance Chinese-style communism and repel Western and Soviet "imperialism" [34]. The introduction of economic reforms in China and the waning of the Cold War have substantially altered Chinese policy in Africa and the rest of the world. Since the early 1990s, commercial interests have taken the lead in guiding Beijing's policy. Its focus is on trade, investment, and energy.

In order to promote these commercial and energy interests, Chinese leaders have emphasized the common bonds they share with African nations. Unlike Western powers, China does not have a colonial history in Africa. In order to distance themselves from this colonial legacy, Chinese leaders have repeatedly called for the non-intervention of foreign powers in the domestic affairs of developing countries. Chinese policy also emphasizes the need to establish a new, just international political and economic order. Finally, human rights are less a restraint on Chinese policy in Africa and elsewhere than on US or European policies. In short, Beijing presents itself as a developing nation that shares similar experiences and aspirations to those of Africa, while playing on African leaders' historic suspicion of Western intentions [35].

Since the early 2000s, China has developed a two-pronged strategy toward energy investments. First, it has pursued exploration and production deals in smaller, low-visibility countries such as Gabon, Equatorial Guinea, and the Republic of Congo. Second, it has gone after the largest oil producers by offering integrated packages of aid [36]. In both Angola and Nigeria, Beijing has offered packages of loans and aid that include funds for Chinese companies to build railroads, schools, roads, hospitals, bridges, and offices.

This Chinese policy – loans and aid in return for oil – has been more successful in Angola than in Nigeria. China has long enjoyed a healthy commercial relationship with Nigeria. China sees Nigeria, which has the largest population in Africa, as a key market for its cheap goods. Accordingly, Chinese companies are actively pursuing trade and investment opportunities in Nigeria. These strong economic and commercial ties are further cemented by frequent high-level visits by top officials to each other's capitals. Despite this close cooperation between Beijing and Lagos, large-scale corruption, lack of transparency, and lack of political and security stability have led to the cancelation of several Chinese projects.

On the other hand, since 2002, Angola has enjoyed a period of sustained peace. This political stability has produced high rates of economic growth, sustained by high government spending and a rapid increase in oil exports. Relations between China and Angola have improved gradually since the 1990s and Beijing emerged as a top trade partner to Luanda by the late 2000s. Meanwhile, Angola has become a major oil supplier to China. Angola has used its oil to secure credit lines from several countries including Portugal, Brazil, Spain, and, most recently, China. From Angola's perspective, the Chinese provide funding for strategic infrastructure projects that Western donors do not fund. Chinese financing offers better conditions than commercial loans, including lower interest rates and longer repayment times. Unlike loans from major international financial institutions such as the World Bank and the International Monetary Fund, Chinese loans are offered with few conditions. Finally, China also offers Angola cheap technology transfer opportunities. These tend to be more suitable and less expensive than those from Europe or the United States, where the technology gap is bigger [37].

In order to promote and strengthen dialogue and cooperation between China and African nations, several institutions have been established including the Forum on China–Africa Cooperation (FOCAC), the China–Africa Development Fund, and the China–Africa Business Council.

One of the earliest institutions was the FOCAC. It is held every three years and was first held in Beijing in 2000 [38]. Three years later, the meeting was held in Addis Ababa in Ethiopia, followed by a China–Africa Summit in 2006 in Beijing and another one in Sharm El-Sheikh, Egypt, in 2009. In this latest meeting Chinese Premier Wen Jiabao announced eight new measures that the Chinese government will take to strengthen China–Africa cooperation in the following three years. These include establishing a partnership to address climate change; cooperation in science and technology; providing $10 billion in concessional loans; opening up Chinese markets to African products; enhanced cooperation in agriculture; deeper cooperation in medical care and health; enhanced cooperation in human resources development and education; and expanded people-to-people and cultural exchanges [39].

Another institution is the China–Africa Development Fund (CADFund). It is a special fund which aims to support Chinese companies in developing cooperation with Africa and entering the African market. The CADFund is designed to guide Chinese companies' investments in Africa and contribute to a higher standard of living for African nations. It was inaugurated in June 2007 after the Chinese government officially approved its establishment with first-phase funding of $1 billion provided by the China Development Bank and will eventually reach $5 billion. The CADFund's operations and activities are based on free market principles [40].

The China–Africa Business Council (CABC) was jointly founded by the United Nations Development Program and the Chinese Ministry of Commerce. It is a non-governmental organization. Non-state-owned businesses are the principal participants and beneficiaries of the CABC. Its aim is to provide practical business tools to facilitate and strengthen commercial ties between China and Africa. It provides information for enterprises to explore local resources, set up joint ventures, and train human resources [41].

This close Chinese–African cooperation, however, has raised serious strategic, commercial, and cultural concerns. Some African leaders fear that the continent will become a powerless victim in a strategic competition between global powers, as it was between 1960 and 1990. Africa, the argument goes, could become a theater for the new scramble for resources between China, Japan, the US, and Europe. African leaders do not want to be forced to choose between these global powers and prefer to maintain close relations with all of them [42].

In addition to these strategic concerns, there has been a growing backlash in parts of Africa against Chinese corporate behavior [43]. Many Africans are concerned over how China operates in the continent, accusing Chinese companies of underbidding local firms and not hiring Africans. Chinese infrastructure deals often stipulate that the majority of the labor force must be Chinese. Furthermore, international observers say that the way China does business – particularly its willingness to pay bribes and attaching no conditions to aid money – undermines local efforts to increase good governance and international efforts at promoting reform.

4.2.3   Central Asia

Central Asia is situated at the crossroads between the East and West and has historically interacted with a variety of cultures and economies. China has traditionally viewed Central

Asia as its personal trading area and a region heavily influenced by Chinese culture. Many of history's most impressive trading centers were located in Xinjiang or west of China's current borders, such as Jarkand, Samarkand, Urumuqi, and Kokand [44]. The trade between China and Central Asia has always been crucial and favored by both sides. The only change today is that the traders have replaced jade, tea, silk, and rhubarb with oil, natural gas, and weapons. While Central Asia was under Russian and Soviet occupation, China was largely excluded from it. Since the early 1990s, however, Beijing has actively resumed its close cultural, commercial, and strategic relations with the region. China recognized the Central Asian states in December 1991 and soon after diplomatic relations were established. Since independence, Central Asian nations' relations with China have substantially progressed, particularly in the areas of security and energy.

Security on China's western borders and its internal security in the Xinjiang region depend upon stable and peaceful relations with its Central Asian neighbors. The Chinese province of Xinjiang is largely dominated by Muslims of Turkic origin. The disintegration of the Soviet Union and the socio-economic and political independence and development of Central Asian states have ignited a rise in nationalism and Islamic identity in Xinjiang. These calls for autonomy or independence have been met by a crackdown by the Chinese authority. The September 11, 2001 terrorist attacks in the United States and the ensuing "war on terrorism" have further strengthened Beijing's campaign against its opponents in Xinjiang. The Chinese government claims that the Muslim rebels are closely connected to the Taliban and Osama Bin Laden.

Indeed, the Chinese government had warned the Central Asian states against supporting, protecting, or training rebels from Xinjiang long before September 11, 2001. Concerned about the rise of Islamic fundamentalism in their own countries, Central Asian leaders have closely cooperated with their Chinese counterparts to contain this common challenge. Within this context, China has supported the US war against the Taliban in Afghanistan and Muslim militants in general. However, strategic rivalry with the United States has restrained such support. Chinese leaders have never been enthusiastic about a large US military presence in their own "backyard."

Against this background the Shanghai Cooperation Organization (SCO) was created in June 2001. Initially, China, Russia, Kazakhstan, Tajikistan, and Kyrgyzstan established the Shanghai Five in 1996 for the purposes of increasing regional cooperation and in order to resolve many of the border disputes between the Member States. In 2001 Uzbekistan joined the Shanghai Five and the organization was renamed the SCO. Later, India, Iran, Mongolia, and Pakistan joined as observer states.

The main goals of the SCO are: strengthening mutual confidence and good-neighborly relations among the member countries; promoting cooperation in politics, trade, economy, science, technology, culture, education, energy, transportation, and environmental protection; maintaining peace, security, and stability in the region; establishing a new international political and economic order [45]. In 2005 the leaders of the SCO Member States called for a "timeline" on the Coalition's, largely US, military presence in the region.

China's close strategic and security cooperation with Central Asia contributes to its energy security. Since the late 1990s, Chinese leaders have negotiated and signed a number of oil and natural gas deals with Central Asian producers. These supplies reach China by pipelines. Pipeline deliveries are favorably viewed in China for at least two reasons: they lock suppliers into long-term relationships; and they are safer than shipments by sea, where terrorists or pirates can interrupt supplies [46].

Considering that it borders on China and has the largest proven oil reserves in Central Asia, Kazakhstan, unsurprisingly, has been the main focus of Chinese energy investment for many years. One of the largest Chinese ventures in Kazakhstan's energy industry is Aktobemunaigaz, a western Kazakhstan production company in which CNPC has amassed an 88.25% stake [47]. China imports oil from Kazakhstan via a pipeline that runs from Atasu in north-western Kazakhstan to Alashankou in China's north-western Xinjiang region. The pipeline was built in three stages, the construction of which lasted most of the 2000s. The commercial operator of the pipeline is a joint venture between CNPC and Kaztransoil [48].

In addition to oil supplies from Kazakhstan, China has negotiated and signed agreements to import natural gas from Turkmenistan. In 2009 the China Development Bank gave $4 billion in credit to the Turkmen state-run gas company Turkmengaz [49]. Turkmenistan agreed to supply 40 billion cubic meters of gas to China annually for a period of 30 years [50]. The pipeline transporting the gas starts from the Turkmen gas fields near the Amu Darya River. It then enters Uzbekistan at Olot, carries on to southern Kazakhstan, and then to Alashankou in China. In addition to gas from Turkmenistan and Uzbekistan, the pipeline carries gas from Kazakhstan's Karachaganak, Tengiz, and Kashagan gas fields [51].

Political cooperation between China and the Central Asian states has improved since 1991 and has been very positive and lucrative for both sides. Chinese expansion in Central Asia slowed down somewhat after September 11, 2001 when the United States became a competitor to China in the region. Despite the fact that the United States has a great deal to offer, Central Asian states understand that the Americans will probably leave some day and the Chinese will always be present due to their geographic proximity and economic interdependence. This makes China a crucial actor in the region and a long-term counterbalance to both the United States and Russia [52].

4.2.4   Refining

The Chinese refining sector is unique because, while it is already the world's second largest after the US one, it still has substantial upside growth potential. The industry is driven by strong demand and more permissive industrial policies than those of Western countries [53]. Chinese policy for refined products calls for self-sufficiency in gasoline, jet fuel, and diesel. These fuels are considered strategic military fuels and the Chinese government is disinclined to allow the country to require regular imports [54]. Since the early 2000s, building new refineries and expanding the capacity of old ones have increased refinery utilization. Sinopec and CNPC have dominated these efforts. In addition, national oil companies from Kuwait, Russia, Saudi Arabia, and Venezuela have entered into joint ventures with Chinese companies to build new refining facilities.

A major issue for the Chinese downstream sector is the lack of adequate refining capacity suitable for heavier Middle Eastern crude oil. The nation's refining industry was built around light, sweet crude supplies from the Daqing and other Chinese fields. It lacks the capacity to process lower-quality, high-sulfur, and high-acidity crudes. The fact that several Middle Eastern producers such as Saudi Arabia, Iran, the United Arab Emirates, and Kuwait have become major oil suppliers of crude oil (mostly heavy) to China is a major driver for the efforts to build and expand capacity and convert refineries from light to heavy crude.

4.2.5   Shipping

China's oil security faces a serious challenge. The nation's indigenous production is increasingly unable to meet growing demand. Subsequently China is growing more dependent on foreign supplies. These imported supplies are delivered to China largely by one of two means: tankers or pipelines. Most of its oil imports come from the Middle East and West Africa by sea. Commercial and national security incentives have persuaded Chinese leaders to initiate a new shipping strategy.

In 2002, Chinese tankers carried less than 4% of China-bound cargoes from the Middle East and none at all from West Africa. According to a new plan initiated in the mid-2000s, Beijing intends to transport 60–70% of it imported oil in government-owned tankers [55]. Other Asian powers such as Japan and South Korea carry a much bigger proportion of their imported oil by national fleets. Furthermore, China has already established a shipbuilding ability. A significant number of very large crude carriers (VLCCs) have been built in China for foreign operators from Norway, Iran, Germany, Japan, Venezuela, and Algeria among others.

Another Chinese concern is that the great majority of these oil tankers pass through the Strait of Malacca. Located between Indonesia, Malaysia, and Singapore, the Strait links the Indian Ocean to the South China Sea and Pacific Ocean. Malacca is the shortest sea route between the Persian Gulf suppliers and the Asian markets – notably China, Japan, South Korea, and the Pacific Rim. It is the key chokepoint in Asia with more than 15 million b/d. At its narrowest point in the Phillips Channel of the Singapore Strait, Malacca is only 1.7 miles (2.7 km) wide, creating a natural bottleneck as well as the potential for collisions, grounding, or oil spills. Piracy is a constant threat to tankers in the Strait of Malacca [56].

Terrorists or pirates could easily block these bottlenecks. In addition, a dispute with the United States could jeopardize China's oil security. Chinese leaders believe that whoever controls the Strait of Malacca could control their country's oil supplies. In recent years China has invested substantial resources in building a "blue-water navy" to respond to these challenges.

Another way to overcome this vulnerability is to import oil via pipelines. In 2006, China inaugurated its first transnational oil pipeline carrying oil from Kazakhstan and Russia. In May 2009, China started the construction of a crude oil pipeline to Russia. This followed an agreement between the two neighbors under which the former agreed to lend $25 billion to two state-run Russian companies ($15 billion to Rosneft and $10 billion to Transneft) in exchange for 300 million tons of Russian oil to be transported via the pipeline over 20 years. The pipeline runs for about 67 kilometers in Russia and 960 kilometers in China and ends at the Daqing oil field [57].

Finally, in March 2009, China signed an agreement with Myanmar to construct parallel oil and gas pipelines. This scheme is designed as an alternative transport route for crude oil from the Middle East and Africa that would bypass the potential chokepoint of the Strait of Malacca.

4.2.6   Strategic Petroleum Reserve (SPR)

Building a national strategic oil reserve is a relatively new idea in China. High-level discussions on the need for an SPR began after the nation became a net importer of oil in 1993. After a long debate, the 10th Five-Year Plan, passed by the National People's Congress in 2001, declared the goal of developing strategic stockpiles. The plan has three phases:

- Phase I (2004–2008) includes construction of one stockpiling facility each in Zhenhai and Aoshan in Zhejiang province, Qingdao in Shandong province, and Dalian in Liaoning province, with a combined storage capacity of 100 million barrels (25 days of net oil imports).
- Phase II (2008–2010) increases storage capacity to 300 million barrels (42 days of net oil imports).
- Phase III (2010–) storage capacity should reach 500 million barrels, although there is no timetable set for this plan [58].

China ultimately plans sufficient strategic petroleum reserves to cover 90 days of imports, which is also the mandatory goal of the IEA. Sinopec, PetroChina, and Sinochem are responsible for the construction of SPR facilities but do not operate them. Rather, the State Oil SPR Office and State Oil Stockpiling Center, both established by the NDRC, are directly responsible for the management and operation of the SPR [59]. In addition to the government-run SPR, China has commercial crude oil storage capacity. Finally, the government plans to create a strategic refined oil stockpile to be operated by a subsidiary of the NDRC.

4.3   Coal

China holds 14% of the world's proven coal reserves, the third largest after the United States and Russia respectively [60]. Production has been on the rise since the late 1990s. About half of this production is used to generate electricity and the other half is used mainly in the industrial sector. China was a major coal exporter up to 2003 when exports peaked and have since fallen markedly. Meanwhile, demand has soared faster than indigenous supply. This growing imbalance between demand and production means that China is growing increasingly dependent on foreign supplies.

Although China is not about to run out of coal, it does face a number of challenges: average mining depth is increasing; resource recovery rates are low; many mines are located in environmentally sensitive areas with limited water resources; the number of mining fatalities remains unacceptably high; and coal transport routes are relatively long and congested [61]. In addition, China has tens of thousands of small local coal mines where inefficient management, insufficient investment, outdated equipment, and poor safety records prevent the full utilization of coal resources.

In recent years the Chinese government has taken some very effective steps to improve the economic and technical efficiency of coal mining, such as shutting down large numbers of small unsafe mines and inefficient power plants. Furthermore, China has welcomed foreign investment in the coal sector. The goal is to modernize existing large-scale mines and to introduce new technologies into the nation's coal industry.

Finally, China's heavy dependence on coal has had negative impacts on the environment. In recent years China has become the largest contributor to carbon dioxide, a leading greenhouse gas, in the atmosphere, although its per capita emissions remain well below those of the world's industrialized nations [62]. In response Beijing has been promoting the use of cleaner coal technology, such as costly power plants that burn at a much higher temperature and have lower emissions, and is looking into even more expensive, cutting-edge technology that could strip out carbon gas from power plants and store the greenhouse gas underground [63].

4.4 Natural Gas

China's natural gas consumption has grown steadily since the late 1990s. However, natural gas is still a relatively minor fuel in the nation's energy mix, which is overwhelmingly dominated by coal. The Chinese government has been promoting natural gas use in order to improve energy diversification and energy efficiency, and to reduce pollution. Under the 10th Five-Year Plan (2001–2005), the government set the target of raising natural gas use to 10% of the energy mix in 2020, which was basically reiterated in the 11th Five-Year Plan (2006–2010).

China was self-sufficient until 2006 when it started importing LNG. In the late 2000s, China had three operating LNG terminals: Dapeng in southern Guangdong province, the Fujian terminal, and the Shanghai terminal. Other LNG terminals are either under construction or being planned [64]. The continuing rise in demand means that China will grow more dependent on imported gas. Natural gas is used primarily in the industrial sector (including the petrochemical industry), power sector, and residential sector [65].

The expansion of natural gas use was restrained largely for two reasons. First, natural gas production requires substantial investments in infrastructure; and, second, natural gas is a far more expensive fuel than coal. Another important character of natural gas production in China is that major gas fields are located inland in the western and north–central parts of the country, far away from major population and industrial centers. Historically, natural gas exploration has been closely linked to the development of oil fields (associated gas). However, as a result of accelerated exploration since the early 2000s, substantial new gas reserves have been discovered, most of them being non-associated gas.

In order to overcome this mismatch between the location of gas production (the western part of China) and where it is mostly needed (the eastern part), the government has promoted the construction of a natural gas transport infrastructure and improved interconnections between regional networks. After the significant discovery in the Ordos Basin in the late 1980s, the government decided to build the first major interregional pipeline, the Ordos–Beijing pipeline, which was completed in 1997. A huge national project, the West–East pipeline, was approved in 2000 and completed in December 2004. This pipeline delivers natural gas from the Tarim Basin in western China to Shanghai in the south-east [66].

4.5 Nuclear Power

Nuclear electric power in China first became available in 1991, when the Daya Bay plant came on line [67]. By 2009, China got about 2% of its power generating capacity from its 11 nuclear reactors. As governments worldwide look at nuclear power as a possible answer to global warming and as a way to diversify their energy mix away from fossil fuels, China has embarked on an ambitious plan to build dozens of new nuclear reactors, bringing the sector's share to 5% of its generating capacity [68]. China's plans already have been felt in world markets. In recent years Beijing has secured contracts for the uranium needed to power its nuclear reactors with a number of countries including Australia, Kazakhstan, and Niger [69].

4.6   Renewable Energy

Environmental sensitivity is now very much a part of energy security. New environmental constraints narrow the range of energy choices. The coal that has powered China's economic growth is also choking its people. Yet, coal use is only one of the nation's air-quality problems. The transportation boom (e.g., the surge in car ownership due to an expanding middle class) poses a growing challenge to China's air quality. Not surprisingly, oil import dependence is projected to soar from 47% in 2005 to 80% in 2030, making China the largest petroleum importing country in the world [70].

In 2007, a joint research project with the World Bank calculated that 750 000 people die prematurely every year in China due to diseases caused by air pollution [71]. Indeed, China is home to 16 of the world's 20 most polluted cities [72]. Moreover, forests and farmland have been eroded to make room for industry and sprawling cities. In response, the Chinese government has taken several initiatives to reduce its heavy dependence on fossil fuels and to reduce environmental hazards.

In December 2005, the NDRC set up a Committee on Petroleum Alternative Energy Strategy Research. In the 11th Five-Year Plan (2006–2010), clear guidelines were provided concerning alternative fuels for replacing coal, oil, and natural gas. The guidelines called for accelerating the introduction of biofuels and promoting methanol by utilizing by-product gases available from industrial processes. In 2006, China released its first national climate change plan with the objective of slowing its growth of greenhouse gas emissions [73].

Since the mid-2000s, China has made tremendous progress in developing renewable sources of energy. In 2005, it became the third largest country in the world (after Brazil and the United States) in producing bio-ethanol for automotive use. China also has one of the fastest-growing markets for wind power in the world [74]. Finally, it is building one of the world's largest photovoltaic projects in the Mongolian Desert [75]. In the auto sector, the Chinese government is giving strong support to companies which are developing electric vehicles.

To sum up, China has big ambitions to shift its energy network away from dirty coal and imported oil. However, a lot of the groundwork has yet to be put in place. Globally, Beijing believes that developed countries should take the lead in reducing their greenhouse gas emissions and providing financial resources, technology transfer, and capacity-building support to developing countries [76].

4.7   Overseas Exploration and Production

The combination of dwindling domestic energy production and booming economy since the early 1980s has forced Chinese national oil companies (NOCs) to explore new energy resources both at home and abroad. As they have gained a better understanding of the international rules and practices, the Chinese NOCs have become global investors and even have become publicly listed on the New York, Hong Kong, and London Stock Exchanges [77].

In the 1990s China embarked on a path of overseas oil and gas investment. The start of the new millennium marked a new era in which Chinese NOCs expanded their investments overseas, strongly supported by the government in Beijing. Zou Chuqu or "going out," as the then Premier Zhu Rongji called it, was the driving force for a broad strategy to pursue energy security. This strategy was accelerated by China's multi-trillion foreign reserves and the shortage of foreign investment in many oil and natural gas producing regions [78]. Stated differently, Chinese state-owned companies have been encouraged to make acquisitions by a central government convinced that the global financial crisis has created an unmatched buying opportunity.

CNPC is leading the charge in overseas upstream petroleum investment and is joined by its publicly listed subsidiary PetroChina. CNOOC is following CNPC in the pursuit of overseas ventures, but has focused more on gas deals. Sinopec is a latecomer and is trying to catch up the other two [79]. Since the early 2000s these Chinese companies have finalized global acquisitions and offered loans for oil deals in several countries including Russia, Kazakhstan, Turkmenistan, Brazil, Ecuador, and Venezuela. The largest overseas takeover to date by a Chinese company occurred in June 2009 when Sinopec agreed to acquire Addax, a Switzerland-based oil company with interests in Africa and the oil frontier of Iraqi Kurdistan [80]. The purchase also demonstrates growing confidence among Chinese energy companies. In the past, they have preferred to strike government-to-government deals and offer loans for oil [81].

As the world's fastest-growing major economy, China cannot afford not to acquire resources overseas. However, Chinese companies' heavy investments suggest two conclusions. First, the drive to finalize investment deals overseas has occasionally ignited competition between Chinese companies. Competition between PetroChina, Sinopec, and CNOOC were evident in several cases. Furthermore, the presence of multiple state companies from the same country can be confusing to host nations. Second, despite the fact that Chinese energy companies have only a short history of managing the political risk of venture into an overseas market, their aggressive pursuit of overseas investments has drawn scrutiny from other governments. China has already had several high-profile setbacks in its attempt to buy into existing energy companies in Kazakhstan, Russia, and elsewhere.

4.8   Conclusion

Given China's size, population, economy, and standing in the international system, the nation's current and future energy outlook and the choices its leaders make will have important economic, geopolitical, and environmental implications not only inside China, but also way beyond its borders. The preceding discussion of Beijing's energy policies suggests four conclusions. First, China's efforts to reduce its dependence on fossil fuels are likely to improve its energy security. However, in the foreseeable future, China is likely to grow more dependent on foreign supplies. Simply stated, the nation's proven reserves cannot meet its large population's needs. Foreign oil and natural gas supplies are crucial to sustain the high economic growth rate.

Second, China's active involvement in global energy markets has generated some suspicion. The evidence, however, suggests that this growing role is driven largely by commercial and economic interests more than by political objectives. Chinese state-owned energy companies do not seek to promote any ideology; rather, they seek to secure energy supplies and maximize profit.

Third, Beijing's efforts to improve its energy security are not mutually exclusive with those of other nations. Chinese overseas investments will lead to more oil and natural gas supplies in the global market. Equally important, Beijing's efforts to reduce pollution would contribute to global efforts to slow climate change. Cooperation between China and other energy consuming and producing nations would serve the interests of all parties and would further stabilize the global energy market.

Fourth, China's role in climate change has increasingly become a topic of great international attention. The nation is the world's largest emitter of greenhouse gases measured on an annual basis, albeit with relatively low emissions per capita [82]. Under the Kyoto Protocol, China is not obligated to cut its greenhouse gas emissions. However, any future climate treaty will be ineffective unless Beijing agrees to make deep cuts. In the negotiations leading to the Copenhagen Global Climate Summit (December 2009), Chinese leaders pledged to cut carbon intensity, a measure of carbon dioxide emissions per unit of gross domestic product, by 40–45% by 2020, compared to levels in 2005 [83]. This announcement, however, fell short of global expectations. Many environmental experts warned that China's plan does not commit it to reducing emissions; rather, the emissions will continue, though at a slower rate.

References

[1] International Energy Agency (2007) *World Energy Outlook*, International Energy Agency, Paris, p. 44.
[2] British Petroleum (2009) *BP Statistical Review of World Energy*, British Petroleum, London, pp. 6, 8, 11, 22, 24, 27, 32, 34, 35.
[3] Energy Information Administration, *Country Analysis Briefs: China*. Available at http://www.eia.doe.gov/emeu/cabs/china/full.html (accessed August 19, 2009).
[4] Klare, M. (2006) Fueling the dragon: China's strategic energy dilemma. *Current History*, **105** (690), 180–185, 181.
[5] Zha, Daojiong and Hu, Weixing (2007) Promoting energy partnership in Beijing and Washington. *Washington Quarterly*, **30** (4), 105–115, 109.
[6] Zha, Daojiong (2006) China's energy security: domestic and international issues. *Survival*, **48** (1), 179–190, 186.
[7] Koyama, K., Institute of Energy Economics, *Perspectives on Asian Oil Demand: Outlook and Uncertainties*. Available at http://eneken.ieej.or.jp/en (accessed February 27, 2007).
[8] Xu, Yi-Chong (2006) China's energy security. *Australian Journal of International Affairs*, **60** (2), 265–286, 272.
[9] Editorial (2008) China establishes new energy agency. *Oil & Gas Journal*, **106** (30), 5.
[10] China National Petroleum Corporation, *CNPC at a Glance*. Available at http://www.cnpc.com.cn/en/aboutcnpc/cnpcataglance (accessed October 7, 2009).
[11] China National Petroleum Corporation, *Our Strategy*. Available at http://www.cnpc.com.cn/en/aboutcnpc/companyprofile/ourstrategy/OurStrategy.htm (accessed October 7, 2009).
[12] China National Petroleum Corporation, *History*. Available at http://www.cnpc.com/cn/en/aboutcnpc/companyprofile/history/default.htm (accessed October 7, 2009).
[13] China Petroleum and Chemical Corporation, *About Sinopec*. Available at http://english.sinopec.com/about_sinopec (accessed October 8, 2009).
[14] Arruda, M.E. and Li, K.-Y. (2004) Framework of policies, institutions in place to enable China to meet its soaring oil, gas demand. *Oil & Gas Journal*, **102** (33), 20–24, 24.
[15] Jaffe, A.M. and Lewis, S. (2002) Beijing's oil diplomacy. *Survival*, **44** (1), 115–134, 117.
[16] Cornelius, P. and Story, J. (2007) China and global energy markets. *Orbis*, **51** (1), 5–20, 8.
[17] Xu, Yi-Chong (2006) China's energy security. *Australian Journal of International Affairs*, **60** (2), 265–286, 271.
[18] Feng, Lianyong, Li, Junchen, Pang, Xionggi, Tang, Xu, Zhao, Lin, and Zhao, Qingfei (2008) Peak oil models forecast China's oil supply and demand. *Oil and Gas Journal*, **106** (2), 43–47, 44.
[19] Rach, N. (2005) China pushes domestic upstream development. *Oil & Gas Journal*, **103** (26), 35–43, 35
[20] Arruda, M.E. and Li, K.-Y. (2004) Legal reforms opening doors to foreign investment in China's oil, gas sector. *Oil & Gas Journal*, **102** (34), 18–20, 18.
[21] Zha, Daojiong (2006) China's energy security: domestic and international issues. *Survival*, **48** (1), 179–190, 181.
[22] Ogutcu, M. (1998) China and the world energy system: new links. *Journal of Energy and Development*, **23** (2), 281–318, 300.

[23] Klare, M.T. (2006) Fueling the dragon: China's strategic energy dilemma. *Current History*, **105** (690), 180–185, 182.

[24] Xin, Ma,(2008) China's energy strategy in the Middle East. *Middle East Economic Survey*, **51** (23). Available online at http://www.mees.com.

[25] Lee, D.K. (1994) Peking's Middle East policy in the post-Cold War era. *Issues and Studies*, **30** (8), 69–94, 70.

[26] Rynhold, J. (1996) China's cautious new pragmatism in the Middle East. *Survival*, **38** (3), 102–116, 103.

[27] Manning, R. (2000) The Asian energy predicament. *Survival*, **42** (3), 73–88, 83.

[28] Yetiv, S. and Lu, C. (2007) China, global energy, and the Middle East. *Middle East Journal*, **61** (2), 199–218, 201.

[29] Goldstein, J. (2004) The Republic of China and Israel, 1911-2003. *Israel Affairs*, **10** (1 & 2), 223–253.

[30] Rynhold, J. (1996) China's cautious new pragmatism in the Middle East. *Survival*, **38** (3), 102–116, 103.

[31] Leverett, F. and Bader, J. (2005-2006) Managing China-US energy competition in the Middle East. *Washington Quarterly*, **29** (1), 187–201, 190.

[32] Goode, E. and Mohammed, R. (2008) Iraq Signs Oil Deal with China Worth Up to $3.5 Billion. New York Times (August 29).

[33] Shichor, Y. (2006) China's Kurdish policy. *China Brief*, **6** (1), 3–6, 3.

[34] Brooks, P. and Shin, J.H. (2006) *China's Influence in Africa: Implications for the United States*, Heritage Foundation, Washington, DC, p. 1.

[35] Taylor, I. (2006) China's oil diplomacy in Africa. *International Affairs*, **82** (5), 937–959, 938.

[36] Hanson, S. (2004) *China, Africa, and Oil*, Council on Foreign Relations, Washington, DC, p. 3.

[37] Vines, A., Wong, L., Weimer, M., and Campos, I. (2009) *Thirst for African Oil: Asian National Oil Companies in Nigeria and Angola*, Chatham House, London, p. 56.

[38] Forum on China-Africa Cooperation, *The Creation of FOCAC*. Available at http://www.fmprc.gov.cn (accessed October 25, 2009).

[39] China News Agency, *Premier Wen Announced Eight New Measures to Enhance Co-op with Africa in the Next Three Years*. Available at http://news.xinhuanet.com/english/2009-11/08/content_12411421.htm (accessed November 8, 2009).

[40] China-Africa Development Fund, *About us*. Available at http://www.cadfund.com/en/column.asp?columnId=51 (accessed October 25, 2009).

[41] China-Africa Business Council, *Brief Introduction to China-Africa Business Council*. Available at http://www.cabc.org.cn/english/introduce.asp (accessed October 25, 2009).

[42] De Lorenzo, M., American Enterprise Institute, *African Perspectives on China*. Available at http://www.aei.org/include/pub_print.asp?pubID=26917 (accessed October 7, 2007).

[43] Gill, B. and Reilly, J. (2007) The tenuous hold of China Inc. in Africa. *Washington Quarterly*, **30** (3), 37–52, 47.

[44] Millward, J. (1998) *Beyond the Pass: Economy, Ethnicity, and Empire in Qing Central Asia, 1759-1864*, Stanford University Press, Stanford, CA, p. 18.

[45] Shanghai Cooperation Organization, *Brief Introduction to the Shanghai Cooperation Organization*. Available at http://www.sectsco.org/html/00026.html (accessed August 21, 2007).

[46] Watkins, E. (2006) China pursues nearby energy. *Oil & Gas Journal*, **104** (14), 28.

[47] Neff, A. (2006) China competing with Russia for Central Asian investments. *Oil & Gas Journal*, **104** (9), 41–46, 44.

[48] Energy Information Administration, *Country Analysis Briefs: Kazakhstan*. Available at http://www.eia.doe.gov/emeu/cabs/kazakhstan/full.html (accessed January 31, 2008).

[49] Moscow Times (2009) Turkmens Get $4 Billion Loan. Moscow Times (June 26).

[50] News Central Asia, *China, Turkmenistan Sign Landmark Gas Deal*. Available at http://www.newscentralasia.net/print/129.html (accessed May 3, 2009).

[51] Editorial (2009) Turkmenistan completes gas line spur to China. *Oil & Gas Journal*, **107** (41), 11.

[52] Swanstrom, N. (2005) China and Central Asia: a new great game or traditional vassal relations? *Journal of Contemporary China*, **14** (45), 569–584, 576.

[53] Collins, G. (2008) China's refining expansions to reshape global oil trade. *Oil & Gas Journal*, **106**, (7), 22–29, 28.

[54] Vautrain, J. (2008) China, India lead growth in Asian refining capacity. *Oil & Gas Journal*, **106** (47), 52–57, 54.

[55] Collins, G. (2006) China seeks oil security with new tanker fleet. *Oil & Gas Journal*, **104** (38), 20–26, 20.

[56] Energy Information Administration, *World Oil Transit Chokepoints*. Available at http://www.eia.doe.gov/cabs/world_Oil_Transit_Chokepoints/Malacca.html (accessed November 24, 2009).

[57] Moscow Times (2009) China Starts Work on Russia Pipeline. Moscow Times (May 19).

[58] Nieh, D., Wu, K., Wang, L., and Fu, S. (2007) Study examines Chinese SPR growth alternatives. *Oil & Gas Journal*, **105** (28), 80–86, 80.

[59] Collins, G. (2007) China fills first SPR Site, faces oil, pipeline issues. *Oil & Gas Journal*, **105** (31), 20–29, 26.

[60] British Petroleum (2009) *BP Statistical Review of World Energy*, British Petroleum, London, p. 32.

[61] International Energy Agency (2009) *Cleaner Coal in China*, International Energy Agency, Paris, p. 19.

[62] Chatham House (2007) *Changing Climates: Interdependencies on Energy and Climate Security for China and Europe*, Chatham House, London, p. 2.

[63] Oster, S. (2009) China Cuts Small Plant. Wall Street Journal (July 31).

[64] Editorial (2009) China begins commissioning third LNG terminal, *Oil & Gas Journal*, **107** (40), 11.

[65] Wu, K., Hosoe, T., Mukherji, V., and Aik, A.Z. (2008) Emerging Asia-pacific LNG markets must sort pricing, supply uncertainties. *Oil & Gas Journal*, **106** (8), 57–63, 60.

[66] Higashi, N. (2009) *Natural Gas in China Market Evolution and Strategy*, International Energy Agency, Paris, p. 7.

[67] Ebel, R.E. (2009) *Energy and Geopolitics in China: Mixing Oil and Politics*, Center for Strategic and International Studies, Washington, DC, p. 59.

[68] Yang, J. (2009) China Plans to Build Advanced Nuclear-Power Plant. Wall Street Journal (October 30).

[69] Cha, A.E. (2007) China Embraces Nuclear Future: Optimism Mixes with Concern as Dozens of Plants Go Up. Washington Post (May 29).

[70] Zhang, Y. and Chew, C.S. *Alternative Fuel Implementation Policy in China and Its Assessment*, Institute of Energy Economics, Japan. Available at http://eneken.ieej.or.jp/en (accessed May 24, 2007).

[71] Dyer, G. (2009) Difficulties of a Different Energy Model. Financial Times (October 13).

[72] Economy, E.C. (2007) The great leap backward? *Foreign Affairs*, **86** (5), 38–59, 39.

[73] Lewis, J.I. (2008) China's strategic priorities in International climate change negotiations. *Washington Quarterly*, **31** (1), 155–174, 155.

[74] Ebel, R.E. (2009) *Energy and Geopolitics in China: Mixing Oil and Politics*, Center for Strategic and International Studies, Washington, DC, p. 61.

[75] Woody, T. (2009) China and U.S. Company Plan a Bib Solar Project. New York Times (September 9).

[76] Harvey, F. and Lamont, J. (2009) China and India to Resist Caps on Carbon. Financial Times (October 22).

[77] Xu, X. (2007) *Chinese NOCs' Overseas Strategies: Background, Comparison, and Remarks*, James A. Baker III Institute for Public Policy, Rice University, Houston, TX, p. 3.

[78] Editorial (2009) China's unswerving appetite for energy. *Petroleum Economist*, **76** (8), pp. 22–24, p. 22.

[79] Wu, K. and Han, S.L. (2005) Chinese companies pursue overseas oil and gas assets. *Oil & Gas Journal*, **103** (15), 18–25, 18.

[80] MacNamara, W. (2009) Addax Takeover to Take Sinopec into Iraq. Financial Times (June 24).

[81] Chazan, G. and Oster, S. (2009) Sinopec Pact for Addax Boosts China's Buying Binge. Wall Street Journal (June 25).

[82] Lewis, J.I. (2009) Climate change and security: examining China's challenges in a warming world. *International Affairs*, **85** (6), 1195–1213, 1195.

[83] Eilperin, J. (2009) China Announces Planned Emissions Cuts. Washington Post (November 26).

5

Persian Gulf

Iran, Iraq, and the six Arab states on the Persian Gulf (Bahrain, Kuwait, Oman, Qatar, Saudi Arabia, and the United Arab Emirates) enjoy several advantages as the world's major oil and natural gas producers. First, together these eight states hold the world's largest proven oil and gas reserves. In 2010 their combined share of the world's reserves is 56.2%, production is 29.6%, and consumption is 6.6%. The figures for natural gas are 40.1%, 13.4%, and 10.2% respectively [1]. No region in the world holds as much proven reserves. The high level of production and low (though growing) level of consumption mean that a substantial proportion of the oil produced in this region is exported to the rest of the world. Meanwhile, the massive natural gas reserves and the relatively small volume of production mean that the region has the potential to play a leading role as a natural gas supplier, once gas deposits are developed.

Second, the cost of production in the Persian Gulf is one of the lowest in the world. Unlike Russia, the Caspian Basin, the North Sea, and the Gulf of Mexico, most oil and gas fields are either onshore or in the shallow waters of the Persian Gulf. This accessibility means that much of the oil and gas production in the Gulf region is less environmentally challenging and cheaper to produce.

Third, the Gulf region has been producing and exporting oil and gas for decades. Generally the energy infrastructure is well developed. Extensive pipeline networks connect the oil and gas fields to marine export terminals and loading platforms on the Persian Gulf. From there the region has easy access to the high seas and global markets. Unlike other producing regions, shipping ports on the Persian Gulf do not experience major storms or freezing.

Finally, traditionally most of the world's spare capacity of oil is concentrated in the Persian Gulf, particularly Saudi Arabia. This spare capacity serves as an insurance policy against any unexpected interruption of supplies due to natural or political reasons [2]. This concentration of spare capacity can be explained by the fact that oil and gas production is dominated by state-owned national companies. Unlike private international oil companies, which aim mainly at maximizing their profits, these state-owned and state-managed national oil companies are driven by both strategic concerns and commercial interests. Saudi Arabia and other Gulf producers maintain spare capacity to ensure short- and long-term stability of global markets.

Given these advantages and despite rising production in Russia, the Caspian Basin, and Africa, the bulk of the increase in world oil output is projected to come from the Persian

Energy Security: An Interdisciplinary Approach, First Edition. Gawdat Bahgat.
© 2011 John Wiley & Sons, Ltd. Published 2011 by John Wiley & Sons, Ltd.

Gulf. According to the International Energy Agency (IEA), the oil and gas resources of the Persian Gulf will continue to be critical in meeting the world's growing appetite for energy. The region's massive hydrocarbon resources are sufficient to meet rising global demand for the next quarter-century and beyond [3]. The United States Department of Energy echoes these sentiments. The Energy Information Administration (EIA) projects that the bulk of the world's oil and gas production will come from the Persian Gulf region [4].

In short, history and geology put Persian Gulf oil and gas producers in the driver's seat. The region has been producing and exporting crude and natural gas for decades and is certain to maintain this policy and status in the future. The projected rise of the Gulf's share in meeting global demand means that major consumers (i.e., China, India, Japan, Europe, and the United States) are likely to grow more dependent on energy supplies from the Persian Gulf. The smooth continuation of this mutual dependence between Gulf producers and major consumers requires close cooperation in addressing several strategic and commercial challenges. Some of these challenges are domestic while others are regional and international.

Most of the Persian Gulf governments have achieved a modest success in initiating and implementing economic and political reform. There is much to be desired in pursuing economic development and political liberalization. Equally important, international sanctions, wars, ethnic and sectarian strives, terrorism, and overall regional instability have negatively impacted the full utilization of the region's hydrocarbon resources.

These domestic, regional, and international challenges have raised doubts about the reliability of oil and gas supplies from the Middle East. In the last few decades policy-makers, media outlets, and think-tanks in Washington, Brussels, Beijing, and Tokyo have frequently called for reducing energy dependence on the Middle East. US officials, more than their European and Asian counterparts, have repeatedly talked about "energy independence" and stopping or reducing the nation's "addiction to oil."

This study argues that such calls are useful for political rhetoric and gaining votes. As an energy analyst asserts, "Presidents may declare an urgent need to cut imports and boost energy independence – no one ever lost political support by seeing evil and blaming foreigners" [5]. In reality and based on projections by US, European, and Asian governments as well as by major international organizations, the world will grow more dependent on oil and natural gas supplies from the Persian Gulf. Furthermore, the region's long history of producing and exporting hydrocarbon fuels suggests that concerns over interruption of supplies from the Middle East are exaggerated. A close scrutiny indicates that, with a few exceptions, the region has proven a reliable producer and exporter of oil and natural gas.

In the following sections I briefly discuss some of the major socio-economic and political challenges threatening stability in the Persian Gulf region. This will be followed by a close examination of energy policy and history in the three main players with the largest deposits – Saudi Arabia, Iran, and Iraq. The analysis suggests that the Persian Gulf oil and gas producers are partners to major consumers in the quest for global energy security. Any interruption of supplies would harm their interests as much as those of consuming regions. The way forward is more cooperation and less confrontation.

5.1   Socio-economic and Political Challenges

One of the most cited reasons for reducing dependence on Middle Eastern oil supplies is the concern that Arab producers might repeat the experience of the 1973–1974 oil embargo.

The events triggered many developments and the consequences of the embargo need to be scrutinized to determine if it can happen again. Stated differently, can oil producers use their petroleum as a political weapon to intimidate and blackmail one or more consumers?

For a long time, the idea of using oil as a political weapon by Arab producers against Western powers that support Israel has been considered. Thus, in the Suez Crisis (1956) and the Six-Day War (1967), limited embargoes were implemented with little effect. In 1956, Syria attacked Anglo-French pipeline interests, most notably putting the Kirkuk–Tripoli pipeline out of operation and stemming the flow of oil from Iraq into the Mediterranean. The United States was able to compensate for the embargo against the United Kingdom and France from its own domestic production. In 1967, the Saudi government ordered Aramco to stop oil supplies to the United States and the United Kingdom during the Arab–Israeli War. The ability of the international oil companies to reroute supplies, however, made the embargo ineffective. In addition, other producers, mainly Iran and Venezuela, ramped up their production and the net impact of the boycott proved to be extremely minor [6]. Furthermore, the Saudi leaders were not fully convinced of the validity of mixing oil with policy. Rather, they preferred to use oil revenues as a "positive weapon" to build up the military and economic strength of the Arab world [7]. Accordingly, Saudi Arabia and other Arab oil producers gave substantial financial aid to Egypt and Jordan from 1967 to 1973.

The outbreak of the 1973 war between Israel on one side and Egypt and Syria on the other, and the political and military support the former received from the United States and other Western powers, changed the political dynamics in the Middle East and prompted Arab oil producers to cut their production and impose restrictions on oil exports to the United States and a few other countries.

On October 17, 1973, 11 days after the Yom Kippur War started, Arab oil ministers met in Kuwait and decided to reduce production by 5% each month (from the previous month) until the total withdrawal of all Israeli forces from the territories occupied during the Six-Day War of 1967 was complete. In the following few days the United States announced a massive new program of military aid to Israel and Arab countries announced a total ban on shipments of oil to the United States, the Netherlands, Portugal, and South Africa in retaliation for their support of Israel [8]. The Arab countries also declared that "Arab-friendly states" would not be affected by that decision. Western oil companies did their best to spread the burden of the oil shortfalls among all countries. Other OPEC countries (i.e., Iran) increased their production.

Between October and December 1973, prices went wild. The companies and countries that were most severely hit by the cuts went looking for crude wherever it could be found, offering prices that a month before would have seemed insane [9]. Finally, in December, OPEC decided to raise the official posted price of benchmark crude – Arabian light – to $11.65 per barrel, which meant a four-fold increase in less than four months; even more shocking, oil prices had skyrocketed by almost a factor of 10 since 1970.

Peace negotiations got under way in Geneva between Egypt, Jordan, and Israel in December and restrictions were gradually lifted as the talks made progress. Little by little, the situation returned to normal, but it was not until March 18, 1974 that a decision was taken by the Arab countries to lift the embargo on shipments to the United States. This selective embargo had come to nothing, and hopes of a total Israeli withdrawal from the occupied Arab territories proved quite unrealistic. Yet, the commotion it provoked had already shattered post-war economic certainties [10].

Despite occasional rhetoric by officials from some oil producing countries, the likelihood of using oil as a political weapon and imposing an oil embargo similar to the 1973 experience is highly unlikely. Not only did the 1973 embargo fail to achieve its political goals, but also it substantially hurt the interests of oil producers. Confidence in the Persian Gulf and the broader Middle East as a reliable source was greatly damaged. Major consuming countries allocated massive financial resources to the exploration and development of oil deposits outside the Middle East. A substantial proportion of these investments went to the North Sea and for the next few decades the supplies from the North Sea satisfied part of European energy consumption, competing with the Persian Gulf. Finally, the increasing globalization of energy markets means that the concept of "selective embargo" or restricting exports to a few consumers is obsolete.

This low probability of another oil embargo, however, does not mean that the non-interruption of oil and gas supplies from the Persian Gulf should be taken for granted. Other socio-economic and political challenges need to be addressed. The list includes sanctions; wars and ethnic/sectarian strife; terrorism; closure of the Straits of Hormuz; political instability; and underinvestment.

5.1.1   International Sanctions

Probably more than any other global power, the United States has taken the lead in imposing unilateral and multilateral sanctions against oil producing countries to force them to change their policies. Libya and Iraq had been under such sanctions for most of the 1990s and early 2000s. Since the 1979 revolution, Iran has been under strict US economic sanctions that ban US companies from investing in the country's energy sector and threaten to penalize companies from other countries if they do business with the Islamic Republic. In recent years, the United States managed to reach a consensus with other major powers to issue UN economic sanctions against Iran due to a dispute over Tehran's nuclear program. These sanctions do not stop the flow of oil and natural gas from the targeted country, but they limit foreign investment and technology transfer. Eventually, economic sanctions prevent producers from reaching their full potential and reduce oil and gas supplies in the global energy markets. In short, it can be argued that consuming countries, particularly the United States, have recently used oil as a political weapon to force a change in political behavior and orientation more often than producing countries have.

5.1.2   Wars and Ethnic/Sectarian Strife

In the last few decades the broader Middle East has experienced a number of wars. These include the 1948, 1956, 1967, and 1973 wars between Arab countries and Israel; the Iran–Iraq War (1980–1988); the Iraqi invasion of Kuwait and the subsequent Gulf War (1990–1991); and the war in Iraq (started in 2003). The region also witnessed several military confrontations between Israel and Hamas in the Gaza Strip and Hezbollah in Lebanon. Finally a number of Middle Eastern countries have suffered from intense ethnic/sectarian strife (namely, Bahrain, Iraq, and Lebanon). These wars and domestic conflicts have multiple negative effects on oil and gas production. Oil and gas infrastructures are attacked and domestic financial resources are diverted from updating and modernizing these infrastructures to resolving these war and ethnic/sectarian conflicts. Finally, this instability scares away foreign investment. International energy companies prefer to operate in stable regions.

5.1.3   Terrorism

Like the rest of the world, Persian Gulf states have been subjected to terrorist attacks. In the last few years, oil and gas installations, pipelines, and tankers have been subjected to numerous terrorist attacks, mainly in Iraq and to a lesser extent in Saudi Arabia. Most oil and gas producers have recently spent millions of dollars on improving the security of their energy installations and infrastructures. The price of oil reflects this perceived threat of terrorism. The so-called "fear premium" refers to the rise in insurance to cover damage from potential terrorist attacks.

5.1.4   Closure of Straits of Hormuz

Most of the oil and natural gas supplies from the Persian Gulf are shipped to the global markets via the Straits of Hormuz. During many of the wars that the region has witnessed, threats to close the Straits were made. In any future military confrontation in the region, such threats cannot be ruled out. In reality, however, the risks to maritime flows of oil seem smaller than is commonly assumed. The most notable example comes from the Iran–Iraq War, during which the two countries attacked shipping in the Persian Gulf to weaken each other's economies. Neither of them was able to succeed. Although commercial shipping in the Gulf initially dropped by about 25% and the price of crude oil spiked, the so-called tanker war did not in the end significantly disrupt oil shipments.

In 1987, the United States offered to protect Kuwaiti ships by letting them fly US flags and providing escorts, which deterred further attacks. Even at its most intense, the tanker war failed to ensnare more than 2 % of the ships traveling through the Persian Gulf [11]. Since then the size and strength of the global tanker fleet have increased markedly, making it even more difficult to disrupt maritime oil shipments.

5.1.5   Domestic Instability

Despite attempts to diversify their economies and to create other sources of income besides oil and gas revenues, the Persian Gulf states continue to be heavily dependent on oil as the main source of national income. This dependence contributes to high levels of unemployment and other socio-economic problems. These problems and the general lack of political liberalization do not necessary mean domestic instability. The royal families in the Persian Gulf have proven resilient to major upheavals such as the rise of Arab nationalism in the 1950s and 1960s and Islamic fundamentalism since the 1980s. Many analysts have underestimated the adaptability of these regimes. At the end of the first decade of the twenty-first century, there is no sign that any of the Gulf monarchies is about to be overthrown. Scenarios of bloody and violent changes in political regimes in Saudi Arabia and other Gulf states which could ruin the social fabric of the country and pose risks to oil exports are highly unlikely. Similarly, the Islamic regime in Tehran seems relatively secure from internal revolution. Certainly, there is much to be desired in terms of economic and political reform, but it seems that the Gulf governments have been able to manage public discontent. Iraq is a special case, where the regime was toppled by foreign powers. Still, significant improvements in security and stability have been made since 2003.

It is difficult to envisage a scenario under which an anti-Western regime takes over any of the Persian Gulf states and voluntarily curtails oil exports. Iran is a classic example: despite a violent and far-reaching revolution that saw the substitution of a Western-friendly government by an anti-Western regime, oil has continued to flow from Iran to the West and the rest of the world, even at the climax of the Islamic Revolution, although at half-capacity. The reduction in Iran's oil capacity was not the result of a deliberate policy or restricting oil exports, but a combination of factors including the Iran–Iraq War, sanctions, and underinvestment due to the unattractive business environment [12].

5.1.6   Underinvestment

Many oil and gas fields in the Persian Gulf have been producing for decades, some have just come on-stream, and others are being developed. In order to compensate for depletion, maintain the current level of production, and add extra capacity, massive capital needs to be invested. According to most resources, the region's hydrocarbon deposits are not in doubt. The development of these resources, however, would largely depend on the availability of the necessary investment. How much international energy companies are allowed into the region's oil and gas sectors varies from one county to another. National oil companies dominate the industry in most of the Persian Gulf states and the broader Middle East. The full participation of foreign investment is restrained by geopolitical considerations, including sanctions and national laws and regulations. In recent years there have been some signs of changing attitudes; however, more is needed to establish a partnership with international energy companies that will bring the required financial resources and advanced technology.

Saudi Arabia, Iran, and Iraq, the three biggest countries in the region, have pursued different strategies in developing their hydrocarbon resources. The similarities and differences between these strategies are discussed in the following sections.

5.2   Saudi Arabia

The kingdom of Saudi Arabia is by far the world's most influential oil producing country and potentially a significant natural gas producer. In 2008 it held approximately one-quarter of the world's proven oil reserves. Its share of world production is 13.1%, while that of consumption is 2.7% [13]. This large difference between the volumes of production and consumption has made the kingdom the largest exporter in the world. Saudi Arabia has always maintained the world's largest spare capacity and over the years has developed an impressive petrochemical industry and refinery capacity both at home and in partnership with international energy companies. Saudi Arabia also holds the world's fourth largest natural gas proven reserves (after Russia, Iran, and Qatar respectively) and has recently sought to develop these massive gas deposits.

Unlike other major producers in the Persian Gulf and OPEC, oil exploration and development in Saudi Arabia have been carried out almost entirely by US companies. In the early 1930s, US oil companies were looking for commercial opportunities overseas. Promising oil reservoirs had been discovered in Iran, Iraq, and Bahrain. This newly discovered hydrocarbon wealth was dominated by European companies, particularly from the United Kingdom. Meanwhile, indigenous leaders were interested in granting concessions to foreign companies in order to strengthen their rising economic and political power.

Under these circumstances, in 1933 King Saud Ibn Abd al-Aziz, the founder of modern-day Saudi Arabia, who was suspicious of European intentions, gave Standard Oil Company of California (Socal, later Chevron) a 60-year exclusive right to explore for oil in an area in eastern Saudi Arabia the size of Texas, New Mexico, and Arizona combined [14]. The California–Arabian Standard Oil Company (CASOC) was formed to exploit the concession. In 1936, Socal sold a half-share in CASOC to the Texas Oil Company (Texaco), and later two more US companies acquired shares: Standard Oil Company of New Jersey (later Exxon) and Standard Oil Company of New York (originally Socony, later Mobil) [15]. A supplementary agreement was signed in May 1938, adding six years to the original agreement and enlarging the concession area. It also included rights in the Saudi government's half-interest in the two neutral zones shared with Iraq and Kuwait.

Early exploratory drilling in Saudi Arabia was not successful. Although the first well was completed in 1935, it was not until March 1938 that oil was struck in commercial quantities in the Dammam structure [16]. First exports of oil took place in 1938 and continued at very modest levels until after World War II. But the event that transformed prospects for the oil industry in Saudi Arabia was undoubtedly the discovery of the Ghawar field in 1948, which proved to be the world's largest, single oil bearing structure [17]. The world's largest offshore field, Safaniya, lies in Saudi Arabian waters of the Persian Gulf. In 1944, CASOC was renamed the Arabian American Oil Company (Aramco).

Unlike other foreign oil companies, Aramco had good relations with the host government, Saudi Arabia, and with the local population. The bitter dispute in the early 1950s between the Iranian authorities and British Petroleum was very different from the smooth cooperation between Aramco and the Saudi government. In 1950 they reached an agreement on a modified system of profit sharing, which introduced the notion of a 50/50 division between the host country and the concessionaire. In 1973, the Saudi government took a 25% stake in Aramco. A year later, this share was increased to 60% and in 1980 it was amicably agreed that Aramco should become 100% Saudi owned, with the date of ownership backdated to 1976 [18]. Prior to the Saudi takeover, Aramco had been the largest single US investment in any foreign country. This friendly and non-confrontational change of ownership helped the two sides to maintain their cordial cooperation. Despite the Saudi takeover of Aramco, US administrators and technicians, side by side with their Saudi counterparts, continued to occupy important positions in the company. Finally, in April 1989, the last American to preside over Aramco, John J. Kelberer, handed over power to its first Saudi boss, Ali al-Naimi, who later became an oil minister.

As the national oil company, Aramco is an important player in planning and executing the country's energy policy. Other institutions include the Ministry of Petroleum and Mineral Resources, which was established in 1960 to execute the general policy related to oil, gas, and minerals. The Ministry supervises its affiliate companies by observing and monitoring exploration, development, production, refining, transportation, and distribution activities related to petroleum and petroleum products [19]. Another institution is the Supreme Council for Petroleum and Minerals, established in 2000. It is comprised of members of the royal family, industry leaders, and government ministers. It has the final word on all energy issues including oversight of the oil and natural gas sectors and Aramco, and drawing up the general strategy [20].

Saudi Arabia produces a range of crude oils: Arabian heavy, Arabian medium, Arabian light, Arabian extra light, and Arabian super light. About 65% to 70% is considered light gravity, with the rest either medium or heavy [21]. Lighter grades generally are produced onshore, while medium and heavy grades come mainly from offshore fields. Most Saudi oil production, except for extra light and super light, is considered "sour," containing relatively high levels of sulfur. Crude comes from the fields given in Table 5.1.

Table 5.1   Oil and natural gas fields in Saudi Arabia.

No.	Field name	Date discovered
1	Dammam	1938
2	Abu Hadriya	1940
3	Abqaiq	1940
4	Qatif	1945
5	Ghawar	1948
6	Fadhili	1949
7	Safaniya	1951
8	al-Wafrah	1954
9	Khursaniyah	1956
10	Khurais	1957
11	Manifa	1957
12	Khafji	1960
13	Hout	1963
14	Fawaris al-Janub	1963
15	Abu Sa'fah	1963
16	Berri	1964
17	Zuluf	1965
18	Jaham	1966
19	Habari	1966
20	Janubi Um-Qdir	1966
21	Kidan	1967 "Gas"
22	al-Lulu	1967
23	Dorra	1967
24	Marjan	1967
25	Karan	1967
26	Jana	1967
27	Jurayd	1968
28	Juraybi'at	1968
29	Shaybah	1968
30	Barqan	1969
31	Marzouk	1969
32	Harmaliyah	1971
33	Mazalij	1971
34	Shutfah	1972
35	El Haba	1973
36	Maharah	1973
37	Abu Jifan	1973
38	Qirdi	1973
39	Rimthan	1974

Table 5.1   (*Continued*)

No.	Field name	Date discovered
40	Kurayn	1974
41	Ramlah	1974
42	Bakr	1974
43	Lawhah	1975
44	Dibdibah	1975
45	Ribyan	1975
46	Watban	1975
47	Suban	1976
48	Sharar	1976
49	Hasbah	1976
50	Jaladi	1978
51	Harqus	1978
52	Wari'ah	1978
53	Jawb	1979
54	Dhib	1979
55	Faridah	1979
56	Hamur	1979
57	Samin	1979
58	Lughfah	1979
59	Maghrib	1982
60	Tinat	1982
61	Jauf	1983
62	Farhah	1983
63	Sahba	1984
64	Hawtah	1989
65	Dilam	1989
66	Raghib	1990
67	Hazmiyah	1990
68	Nuayyim	1990
69	Hilwah	1990 "Gas"
70	Ghinah	1990
71	Kahf	1991 "Gas"
72	Midyan	1992
73	Wajh South	1993 "Gas"
74	Umluj	1993 "Gas"
75	Umm Jurf	1993
76	Nisalah	1993
77	Layla	1994
78	Abu Markhah	1994
79	Abu Rakiz	1995
80	Burmah	1995
81	Usaylah	1996
82	Shiblah	1996
83	Mulayh	1996
84	Abu Shidad	1996

(*Continued*)

Table 5.1 (*Continued*)

No.	Field name	Date discovered
85	Khuzama	1997
86	Waqr	1997 "Gas"
87	Hamma	1998
88	Wudayhi	1998 "Gas"
89	Sham'ah	1998 "Gas"
90	Sidr	1998
91	Kahla	1998 "Gas"
92	Shaden	1999 "Gas"
93	Niban	1999
94	Ghazal	2000
95	Manjurah	2000 "Gas"
96	Jufayn	2001
97	Warid	2002
98	Tukhman	2002
99	Yabrin	2003
100	Awtad	2003
101	Abu Sydr	2004
102	Midrikah	2004
103	Duayban	2005
104	Halfa	2005
105	Muraiqib	2005
106	Zimlah	2006 "Gas"
107	Kassab	2006 "Gas"
108	Nujayman	2006 "Gas"
109	Mabruk	2007
110	Dirwazah	2007

Source: Ministry of Petroleum and Mineral Resources, *Fields*.
Available at http://www.mopm.gov.sa/mopm/detail.do?
content=fields (accessed September 13, 2009).

As has been mentioned above, Ghawar is the major oil structure in Saudi Arabia and by far the largest in the world. It is the main producer of Arab light and contains several fields. It also contains huge fields of natural gas deep beneath the oil formations. These deposits had been actively sought since industries and utilities in the Eastern Province depend on associated gas for power and feedstock. Oil production costs at Ghawar are the lowest in Saudi Arabia. Safaniya is by far the largest offshore field in the world. It was found in 1951 by Texaco. Deposits at Safaniya are recoverable at relatively low cost, but the oil is heavy with sulfur (not high quality) [22].

The Saudi–Kuwait Divided Zone or the "Neutral Zone," which was left undefined in 1922, is shared between the two countries. Japan's Arabian Oil Company (AOC) operated the two Saudi offshore fields of Khafji and Hout. But in February 2000, AOC lost the concession when Japan refused to invest in development projects requested by the Saudis. As a result, Aramco took over operation of the fields.

Saudi oil is exported via three primary terminals: Ras Tanura, the world's largest offshore oil loading facility, Ras al-Ju'aymah (both on the Persian Gulf), and Yanbu' terminal on the Red Sea. The Saudi government owns and operates a number of shipping companies including the National Shipping Company of Saudi Arabia (NSCSA). Founded in 1979, NSCSA owns more than a dozen very large crude carriers (VLCCs) and uses them in the transportation of crude oil, chemicals, and liquefied petroleum gas (LPG) [23]. Another shipping company, Vela, was established in 1984 to provide marine transportation for refined products and crude oil produced by Aramco. It began as a small company operating with only four crude oil tankers, but over the years has acquired more tankers [24].

In addition, Aramco operates a number of pipelines, most notably the Petroline to transport Arabian light and super light from the Abqaiq refineries in the Eastern Province to Yanbu' on the Red Sea for export to European markets. This pipeline is considered a potential alternative to Gulf port facilities if exports were to be blocked from passing through the Straits of Hormuz in the Persian Gulf. The Abqaiq–Yanbu' liquid natural gas pipeline, which serves Yanbu's petrochemical plants, runs parallel to the Petroline. Two major pipelines ceased to operate due to regional conflicts – the Trans-Arabian pipeline (*Tapline*) from Qaisumah to Sidon in Lebanon, and the Iraqi pipeline across Saudi Arabia, which runs parallel to the Petroline.

Since the discovery of oil, the following dynamics have shaped the kingdom's oil policy:

- The heavy dependence on oil revenues. This means that securing adequate revenues will always be critical for social, economic, and political stability.
- The large share of Saudi oil in the world petroleum market. This implies maintaining a high level of production and large volume of exports to the major markets – Asia, Europe, and the United States.
- The leading role of Saudi Arabia among oil exporting countries. This has prompted the kingdom to seek consensus, occasionally play the role of swing producer, and keep a large volume of spare capacity.
- The challenge of economic diversification. Saudi Arabia, like many other oil producers, depends heavily on oil revenues as the main source of national income. The need to diversify from oil and create other sources of income are the main drivers for investment in hydrocarbon-related industries such as petrochemicals.
- The vital importance of security. Oil installations and infrastructures are easy targets. Protecting them from external threats or terrorist attacks has always been considered a core national security objective [25].

Saudi leaders have always considered their country as the world's central oil bank. The kingdom's massive hydrocarbon reserves justify such a perception and make it in Riyadh's self-interest to stabilize global oil markets. As for other commodities, prices of oil are determined largely by a balance between supply and demand. In recent decades, with a few exceptions, Saudi Arabia has played the role of swing producer – increasing production to make up for any shortages or decreasing production to prevent a price collapse. Against this background, the Saudi policy has been to avoid too low or too high prices and to promote stability at a "reasonable level."

A decline in prices benefits oil consuming countries. It acts in the same was as a tax cut, increasing consumer disposable income. This allows for looser monetary policy and hence lower interest rates with lower inflation and stronger economic growth than would otherwise be the case. On the other hand, sharply higher oil prices have been a major cause of recessions in the United States and other Western economies in the last few decades and are believed to have contributed to the global economic meltdown in the late 2000s.

For Saudi Arabia and other oil producers, excessively low prices have harmful political and economic ramifications. Considering that oil revenues are the main source of national income in Saudi Arabia and other oil producers, very low prices mean shrinking financial resources and reduced public spending, which can lead to social and political upheaval. Meanwhile, too high prices are not in the best interest of major oil producers such as Saudi Arabia. One of the most salient disadvantages could be the development of alternative or competing energy sources, which could undermine the importance of petroleum. In this respect, it is known that when petroleum loses its competitive edge, it will be difficult to recover it even if prices subsequently decline. Another disadvantage of high prices is the exploration and development of oil in high-cost areas such as the North Sea and Gulf of Mexico, which in turn leads to an increase in supply and exerts downward pressures on prices. Third, excessively high prices could have a negative impact on the world economy, which in turn weakens demand in the long run and causes producing countries to lose their credibility.

Within this context, Saudi leaders were not satisfied with the excessively high prices in the early 2000s up to 2008 and the collapse that followed. Saudi Arabia sponsored a summit of OPEC members in Riyadh in November 2007 and a meeting between major consumers, producers, and international oil companies in Jeddah the following year (June 2008). The goal was to reach a consensus on stabilizing prices and markets. In late 2008, King Abdullah of Saudi Arabia called for a price around $75 per barrel, suggesting that such a price would be fair to both consumers and producers [26]. In order to achieve this goal and to assert its central role and leadership, Saudi Arabia increased its oil production capacity to a record 12.5 million b/d in 2009 despite the severe decline in prices in late 2008 and early 2009 [27]. This capacity expansion has helped offset losses from Nigeria, Iran, Iraq, and Venezuela.

In addition to the sizable contribution that oil revenues make to the Saudi gross domestic product (GDP), government revenues, balance of payments, and exports, oil and gas also play a major role in establishing other industries and related services, mainly petrochemicals, electricity, water desalination, and energy-intensive heavy industries. Since the mid-1970s the kingdom has invested substantial resources in establishing and developing a sophisticated petrochemical industrial base. The development of this industry is closely associated with the country's massive natural gas deposits.

Saudi gas is generally rich in ethane, the petrochemical industry's most versatile and favorable feedstock. For a long time natural gas was either reinjected into the oil fields or was simply flared and wasted. For the last few decades, the Saudi authorities have sought to avoid wasting this valuable source. Instead of drawing up a natural gas exporting strategy, Saudi Arabia has embarked on an intensive program to use its gas as a feedstock in its growing petrochemical industry. This industry enjoys two important advantages: abundant and cheap hydrocarbon supplies and proximity to the burgeoning Asian markets. The country's main petrochemical centers are in Jubail and Yanbu'.

The main enterprise in charge of the petrochemical industry is the Saudi Basic Industries Corporation (Sabic). It was founded in 1976 first as a completely government-owned company and then, in 1984, 30% of Sabic's ownership was offered to Saudi and Persian Gulf investors [28]. The underlying reason for creating Sabic was to use the country's associated gas, a by-product of oil extraction, to produce value-added commodities such as chemicals, polymers, and fertilizers. Developing such industries would serve two purposes: diversifying the economy and contributing to export volumes. Since the mid-1980s Sabic petrochemical production has increased several fold. In 1985, production was 6.3 million metric tons (mmt), by the end of 2008 it had reached 56 mmt, and by 2020 Sabic plans to produce 135 mmt per year [29].

In addition to the petrochemical industry, the Saudi authorities, led by Aramco, have invested heavily in developing large and growing refinery projects, both at home and overseas. Aramco has taken the lead in some of these projects and has partnered other international companies in others. Some of the largest refinery schemes include the Rabigh refinery on the Red Sea coast in partnership with Sumitomo Chemical of Japan, the complex in Yanbu' in partnership with ExxonMobil, also on the Red Sea coast, and the facility at Jubail in partnership with Shell on the east coast. Aramco also built, operates, and is negotiating refineries in several countries including China, Japan, Philippines, South Korea, and the United States. The kingdom's refining capacity is projected to increase by 50% from its 2006 level, reaching approximately 6 million b/d [30]. The goal is to meet domestic consumption and the fast growing global demand, particularly in Asian markets.

Before 1970, the kingdom's energy industry was dominated by a single fuel – oil. Although gas associated with crude oil was abundant, it was mostly disposed of by flaring as a useless and worthless by-product of oil production. The changes in global energy markets in the mid-1970s prompted the Saudi government to modify its policy on natural gas. Riyadh instructed Aramco to implement the Master Gas System (MGS), which began functioning in 1981. The MGS feeds petrochemical plants and provides fuel to power utilities, including electrical and seawater desalination plants. It comprises gathering systems, processing plants, fractionation plants, storage facilities, transmission pipelines, and export terminals.

This official interest in utilizing the kingdom's vast gas resources has fueled domestic demand. Since the mid-1980s gas consumption has risen at a remarkable rate of 7.2% and is projected to grow on average by 3.7% up to 2025 [31]. Parallel to the increase in consumption, the volumes of both proven reserves and production have substantially increased [32]. Tables 5.2 and 5.3 summarize some of the main projects.

In order to further accelerate the development of the country's natural gas deposits, in the early 2000s the Saudi authorities negotiated and signed agreements with a number of international energy companies including Russia's LUKoil, China's Sinopec, Italy's ENI, Spain's Repsol, France's Total, and the Netherlands'/UK's Royal Dutch/Shell [33]. Finally, in November 2006, the Petroleum Ministry and Aramco announced a $9 billion strategy to expand gas reserves between 2006 and 2016 through new discoveries [21].

Saudi Arabia's massive hydrocarbon resources have made it an attractive target for terrorists seeking not only to deal a heavy blow to the country's economy, but also to severely disrupt global energy markets with significant ramifications to the international system. Since the early 2000s a number of attempted terrorist attacks on the kingdom's energy installations and infrastructure have been reported [34]. The Saudi authorities have pursued at least two strategies to protect their energy infrastructure. First, the kingdom has set up a 35 000-strong security force trained well in the use of surveillance equipment, countermeasures, and crisis management [35]. These forces are in addition to Aramco's own 5000 special security force [36]. Second, Saudi authorities maintain multiple options for transportation and export in its oil system, in part as a form of indirect security against any one facility being disabled.

To sum up, given its massive reserves, production, and exports, Saudi Arabia plays a leading role in oil policy. Saudi leaders have generally perceived their country's energy security to be in

Table 5.2 Proposed refinery additions and expansion in Saudi Arabia (2005–2013).

Project	Announced completion date	Capacity increase (million b/d)	Total capacity (million b/d)	Aramco partners	Notes
Domestic					
PetroRabigh JV (expansion)	2008 (Q4)	60	460	Sumitomo Chemical (Japan)	Integrated oil refinery and petrochemical complex. Upgrade to shift product mix away from low-value heavy products toward gasoline and kerosene. Second upgrade (2010) to bring capacity up to 825 000 b/d has been proposed
Jubail JV	2012 (Q4)	400	400	Total	Export-oriented, heavy conversion
Ras Tanura	2012	400–440	400–440	None	For domestic consumption. Part of existing Ras Tanura refining complex, which will have a total capacity of approximately 950 000 b/d
Yanbu' JV	2013	400	400	Conoco-Phillips	Export-oriented, heavy conversion
Jizan	TBD (2012–2013)	250–400	250–400	TBD	Proposed. Reported to go to bid in 2008
Yanbu' (expansion)	2010–2011	100	330	None	Proposed. Reported to go to bid in 2008
Overseas					
Fujian, China	2007	80	240	Sinopec (50%) ExxonMobil (25%)	First of two facilities in China. Combined refinery and petrochemical plant
Fujian, China (expansion)	2009 (Q1)	160			
Motiva–Port Arthur, TX (expansion)	2010	325	600	Shell	To be the largest US refinery

Sources: Dow Jones, Reuters, Oil Daily, Saudi Aramco, Global Insight, MEES, PFC. Also, Energy Information Administration, *Country Analysis Briefs: Saudi Arabia*. Available at http://www.eia.doe.gov/emeu/cabs/Saudi_Arabia/full.html (accessed August 13, 2008).

Table 5.3 Upstream natural gas projects in Saudi Arabia, 2004–2012.

	Field	Area	Projected capacity addition (bcf/d)	Expected on line	Notes
Non-associated	Karan	Offshore Khuff region	1.50	2012	Saudi Arabia's largest gas project currently in development. Recently increased production expectations by 0.5 bcf/d
	Ghazal	Onshore	0.13	2008	Producing approximately 0.27 bcf/d
	Midrikah	Onshore	Unknown	TBA[a]	
	Fazran	Onshore	Unknown	TBA	In testing
Associated	Qatif/Abu Safa'a	Onshore	0.40	2004	
	Haradh	Onshore	0.25	2005	
	Khurais	Onshore	0.30	2009	
	Khursaniyah	Onshore	0.30	2008	
	Manifa	Offshore	0.12	2011	

[a] To Be Announced.

Sources: Dow Jones, Reuters, Oi! Daily, Saudi Arair/co, Global Insight, MEES. Also, Energy Information Administration, *Country Analysis Briefs: Saudi Arabia*. Available at http://www.eia.doe.gov/emeu/cabs/Saudi_Arabia/full.html (accessed August 13, 2008).

line with regional and international stability. Riyadh has been an active participant in all major regional and international oil organizations and has sought to coordinate its energy policy with other producers. Saudi leaders have also sought to sponsor a dialogue between producers and consumers. The goal is to move the dynamics away from confrontation and closer to cooperation. Within this context, the Saudi authorities have always advocated moderate oil prices and maintained strong oil ties with all major markets (i.e., Asia, Europe, and the United States). Other major energy producers – Iran and Iraq – have been less fortunate.

5.3 Iran

The Islamic Republic of Iran was the first country in the Persian Gulf where oil was discovered. Iran holds massive hydrocarbon deposits. It has the world's second largest proven oil reserves (after Saudi Arabia), is the world's fourth largest crude producer (after Saudi Arabia, Russia, and the United States), and is OPEC's second largest producer and exporter after Saudi Arabia. The figures for natural gas are equally impressive. The country holds the world's second largest natural gas proven reserves (after Russia) and is the world's fourth largest gas producer (after Russia, the United States, and Canada). Tehran also enjoys a strategic location close to the lucrative energy markets of Asia and close to Europe with easy access to the high seas.

Despite these significant geological and geographical advantages, Iran's current status has remained well below its actual potential. Several political and economic factors have hindered

Iran and slowed down the full utilization of its massive hydrocarbon resources. These include the broad upheaval that accompanied the 1979 revolution and the subsequent Iran–Iraq War (1980–1988), economic sanctions (imposed since 1979), a high population growth rate, and economic mismanagement. The combination of all these factors has negatively impacted economic performance, particularly the energy sector – the backbone of national economy. Nonetheless, since the early 2000s the Iranian economy has experienced a period of economic growth and ambitious oil and gas projects have been planned. In the absence of major internal or external crises, the Islamic Republic can substantially develop its hydrocarbon resources, boost its oil and gas exports, and improve energy security both at home and abroad [37].

The discovery and development of oil in Iran reflect three distinctive characteristics: Iran was the first country in the Persian Gulf and the Middle East where oil was found; the relations between the Iranian authorities and international oil companies in charge of petroleum exploration and development operations within the country were, generally speaking, characterized by mistrust and tension; and US involvement in the Iranian hydrocarbon industry started in the mid-1950s, more than four decades after oil was discovered.

In 1901 the Shah of Iran granted a concession to William Knox D'Arcy, a British adventurer, to find, exploit, and export petroleum anywhere in Iran, except for the five northern provinces (Azerbaijan, Gilan, Mazanderan, Astrabad, and Khorassan), which were excluded as a result of Russian influence. Oil was first struck in 1908 in Masjid-i-Suleiman, on the site of an ancient fire temple [38]. The Anglo-Persian Oil Company was formed in 1909; renamed the Anglo-Iranian Oil Company in 1935, the company is now known as British Petroleum (BP). The United Kingdom was particularly interested in discovering oil in Iran for at least three reasons: (1) there were no indigenous oil deposits in the United Kingdom nor in any part of the British Empire, as known at the time; (2) in 1913, shortly before World War I began, the Admiralty, then headed by Winston Churchill, decided to shift from coal to oil as fuel for the Royal Navy [39]; and (3) control of Iranian oil would strengthen a British presence and influence in the Middle East and deter the threat of German or Russian expansion in that area. Taking these reasons into consideration, the UK government bought a controlling interest in the Anglo-Persian Oil Company. Accordingly, Iranian oil facilities were rapidly expanded during World War I and by the early 1950s were still the best developed in the Persian Gulf region.

In spite of the continued expansion of the Iranian oil industry, tension and suspicion between the Anglo-Iranian Oil Company and the authorities in Tehran built up to a showdown in the early 1950s. The Iranian grievances focused on three areas: (1) the monopoly position enjoyed by the company; (2) the close relationship between the company and the UK government; and (3) dissatisfaction with the financial terms of the concession between the company and the Iranian government. Developments between the early 1930s and the 1950s had underscored these Iranian complaints and contributed to the rise and sharpening of nationalism in Tehran. In 1932, the Iranian government decided to cancel the concession it had awarded to the Anglo-Persian Oil Company three decades earlier. After lengthy negotiations, the two sides signed a new concession in 1933 with more favorable provisions to Iran. Following this new agreement, the Iranians were satisfied with the steadily increasing income they were receiving and for a number of years had good relations with the company.

However, World War II introduced new dynamics. In 1941, the United Kingdom and the Soviet Union occupied Iran and forced Shah Reza, who was sympathetic to Germany, to abdicate in favor of his son. Several months later, Tehran, London, and Moscow signed a tripartite treaty of alliance. In 1948 and 1949, negotiations between the Iranian government and the Anglo-Iranian Oil Company produced an agreement supplementary to the 1933 concession, offering improved financial terms to Iran, though allowing little scope for the assertion of Iranian sovereignty over its oil resources [40].

The Iranian Majlis (Parliament) rejected the agreement and passed the Nationalization Bill in April 1951, providing for the creation of a National Iranian Oil Company (NIOC), to which the assets of the Anglo-Iranian Oil Company were to pass and which would be the government agency responsible for running all aspects of the Iranian oil industry. This movement was led by Dr. Muhammad Mossadeq, who became prime minister. In response to these changes, Iranian oil production fell steeply and alternative oil resources were rapidly developed in the Persian Gulf, particularly in Kuwait, Saudi Arabia, and Iraq. The next two years witnessed an economic confrontation between London and Tehran, with Washington fearing that popular discontent and economic straits might lead Iran toward communism. These economic and political uncertainties came to an end with a coup d'état, supported by the US government, that led to the arrest of Mossadeq and the installation of a friendlier government in Tehran. Later, an agreement was signed between the new Iranian government and the Iranian oil participants, which was generally known as the Consortium.

The Consortium was initially made up of several international oil companies: British Petroleum, 40%; Royal Dutch/Shell, 14%; Gulf, Mobil, Standard of New Jersey, Standard of California, and Texaco, each 7%; Compagnie Française des Petroles, 6%; and Iricon Agency, comprising eight US independent oil companies, 5% [41]. The terms of the agreement were in line with the offers that had been made to, and refused by, Mossadeq at various times after the Act of nationalization. Exploitation and marketing rights were assigned to the Consortium, but it was recognized that NIOC remained the sole owner of all fixed assets. Thus, in principle, the 1954 agreement recognized Iranian nationalization of the oil industry and provided what amounted to a 50/50 share of profits. In 1966 the Consortium relinquished 25% of its agreement area, and in 1973 NIOC assumed control over oil production and refining in the whole agreement area.

To sum up, the process of expanding Iranian control over and ownership of its oil resources took place earlier and was more confrontational than the experiment in other Persian Gulf producers. In spite of Iran's failure in the early 1950s to completely control its hydrocarbon wealth, by 1973 Tehran took charge. Still, NIOC allocated areas for exploration and development under joint arrangements with international oil companies and granted them service contracts.

Since its establishment during the Mossadeq crisis, NIOC has faced two often contradictory demands. On the one hand, oil is a fungible economic commodity that must be traded to be valuable. NIOC must sell in the international market to generate revenues for the Iranian national treasury. The state decides how to spend these funds based on both strategic and commercial interests. On the other hand, oil and natural gas are not only economic commodities, but also rather seen as symbols of Iran's national strength and pride [42]. In this regard, NIOC serves as the national custodian of the country's most prized commodities. Due to Iran's confrontational encounters with international oil companies in the early decades of oil production, mistrust of foreign investment and a strong resource nationalism are more present than elsewhere in the region [43]. NIOC manages Iran's oil and natural gas sectors.

As of the late 2000s, Iran has approximately 40 producing fields (27 onshore and 13 offshore), with the majority of crude oil reserves located in the south-western Khuzestan region near the Iraqi border. Production reached a peak of 6 million b/d in 1974, but the authority has not been able to maintain this level since the 1979 Revolution due to war, sanctions, mismanagement, and high rates of natural decline and depletion. One of the most recent and largest oil discoveries is the Azadegan field. It contains 26 billion barrels of proven crude oil reserves, but is geologically complex and its oil difficult to extract. The first exploration well was drilled in the field in 1976, but its discovery was finalized only after drilling a second well in 1999. The field is located west of Ahvaz close to the Iraqi border and is considered one of the biggest discoveries in the world in the last 30 years [44].

These massive deposits are not fully translated into exports. Iran's oil consumption is one of the highest in the world on a per capita basis [45]. Two developments have contributed to the surge in the country's petroleum consumption: (a) rapid population growth, which has more than doubled since the 1979 revolution; (b) heavy subsidization, where, according to the International Monetary Fund, the Iranian government lost an estimated $32 billion (11% of GDP) by underpricing crude oil and its derivatives in the domestic market in 2007–2008 [46]. The rationale behind this heavy subsidization was based mainly on two major theoretical considerations. The first was that Iranians, being the ultimate owners of their hydrocarbon reserves, should be allowed to purchase their products at "reasonable prices." And the second justification was the assumption that low energy prices would serve as a catalyst to industrialization, theoretically compensating for technological and infrastructural deficiencies vis-à-vis international competition.

These assumptions have not worked the way they were intended to. Instead, the very cheap gasoline prices have discouraged efficiency and conservation and encouraged a brisk smuggling trade as Iranians buy millions of gallons of fuel at the subsidized price and truck them into neighboring Pakistan, Turkey, Afghanistan, and Iraq for sale at market rates [47]. Confronting this dire situation, the Iranian authorities have imposed a petrol rationing system since the summer of 2007, under which people are allowed to buy a fixed amount of gasoline at a subsidized cheap price, with extra fuel sold at a higher price [48].

In addition to rationing consumption, Tehran has sought to add more refining capacity. The severe shortage of gasoline and other petroleum products has been recognized for several years. In February 1999 the Majlis passed an important law allowing foreign investors up to 45% equity participation in Iranian oil refineries [49]. Also in the same year, the National Iranian Oil Refining, Production, and Distribution Company (NIORPDC) was set up by the oil ministry to supervise the nation's nine operating refineries (Table 5.4).

In order to address this severe shortage of refineries, the Iranian authorities have embarked on a large refining expansion and upgrading program as well as building new refineries. Iran is expanding capacity in the Arak, Isfahan, and Abadan plants [50]. It also plans to build seven new refineries across the country. These are Khuzestan, Persian Gulf Star, Shahriar, Anahita, Hormoz, Caspian, and Pars [51]. The government claims that when these refineries are built, Iran will not need any gasoline imports and indeed will become an exporter [52]. The full implementation of this plan and the timely construction of these new plants and adding capacity to the existing refineries would require huge investments. In 2007, Oil Minister Kazem Vaziri-Hamaneh announced that the refining industry needed $15 billion for its development [53]. A year later (2008) another Oil Minister, Gholam Hossein Nozari, echoed the same sentiments, saying that Iran needed some $500 billion of oil industry investment over the next 15 years [54]. The flow of these badly needed investments has been largely blocked by US economic sanctions, now more than three decades old.

Table 5.4 Iran crude refining capacity, 2009

Refinery	Thousand b/d
Abadan	350
Isfahan	284
Bandar Abbas	232
Tehran	220
Arak	170
Tabriz	100
Shiraz	40
Kermanshah	25
Lavan Island	30
Total	1451

Source: Energy Information Administration, *Country Analysis Briefs: Iran*. Available at http://www.eia.doe.gov/ emeu/cabs/Iran/full.html (accessed February 4, 2009).

The United States has actually maintained sanctions against Iran since 1979, following seizure of the US Embassy in Tehran. Economic sanctions against Iran became more exclusive in the mid-1990s with the signing of several executive orders and the enacting of the Iran–Libya Sanctions Act (ILSA). In 1995, President Clinton issued two executive orders that established a total ban on trade with Iran. The first order prevents US companies from supervising, managing, and financing projects relating to the development of Iran's oil and gas resources. The second executive order stated that Americans may not trade in Iranian oil, finance, broker, approve, or facilitate such trading, or finance or supply goods or technology that would benefit the Iranian petroleum sector.

To tighten the embargo further, Congress unanimously passed ILSA, and President Clinton signed it into law in August 1996. ILSA mandates the president to impose two sanctions selected from a menu of six on any US or foreign personnel who invest $20 million or more in an Iranian project (lowered in August 1997 from $40 million) if the investment directly and significantly contributes to the enhancement of Iran's ability to explore for, extract, refine, or transport by pipeline its oil and natural gas. In other words, the legislation was designed to force foreign companies into choosing to do business with either Iran or the United States [55].

Following the election of President Muhammad Khatami in 1997, the Clinton administration took several steps aimed at reducing tension with Tehran. In July 1999, new regulations were issued to allow some exports of food, medicines, and medical equipment to Iran under specific conditions. In April 2000, the ban on the import of caviar, pistachio nuts, and carpets was lifted. This move was largely symbolic, since these exports account for only a fraction of Iran's foreign exchange earnings, which come mainly from oil sales.

To sum up, by the late 2000s, US sanctions against investing in Iran's hydrocarbon resources imposed since 1995 were still in place. It can be argued that at least initially these sanctions have slowed down foreign investment in Iran, but they have not succeeded in deterring many IOCs from signing expensive deals with Tehran. Put differently, it is likely that without US sanctions Iran would have been more successful in attracting foreign investment to develop its oil and gas fields. The few steps that President Clinton took before leaving office were largely symbolic and had no serious impact on the substance of either the bilateral relations between the two countries or the US sanctions on Iran's energy industry.

President George W. Bush maintained the US sanctions against Iran. Furthermore, due to a dispute over the country's nuclear program, the United Nations Security Council (UNSC) imposed two sets of sanctions on the Islamic Republic. In December 2006 the UNSC ordered countries to stop supplying Iran with materials that could contribute to its nuclear and missile program, and froze the assets of some Iranian firms and individuals related to those programs. In March 2007 the UNSC broadened the previous sanctions to include a ban on all Iranian arms exports. It also asked countries to restrict financial aid and loans to Tehran and froze the assets of 28 additional officials and institutions linked to Iran's nuclear and missile programs [56]. In June 2010 another set of sanctions was approved.

This US-led financial clampdown has made it harder for Iran to raise loans, obtain foreign currency, or hold offshore assets. Greater European participation has also made it much harder for Iran to simply shift oil transactions out of dollars and into euros. In the late 2000s several major IOCs suspended or reduced their engagement in Iran's oil and gas industry. In 2004, Japan's Inpex Company agreed to invest 75% of a $2 billion plan to develop Azadegan. But two years later, Iran cut Inpex's share to 10%, complaining that the firm was delaying the project – apparently under pressure from Washington. Similarly, in 2008 Royal Dutch Shell and Spain's Repsol withdrew from a $10 billion plan to develop phase 13 of South Pars [57]. This reluctance by some of the IOCs to pursue commercial opportunities in Iran underscores the hard choices they have to make between developing Iran's hydrocarbon resources or participate in punishing Iran for dispute over its policy.

In order to avoid complete isolation and to pursue other investment opportunities, Tehran has signed several agreements with Chinese (China National Petroleum Corporation and Sinopec), Russian (Gazprom), Malaysian (SKS Ventures), and some European companies (Elektrizitats-Gesellschaft Laufenburg of Switzerland).

It is important to point out that sanctions are not the only reason why IOCs have been hesitant to invest in Iran. Another major reason is the unattractive financial and legal terms Iran has offered in the last two decades. Following the 1979 revolution, a new constitution was established. Reflecting one of the issues at stake in the revolution itself, the new constitution stipulates that foreign ownership of the country's natural resources was illegal. Specifically, it prohibits the granting of petroleum rights on a concessionary basis or direct equity stake. The petroleum industry was nationalized, the government expropriated the assets of foreign petroleum companies operating in the country, and concession agreements were broken. Eight years later, when the attitude toward foreign investment started to change, the Petroleum Law of 1987 was promulgated. It permits the establishment of contracts between the Ministry of Petroleum and state companies on one side and local and foreign natural citizens and legal entities on the other [58]. Since reopening to the IOCs, Iran has employed the buy-back framework for upstream contracts. The adoption of this principle represents a radical break with the past and was a courageous move that carried huge political risks for those who espoused it.

The buy-back model is better explained in comparison to other formulas that have been utilized between oil producers and foreign companies. These include the joint venture (JV), the production-sharing agreement (PSA), and the technical services contract (TSC). JVs are arrangements under which a state enterprise and a foreign firm invest stated amounts of capital that can take various forms, including funds, intellectual property, or physical assets and rights to land. The partners share the risk and the reward of the venture in proportion to the capital contributed. Under PSAs, the foreign contractor is reimbursed for exploration, development, and operating costs by way of a certain share of production [59]. In other words, the foreign partner holds no equity stake but does maintain a firm legal entitlement to a certain percentage of oil output volume. Finally, TSCs involve a simple cash payment for services rendered. These types of contracts entail little risk, if any, for the foreign firms. Subsequently, the rewards do not rise should the discovery be substantial or oil prices increase [60]. In addition, TSCs tend to be awarded for short periods.

The buy-back formula does not violate the Iranian constitution. It is a service contract under which one or more parties are contracted by the Ministry of Petroleum to carry out necessary exploration and development work on a field that, once completed, reverts wholly to the Ministry. Thus, the foreign company is neither a partner nor a concessionaire, but acting in the role of a hired contractor servicing the national company. The buy-back model demands that the foreign partner provides all the investment capital for exploration and, in return, is paid a predetermined rate or return on capital invested. This is paid in kind after the production of the first commercial oil. Buy-back contracts generally are designed by the Iranian negotiators to last five to seven years and are thus fairly short term in the context of the traditional upstream contract [61].

Although there have been several agreements signed between Iran and IOCs based on buy-back formulas, the model has been criticized particularly for three reasons. First, it is almost risk free for the IOCs, since they are guaranteed a return on their investment. Second, the brief duration of the contracts does not provide incentives for the IOCs to introduce and fully utilize their advanced technology and management skills. Third, the fixed rate of return leaves IOCs with little incentive to maximize the profit from oil and gas fields. In short, the criticism focuses on the disconnection between performance and profit and the absence of long-term relationships between the Iranian oil authority and IOCs [62].

With these shortcomings taken into consideration, there have been intense debates in Tehran and negotiations with IOCs regarding modification of the buy-back model. New contracts signed with IOCs link performance to payments and provide for long-term relations with foreign companies. The goal is to optimize oil and gas field utilization. A consensus is growing in Iran that, without modifying the buy-back formula, the country could fall short of its expectations in terms of attracting the necessary foreign investment. This is particularly important to offset how some foreign investors see Iran as a political risk due to US and UN sanctions.

This buy-back formula has served as the basis for negotiations between IOCs and the Iranian authorities to develop the country's oil and gas deposits. By far the most important natural gas structure is the "super-giant" South Pars. Originally Shell discovered the field in 1966 and a few years later (1971) Shell discovered South Pars' extension, the North Field in Qatar [63]. The Iranian portion is estimated to contain some 14 trillion cubic meters of gas reserves and 18 billion barrels of gas condensates, making it the country's main gas reservoir and one of the largest energy structures in the world [64]. The Iranian authorities are working in partnership with IOCs to develop the field in 25 phases. South Pars and most other gas fields are located primarily in the shallow waters of the Persian Gulf [65].

Iran's gas production and its place on the world gas stage do not accurately reflect its declared proven reserves [66]. The country's share of the world's proven reserves is 16%

while its share of production is only 3.8%. This discrepancy is due largely to lack of interest in developing gas deposits till the early 1990s. Traditionally, Iran has been an oil producer and exporter for most of the past century. For a long time gas received a low priority because oil offered better and higher foreign exchange returns. Nevertheless, gas exports began on a large scale in 1970 to the former Soviet Union [67]. These exports continued for several years until they were stopped due to price and payment difficulties.

Since the early 1990s, official attention has focused on developing gas resources. This was partly in response to the skyrocketing of domestic oil consumption, which cut into crude exports and threatened to deprive the state of a substantial proportion of annual revenues. The Iranian government responded by launching a gasification program in 1992. Under this program, hundreds of cities and towns were given access to gas. Since then gas consumption and production have witnessed a steady increase. This rise in production serves at least four objectives: (1) to satisfy growing domestic demand; (2) to use as a source of foreign revenues by exporting gas to neighboring markets; (3) to inject into depleted oil fields to maintain reservoir pressure and allow higher recovery; and (4) to free more oil for export.

In recent years Iran has developed an ambitious plan to export its gas, both via pipelines and by tankers as LNG, to several countries. Some of these projects are already on line, others are either under construction or being negotiated. The list includes Armenia, Azerbaijan, India, Oman, and Turkey. In addition, the Islamic Republic is eager to export gas to the lucrative European market by either contributing to pipelines such as Nabucco or building a pipeline via Turkey. The scale of Iran's export ambitions was highlighted in 2006 by Reza Kasaei Zadeh, Deputy Oil Minister and General Director of the National Iranian Gas Company: "Iran's policy is to achieve 8–10% of the world's gas trade and its by-products within 20 years" [68].

Despite these ambitious goals, the timely full utilization of Iran's natural gas deposits is restrained by a number of overlapping factors:

1. Iran has continued to be under unilateral and multilateral economic sanctions. Several governments and energy companies are reluctant to economically and politically engage with Iran until these sanctions are lifted.
2. Natural gas projects are expensive and require the active participation of foreign investment. Iran has succeeded in signing several agreements with a number of IOCs. Still, much more is needed. It is expected that once the country's stance on the international stage is improved, more foreign investments and foreign technology will be available.
3. The buy-back formula needs to be modified and provide more attractive financial terms for IOCs.
4. More pipelines need to be built to connect Iran's gas fields and terminals to global markets.

To sum up, oil revenues provide a substantial proportion of Iran's national income and export earnings. Iran is not yet a major gas exporter, but its gas production satisfies more than half of its domestic consumption and thus is crucial in releasing oil for export. The nation's economic development and political stability depend on maintaining and upgrading oil and gas production. Stated differently, Iran's energy security depends on maintaining and expanding its volumes of production and exports. Iran's OPEC Governor Mohammad Ali Khatibi argues that the oil and gas industries need "peace and stability to invest in production and refining" [69].

In addition to the necessity for regional and international peace and stability, the Iranian authorities have to establish a more favorable political and commercial environment to encourage private and foreign investment in both upstream and downstream energy sectors. High levels of subsidy on oil product prices have led to excessive growth in domestic consumption and smuggling, eroding the crude export surplus and challenging the delivery of products to the economy. Development of the gas potential has been delayed because of growing opposition within Iran to developing gas for export purposes and the poor financial return obtained from gas sales at subsidized prices to the domestic power sector. The existence of sanctions has encouraged a degree of increasing self-reliance within the sector, but a business climate that offers limited comfort to private investment has meant that the potential to develop the hydrocarbon deposits has not been fully realized [70].

5.4 Iraq

Despite the fact that Iraq has massive proven oil reserves (the world's third largest after Saudi Arabia and Iran) and that petroleum was discovered in Iraq earlier than in most other Middle Eastern producers, the country's hydrocarbon resources are largely underexplored and underdeveloped. Oil was first struck in commercial quantities at Naft Khana adjacent to the Iranian frontier in 1923 [71]. Shortly after that, the Turkish Petroleum Company (TPC) was confirmed in its concession covering most of Iraq. The shares in this firm were divided between British, Dutch, and French partners. In 1927, a giant oil structure was discovered in Kirkuk. These promising oil resources attracted US firms. Accordingly, in the late 1920s the composition of the TPC was changed. Five American oil companies (Standard Oil of New Jersey, Standard Oil of New York, Gulf Refining Company, Atlantic Refining Company, and Pan American Petroleum and Transport Company) acquired shares [72], and the new firm was renamed the Iraqi Petroleum Company (IPC). Thus, US oil companies had participated with their European counterparts in developing Iraqi oil resources at an early stage.

The demise of the monarchy in 1958 and the rise to power of nationalistic and leftist regimes had drastically altered the relations between the Iraqi government and IPC. An important step in this direction was the issue of Public Law 80 of 1961 under which the Iraqi government seized approximately 99% of the concession territory of IPC and its affiliates. A few years later, a state-owned company – the Iraq National Oil Company (INOC) – was established. Finally, in 1972 the Iraqi government nationalized IPC, and by 1975 the holdings of various private companies working in Iraq were completely transferred to INOC. Since then, INOC has dominated the country's energy sector with a minimum role preserved for IOCs.

In the following years, Iraq's oil production reached its peak, but these favorable conditions did not endure. Since 1980, the Iraqi oil industry has been a victim of two wars, a prolonged and comprehensive economic sanctions regime, and serious looting, damage, and unrest. Consequently, the country's level of production has deteriorated, reflecting these unfavorable political and military developments. Since the late 1990s, Iraqi oil production has risen in response to several resolutions passed by the UNSC. The fluctuations in Iraqi oil production illustrate the impact of wars and political problems. The peak of 1979 (3.5 million b/d) was reached before the crises of the following two decades. The huge drop in 1981 was in response to the war with Iran in which the two sides attacked each other's oil installations.

The substantial increase in production in 1989 coincides with the only year when Iraq was not at war with any of its neighbors.

However, this relative peace did not last long. The Gulf War and economic sanctions took a heavy toll on Baghdad's oil industry. Both upstream and downstream installations were targets for international alliance's missiles and bombs. Furthermore, since oil was the main source of foreign currency revenue, it became the main focus of the sanctions. Iraqi oil exports were restricted to supplying Jordan with limited quantities. The rise in production in 1997 was the result of UNSC Resolution 986 (also known as oil for food). In April 1995, the UNSC had passed Resolution 986, which allowed limited Iraqi oil exports for humanitarian and other purposes. Iraq actually began exporting oil under Resolution 986 in December 1996. With Iraq steadily increasing its oil export revenues, the UNSC passed Resolution 1284 in December 1999 to remove any limits on the amount of oil Iraq could export. The political and security turmoil that followed the toppling of Saddam Hussein's regime in 2003 dealt another heavy blow to the country's oil production and infrastructure.

These continuous political crises meant that few financial resources were available to modernize Iraq's energy infrastructure. The country is heavily underexplored and underdeveloped. Even worse, it is not known how badly the oil wells were damaged by the wars and sanctions.

Political and military crises have not only restrained Iraq's level of production, but also seriously reduced its ability to ship its crude to the international markets. Iraq is almost a landlocked state with a narrow outlet to the Persian Gulf, where it has three tanker terminals: Mina al-Baker, Khor al-Amaya, and Khor al-Zubair. The infrastructures for these terminals were almost completely destroyed in the war with Iran and the Gulf War. Since then, substantial repairs have been done. Nevertheless, these terminals do not have enough facilities for the massive Iraqi oil exports. The country has had to rely on pipelines that traverse other countries. These include Israel/Palestine, Syria, Saudi Arabia, and Turkey.

Shortly after oil was discovered in commercial quantities in Iraq, the IPC built a pipeline to Haifa in British-controlled Palestine and another pipeline to Tripoli in Lebanon with a sideline forking from Homs to Banias in Syria [73]. The establishment of Israel and the resulting Arab boycott caused the shutdown of the Haifa pipeline. The civil war in Lebanon (1979–1989) raised security concerns regarding the pipeline to Tripoli, so the pipeline was closed for a number of years. During the Iran–Iraq War, Syria supported Tehran and in 1982 closed the Tripoli–Banias pipeline to deny Baghdad oil exports and revenues. This pipeline remained closed until late 2000. Anticipating troubles in its relations with Syria, Iraq decided to take protective measures. It built two pipelines to the Mediterranean via the territory of Turkey (Kirkuk–Ceyhan) in the 1970s and 1980s. The pumping stations and other facilities for these pipelines were destroyed during the Gulf War, and since then massive repairs have been done.

In addition to these pipelines to the Mediterranean, Iraq constructed two pipelines (in 1985 and 1990) to the Red Sea through Saudi Arabia. Both were owned entirely by Iraq but were shut down by Saudi decision in August 1990, when Iraq attacked Kuwait. In June 2001, Saudi Arabia decided to expropriate the two pipelines. The reason, according to Riyadh, was "the continued Iraqi threats of aggression in the years after the occupation of Kuwait and the Gulf war" [74]. Finally, in order to optimize export capabilities, Iraq constructed a reversible "Strategic Pipeline" in 1975. This pipeline consists of two parallel lines. The North–South system allows for export of northern Kurkuk crude from the Persian Gulf and for southern Rumaila crude to be shipped through Turkey.

In the aftermath of the 2003 war in Iraq most of these pipelines had been a favorite target for sabotage by insurgents for at least two reasons. First, a successful attack on pipelines leads to interruption of the flow of oil supplies and costs a lot of money to repair. Second, pipelines run for thousands of miles, so these long distances complicate the protection of each pipeline from where it originates all the way to where it delivers. In short, pipelines are relatively easy targets and attacking them causes economic disruption. Working with the Iraqi government, the United States has sought to improve pipeline security and repair procedures. The United States funded the Pipeline Exclusion Zones (PEZs), a security measure around each oil pipeline that provides protective fences and concertina wire, as well as gates and guardhouses [75]. These measures have contributed to improved security.

The distribution and concentration of Iraqi oil further complicate the country's ethnic and sectarian divisions. Most known hydrocarbon resources are concentrated in the Shi'ite areas of the south and the ethnically Kurdish north, with few resources in control of the Sunni minority. Approximately 70–80% of the country's proven oil reserves are in the south while about 20% are in the north near Kirkuk, Mosul, and Khanaqin. Iraq's proven reserves are housed in some 80 oil fields. However, there is a concentration of reserves in a few fields. The seven largest fields – Rumaila South and North, Kirkuk, East Baghdad, Majnoon, West Qurna, and Zubair – contain two-thirds of total reserves [76].

Despite improvements in recent years, the refining sector has not been able to meet domestic demand for most refined products. As a result, Iraq relies on imports for about one-quarter of the petroleum products it uses. To alleviate product shortages, Iraq's 10-year strategic plan for 2008–2017 set a goal of increasing refining capacity from 600 000 to 1.5 million b/d [76] as in Table 5.5.

The development of the natural gas sector in Iraq is similar in many ways to other Middle Eastern countries. Most Iraqi gas is associated gas and the exploration and development operations of natural gas lagged behind those of the oil sector. Initially, most of the gas was flared due to the lack of sufficient infrastructure to utilize it for consumption and export.

The production of gas started with the commencement of oil production in the late 1920s. It witnessed a marked increase during the 1970s in parallel with a similar increase in oil

Table 5.5 Existing refineries in Iraq, 2010.

Refinery	Location	Capacity (b/d)	Notes
Baiji	North–Central Iraq	310 000	Improvements in operational issues
Basrah	Near Basrah	150 000	Considering adding 70 000 b/d distillation tower
Daura	Baghdad	110 000	Considering adding 70 000 b/d distillation tower
Mosul–Qaiyarah, Kirkuk, Khanaqin, K3–Haditha	Scattered	<10 000 each	Topping plants making low-grade diesel and kerosene
Muftiah, Najaf, Maysan, and Nassiriyah–Samawah	Scattered	<10 000 each	Topping plants making low-grade diesel and kerosene

Source: Energy Information Administration, *Country Analysis Briefs: Iraq*. Available at http://www.eia.doe.gov/emeu/cabs/Iraq/full.html (accessed June 4, 2009).

production. Wars and economic sanctions inflicted severe damage on gas treatment, processing, and compression facilities [77]. Gas is widely used in power generation and in injecting into oil fields to enhance recovery efforts.

Iraq's 10-year strategic plan for 2008–2017 set a goal of almost doubling natural gas production. In order to reduce flaring, the state-owned South Gas Company signed an agreement with Shell in September 2008 to implement a 25-year project to capture flared gas and provide it for domestic use, with any surplus sent to an LNG project for export. In late 2000s, the Iraqi authorities had expressed interest in exporting gas to Kuwait, Syria, Lebanon, Turkey, and Europe via the proposed Nabucco pipeline.

For a long time, both the Iraqi oil and natural gas sectors were supervised by INOC in cooperation with the Oil Ministry and other state institutions. The creation, evolution, and performance of INOC reflect the conflicts between pursuing commercial interests by the country's engineers and other energy professionals on one hand, and political considerations imposed by Iraqi leaders on the other hand.

In August 1964, Law 11 set up the first INOC. The objective was to exercise all aspects of the oil industry, inside and outside of Iraq, in all its phases including exploration, production, transport, refining, storage, distribution, and marketing of all hydrocarbons, their products and by-products, as well as the manufacturing of equipment. INOC was allowed to create subsidiaries on its own or in partnerships and to join established companies. The Law stipulated that INOC's capital be purely governmental in keeping with the principle of sovereignty over mineral resources of the state that are natural monopolies. The Law also emphasized that the company shall adhere to the general oil policy of the state and shall be attached to the Ministry of Oil [78].

In 1967, Law 123 was enacted, annulling Law 11 of 1964 and establishing a second version of INOC. The need to expand INOC's role and responsibilities was the main reason behind this change. In the following years INOC demonstrated a high degree of success including signing service agreements with several IOCs and starting oil exploration and development on its own [79]. Despite this success, Law 101 of 1976 for the "Organization of the Oil Ministry" brought drastic changes to the structure and role of INOC. Essentially, Law 101 underscored the Oil Ministry's authority over INOC and limited any independence that it had. In effect, the Oil Ministry took charge of the management of all aspects of the oil sector including exploration, drilling, production, refining, gas processing, transport, and marketing.

In spite of this intensive political interference, INOC continued its relative success in the late 1970s, which was interrupted by the outbreak of the Iran–Iraq War in 1980 and the political turmoil that characterized the following three decades. However, in the mid-1980s Saddam Hussein promulgated several decrees transforming the upstream and downstream subsidiaries that INOC had created and managed into administratively and financially independent enterprises reporting to the Oil Ministry. Finally, Decree 267 of 1987, "Merging INOC with the Oil Ministry," completely dismantled IONC and stipulated that the oil minister will replace its board of directors [80].

Frustrated with the slow development of the oil sector, the Iraqi government decided in July 2008 to establish a National Council for Reconstruction and Development to accelerate, among other things, the approval of oil upstream contracts [81]. The following year (2009), the government approved a draft law creating, or recreating INOC. The lack of consensus among Iraqi policy-makers on the role of foreign investment and on how much autonomy the Kurdish authority should have has contributed to the slow progress in the country's oil sector.

Seeking to weaken the international sanctions regime, Saddam Hussein negotiated and signed agreements with a few Russian, Chinese, and European oil companies. However, due to political and security instability the implementation of these agreements never materialized. Driven by the need to modernize the country's energy infrastructure and the lack of financial resources, the Iraqi government started negotiating with IOCs in the mid-2000s. Initially Baghdad refused to award any PSAs and insisted that the contracts would be awarded on a technical service basis (TSB). This means, in effect, that the IOCs would be paid fees for their services and not allowed any equity in the reserves, as they prefer. The Iraqi oil authorities adopted the TSB model to increase production over a short period of time as well as to ensure that it had sufficient public and parliamentary support to sign upstream contracts with IOCs and also to silence the critics among the politicians in Baghdad who accused them of not moving fast enough to reach agreements with IOCs and increase production. In September 2008, the Iraqi Cabinet approved a $3 billion oil service contract with the Chinese National Petroleum Company (CNPC), restoring a deal originally signed in 1997 [82].

While the Iraqi authorities in Baghdad were negotiating with some of the world's biggest oil companies, many smaller ones decided to bet on Kurdistan, the more stable semi-autonomous region in the north of the country. Their decisions were risky, because Baghdad retaliated by prohibiting those companies from bidding for work in the southern areas it controlled. In May 2009 the Kurdistan Regional Government (KRG), the official ruling body of a federated region in northern Iraq that is predominantly Kurdish, reached a deal with Addax, a small Swiss oil company, and DNO, a Norwegian company, to export 100 000 b/d of its oil through the Iraq–Turkey pipeline. Baghdad agreed, ending a stalemate that threatened the development of oil deposits in the Kurdish area [83]. The KRG also signed oil production sharing, development, and exploration contracts with other foreign firms.

This confusion between the authority of the Iraqi government and that of the KRG is due to the ambiguity of the constitutional articles governing oil and gas assets. The discrepancies focus on Article 108 and Article 109. The former states that "Oil and gas are the property of the people of all the regions and governorates." The latter states that "The federal government will manage the production of oil and gas fields in cooperation with the governments of the producing regions or governorates providing that the revenues will be distributed in a manner compatible with demographical distribution all over the country" [84].

This ambiguity has further complicated the issuing of an oil law. In May 2006, Oil Minister Husain al-Shahristani formed a committee of three prominent Iraqi oil experts to draft the oil law. The draft, completed in August 2006, was adopted by the Ministry of Oil and submitted to the prime minister, who in turn appointed a ministerial subcommittee to review it. The committee represented a spectrum of interests, including federal representatives and officials from the KRG. The lack of agreement between the federal government and the KRG prompted the Kurdish Parliament to pass its local oil and gas law in August 2007 [85]. This is based primarily on the KRG interpretation of the constitution, which omits petroleum from the list of exclusive powers of the federal authorities and views the role of the federal government as merely administrative [86].

To conclude, like Saudi Arabia, Iran, and other Middle Eastern producers, Iraq is heavily dependent on oil export revenues. But probably more than any other country in the region, Iraqi oil and natural gas deposits are largely underdeveloped and underexplored. Iraq's energy security is tied to the timely utilization of its massive hydrocarbon resources. The development of these resources requires major investments. These private and foreign investments will not materialize without a sense of political stability and security as well as a credible and transparent management of the energy sector.

5.5   Conclusion: The Way Forward

Two broad conclusions can be drawn from the above detailed examination of oil and gas policies in Saudi Arabia, Iran, and Iraq. First, despite efforts to diversify their economies and create viable economic alternatives, these three states and other Middle Eastern producers remain deeply dependent on oil export revenues. Their economic prosperity and political stability are tied to the steady flow of oil exports and oil revenues. The implication is that they have great interest in the stability of global energy markets and the international system. Economic and political upheavals have significant ramifications on their stability. In short, they share the same economic and strategic interests as consuming countries.

Second, despite recent discoveries and rises in oil and gas production in Africa, the Caspian Sea, Russia, and other regions, the Persian Gulf still occupies the driver's seat. The region's substantial geological and geographical advantages suggest that the world is growing more dependent on Middle Eastern supplies. American, Asian, and European economies lack sufficient indigenous resources and they need Middle Eastern oil and gas to sustain their high standard of living. The implication is that the calls for reducing dependence on the Middle East are unrealistic and, indeed, misleading.

The growing interdependence between Middle Eastern oil and gas producers and global consumers underscores their mutual interest in economic and political stability. A policy based on improving regional security, containing domestic tension, and creating the right environment for investment seems to be the right course. The region needs more cooperation and less confrontation.

References

[1] British Petroleum (2010) *BP Statistical Review of World Energy*, British Petroleum, London, pp. 6, 8, 22, 24.

[2] Khadduri, W. (1996) Oil and politics in the Middle East. *Security Dialogue*, **27** (2), 155–166, 157.

[3] International Energy Agency (2005) *World Energy Outlook*, International Energy Agency, Paris, p. 46.

[4] Energy Information Administration (2009) *International Energy Outlook*, United States Government Printing Office, Washington, DC, pp. 28 and 38.

[5] Adelman, M.A. (2004) The real oil problem. *Regulation*, **27** (1), 16–21, 19.

[6] Horsnell, P. (2000) *The Probability of Oil Market Disruption: With an Emphasis on the Middle East*, James A. Baker III Institute for Public Policy of Rice University, Houston, TX, p. 18.

[7] Stork, J. (1975) *Middle East Oil and the Energy Crisis*, Monthly Review Press, New York, p. 212.

[8] Parra, F. (2004) *Oil Politics: A Modern History of Petroleum*, I.B. Tauris, London, p. 180.

[9] Skeet, I. (1988) *OPEC: Twenty-Five Years of Prices and Politics*, Cambridge University Press, Cambridge, p. 100.

[10] Maugeri, L. (2006) *The Age of Oil: The Mythology, History, and Future of the World's Most Controversial Resource*, Praeger, London, p. 114.

[11] Blair, D. and Lieberthal, K. (2007) Smooth sailing: the world's shipping lanes are safe. *Foreign Affairs*, **86** (3), 55–68, 58.

[12] Fattouh, B. (2007) *How Secure Are Middle East Oil Supplies?* Oxford Institute for Energy Studies, Oxford, p. 12.

[13] British Petroleum (2009) *BP Statistical Review of World Energy*, British Petroleum, London, pp. 6, 8, 11.

[14] Bronson, R. (2006) *Thicker than Oil: America's Uneasy Partnership with Saudi Arabia*, Oxford University Press, Oxford, p. 18.

[15] Lenczowski, G. (1960) *Oil and State in the Middle East*, Cornell University Press, Ithaca, NY, p. 24.

[16] Ministry of Petroleum and Mineral Resources, *Background about Saudi Oil*. Available at http://www.mopm.gov.sa/mopm/detail.do?content=history_oil_and_gas (accessed September 13, 2009).

[17] McLachlan, K. (1980) Oil in the Persian Gulf area, In *The Persian Gulf States: A General Survey* (ed. A.J. Cottrell), Johns Hopkins University Press, Baltimore, MD, pp. 195–224.

[18] Aramco, S., *Our Story*. Available at http://www.SaudiAramco.com (accessed September 19, 2009).

[19] Ministry of Petroleum and Mineral Resources, *About Ministry – Introduction*. Available at http://www.mopm.gov.sa/mopm/detail.do?content=ministry_structure (accessed September 13, 2009).

[20] Supreme Council for Petroleum and Mineral Affairs, *Introduction*. Available at http://www.saudinf.com/main/c551.htm (accessed September 19, 2009).

[21] Energy Information Administration, *Country Analysis Briefs: Saudi Arabia*. Available at http://www.eia.doe.gov/emeu/cabs/Saudi_Arabia/full.html (accessed August 13, 2008).

[22] Shammas, P. (2000) Saudi Arabia: petroleum industry review. *Energy Exploration and Exploitation*, **18** (1), 1–86, 25.

[23] National Shipping Company of Saudi Arabia, *About us*. Available at http://www.nscsa.com/printable/aboutus.php (accessed September 13, 2009).

[24] Vela, *History*. Available at http://www.vela.ae/-History-Vela8.html (accessed September 13, 2009).

[25] Mabro, R. (2001) Saudi Arabia's oil policies. *Middle East Economic Survey*, **44** (49). Available at http://www.mees.com/news/a44n49d01.htm (accessed December 3, 2001).

[26] Hoyos, C. (2008) OPEC Eyes Production Cut as Saudis Signal Oil Price Target. Financial Times (December 1).

[27] Hoyos, C. (2009) Saudis Wield Influence with Oil Expansion. Financial Times (September 6).

[28] al-Sa'doun, A. (2006) Saudi Arabia to become major petrochemical hub by 2010. *Oil & Gas Journal*, **104** (1), 52–56, 53.

[29] Saudi Basic Industries Corporation, *Development*. Available at http://www.Sabic.com/corporate/en/ourcompany/corporateprofile/default.aspx (accessed September 20, 2009).

[30] Fletcher, S. (2006) Al-Naimi: oil industry needs better data to plan growth. *Oil & Gas Journal*, **104** (6), 24–26, 26.

[31] al-Falih, K. (2004) Saudi Arabia's gas sector: its role and growth opportunities. *Oil & Gas Journal*, **102** (23), 18–24, 19.

[32] Aitani, A. (2002) Big growth ahead seen for Saudi gas utilization. *Oil & Gas Journal*, **100** (30), 20–27, p. 21.

[33] Mabro, R. (2002) *Saudi Arabia's Natural Gas: A Glimpse at Complex Issues*, Oxford Institute for Energy Studies, Oxford, p. 2.

[34] Gavin, J. (2004) Saudi oil comes under threat. *Petroleum Economist*, **71** (7), 3–4, 3.

[35] England, A. (2007) Saudis Set Up Force to Guard Oil Plants. Financial Times (August 26).

[36] Webb, S. (2007) Saudi Builds Security Force of 35,000 to Guard Oil. Washington Post (November 16).

[37] Maleki, A. (2007) Energy supply and demand in Eurasia. *China and Eurasia Forum Quarterly*, **5** (4), 103–113, 106.

[38] National Iranian Oil Company, *Brief History of Iran Oil Nationalization*. Available at http://www.nioc.ir/English/brief_history/index.html (accessed September 13, 2009).

[39] Lenczowski, G. (1960) *Oil and State in the Middle East*, Cornell University Press, Ithaca, NY, p. 124.

[40] McLachlan, K. (1980) Oil in the Persian Gulf area, In *The Persian Gulf States: A General Survey* (ed. A.J. Cottrell), Johns Hopkins University Press, Baltimore, MD, pp. 195–224, p. 202.

[41] Lenczowski, G. (1960) *Oil and State in the Middle East*, Cornell University Press, Ithaca, NY, p. 126.

[42] Brumberg, D. and Ahram, A. (2007) *The National Iranian Oil Company in Iranian Politics*, James A. Baker III Institute for Public Policy at Rice University, Houston, TX, p. 4.

[43] Marcel, V. (2006) *Oil Titans: National Oil Companies in the Middle East*, Brookings Institution, Washington, DC, p. 42.

[44] Energy Information Administration, *Country Analysis Briefs: Iran*. Available at www.eia.doe.gov/emeu/cabs/Iran/full.html (accessed February 4, 2009).

[45] Samsam Bakhtiari, A.M. and Shahbudaghlou, F. (2000) Energy consumption in the Islamic Republic of Iran. *OPEC Review*, **24** (3), 211–233, 221.

[46] International Monetary Fund (2008) *Islamic Republic of Iran: Selected Issues*, International Monetary Fund, Washington, DC, p. 26.

[47] Spindle, B. (2007) Soaring Energy Use Puts Oil Squeeze on Iran. Wall Street Journal (February 20).
[48] Smyth, G. (2007) Iranians Lose Access to Unlimited Cheap Fuel. Financial Times (March 8).
[49] Shammas, P. (2001) Iran: review of petroleum developments and assessments of the oil and gas fields. *Energy Exploration and Exploitation*, **19** (2 & 3), 207–260, 246.
[50] Editorial (2009) Report updates Iran's refinery project status. *Oil & Gas Journal*, **107** (3), 62–63, 63.
[51] Mehr News, *Construction of seven refineries moving ahead*. Available at http://www.mehrnews. com/en/NewsPrint.aspx?NewsID=800673 (accessed December 14, 2008).
[52] Mehr News, *Iran to join world petroleum exporters by 2011*. Available at http://www.mehrnews.com/ en/NewsPrint.aspx?NewsID=529397 (accessed August 4, 2007).
[53] Payvand, *Iran's oil refining industry needs $15 billion investment: Oil Minister*. Available at http://www. payvand.com/news/07/feb/1215.html (accessed February 17, 2007).
[54] Karimi, N., Associated Press, *Iran looks to tap key oil field with homegrown crews*. Available at http://news.yahoo.com/s/ap/20080511/ap (accessed May 11, 2008).
[55] United States Department of the Treasury, *Iran: What you need to know about U.S. economic sanctions*. Available at http://www.treasury.gov (accessed September 13, 2009).
[56] Donovan, J., Payvand, *Iran: top U.S. official says financial clampdown is working*. Available at http://www.payvand.com/news/07/oct/1166.html (accessed October 18, 2007).
[57] Fifield, A. (2008) Shell and Repsol Drop Iran Gas Project. Financial Times (May 11).
[58] Gaddy, D. (2001) Iran expands Middle East influence. *Oil & Gas Journal*, **99** (10), 74–81, 78.
[59] Seymour, I. (2000) Kuwait's upstream oil opening in the context of parallel development elsewhere. *Middle East Economic Survey*, **43** (1), 24–26.
[60] Bindemann, K. (1999) *A Little Bit of Opening Up: The Middle East Invites Bids by Foreign Oil Companies*, Oxford Institute for Energy Studies, Oxford, p. 6.
[61] Fesharaki, F. and Varzi, M. (2000) Investment opportunities starting to open up in Iran's petroleum sector. *Oil & Gas Journal*, **98** (7), 44–52, 48.
[62] Brexendorff, A. and Ule, C. (2004) Changes bring new attention to Iranian buyback contracts. *Oil & Gas Journal*, **102** (41), 16–21, 18.
[63] Shammas, P. (2001) Iran: review of petroleum developments and assessments of the oil and gas fields. *Energy Exploration and Exploitation*, **19** (2 and 3), 207–260, 242.
[64] Pars Oil and Gas Company, *South Pars gas field*. Available at http://www.pogc.ir/Default.aspx?tabid=136 (accessed September 13, 2009).
[65] Omidvar, H. (2007) Iran details LNG liquefaction plans. *Oil & Gas Journal*, **105** (21), 70–71, 70.
[66] Ghorban, N. (2006) Monetizing Iran's gas resources and the debate over gas export and gas-based industries options. *Middle East Economic Survey*, **49** (28), 26–28.
[67] Takin, M. (1999) Iranian gas to Europe? *Middle East Economic Survey*, **42** (15), 18–20.
[68] Forbes, A. (2007) The struggle to market. *Petroleum Economist*, **74** (5), 4–5, 4.
[69] Fars, *Iran calls for regional security to tame soaring oil prices*. Available at http://www.farsnews. com/English/printable.php?nn=8704011091 (accessed June 21, 2008).
[70] Audinet, P., Stevens, P., and Streifel, S. (2007) *Investing in Oil in the Middle East and North Africa: Institutions, Incentives and the National Oil Companies*, World Bank, Washington, DC, p. 70.
[71] Longrigg, S.H. (1968) *Oil in the Middle East: Its Discovery and Development*, Oxford University Press, Oxford, p. 140.
[72] Karlsson, S. (1985) *Oil and the World Order: American Foreign Oil Policy*, Barnes & Noble, Totowa, NJ, p. 68.
[73] Lenczowski, G. (1995) Major pipelines in the Middle East: problems and prospects. *Middle East Policy*, **3** (4), 40–46, 41.
[74] Editorial (2001) Saudi Arabia seizes IPSA pipeline. *Middle East Economic Survey*, **44** (25), 20–22.
[75] Energy Information Administration, *Country Analysis Briefs: Iraq*. Available at http://www.eia.doe.gov/ emeu/cabs/Iraq/full.html (accessed June 4, 2009).
[76] Shafiq, T. (2009) Iraq's oil prospects face political impediments. *Oil & Gas Journal*, **107** (3), 46–49, 46.
[77] Ghadhban, T.A. and al-Fathi, S. (2001) The gas dimension in the Iraqi oil industry. *OPEC Review*, **25** (1), 27–52, 33.
[78] Editorial (2005) Stipulations establishing INOC. *Oil & Gas Journal*, **103** (34), 19.
[79] Shafiq, T. (2005) Iraqi oil – 2: future grounded in old INOC. *Oil & Gas Journal*, **103** (34), 18–21, 20.
[80] Husari, R. (2009) Iraq National Oil Company, an historical and political perspective. *Middle East Economic Survey*, **52** (38), 20–22.

[81] Khadduri, W. (2008) Iraq oil development: lack of direction and decision-making. *Middle East Economic Survey*, **51** (34), 23–25.

[82] Radio Free Europe, *Iraq Offers up Giant Oil Fields to Foreign Firms*. Available at http://www.rferl. org/articleprintview/1365542.html (accessed January 1, 2009).

[83] Hoyos, C. (2009) Baghdad May Yet Fulfill Its Potential. Financial Times (May 26).

[84] Shafiq, T. (2006) Oil industry implications of Iraq's constitutional articles 108, 109 & 111 governing oil and gas assets: a call to reconsider. *Middle East Economic Survey*, **49** (23), 22–24.

[85] Shafiq, T. (2009) Impediments abound to exploiting Iraq's vast petroleum resource. *Oil & Gas Journal*, **107** (4), 31–36, 32.

[86] Khadduri, W. (2007) Perspectives on Iraqi oil and politics. *Middle East Economic Survey*, **50** (36), 20–22.

6

Africa

Africa is a continent of 54 countries with an estimated mid-2000 population of 805 million people. The continent's vast hydrocarbon resources are crucial to the economic well-being of these people as well as to global energy security. In 2010 Africa's proven oil reserves are estimated at 127.7 billion barrels (9.6% of the world's total), production 9705 million b/d (12.0% of the world's total). The figures for natural gas are 14.76 trillion cubic meters of proven reserves (7.9% of the world's total) and 203.8 billion cubic meters of production (6.8%) [1]. These figures indicate that Africa holds massive oil and natural gas reserves and is already making a significant contribution to global energy markets.

African oil producers enjoy several geological and geopolitical advantages. First, unlike other producers such as the United States, Mexico, China, and the North Sea where production has already peaked or started declining, most African reservoirs are largely untapped. Upon development, these African reservoirs have the potential to make a significant contribution to the world's oil supply. Indeed, a significant proportion of the increase in world oil production and proven reserves in the last two decades has come from Africa. Second, most of Africa's oil is high-quality, low-sulfur oil. This crude is highly prized by refiners in Europe and North America because it yields far more lucrative refined products than oil from other regions. Third, the location of most oil fields is convenient for shipment to major consuming regions in Asia, Europe, and North America. Finally, compared to other producing regions, most African producers offer attractive fiscal terms to international oil companies (IOCs). Most of the companies operate under favorable production-sharing contracts that allow even the most challenging fields in deep water to be developed at a good financial return. Despite growing competition from national oil companies (NOCs), foreign investment is generally welcomed in most African countries in both upstream and downstream sectors.

These characteristics (substantial reserves, high quality, easy transportation, and attractive investment environment) have laid the foundation for a partnership between major US, European, and Asian energy companies and African producers [2]. This cooperation is the main reason for the continent's rising oil production in recent decades.

As concerns mount about global energy security, North African producers find themselves in a prime position and attracting much attention [3]. Algeria is already a well-established oil and natural gas producer and exporter, though its oil production has declined in the last few

Energy Security: An Interdisciplinary Approach, First Edition. Gawdat Bahgat.
© 2011 John Wiley & Sons, Ltd. Published 2011 by John Wiley & Sons, Ltd.

years, Egypt's gas sector has received substantial interest and investment from international energy companies and the country is on its way to becoming an important gas exporter. Since the lifting of economic sanctions in the early 2000s, Libya has gradually resumed its leading stance as a major African oil and gas producer and exporter.

North Africa is one of the few regions in the developing world where oil companies have full access to reserves and with the participation of the state-owned companies. Indeed, foreign oil companies have always formed the backbone of the North African oil industry. This has been further reinforced since the late 1990s when their share of output continued to rise. This is particularly interesting given that Algeria, Egypt, and Libya have generally followed a model where the state plays a dominant role in the economy. Oil contracts and fiscal terms vary considerably across the North African producers. The production-sharing agreement (PSA) remains among the most common types of agreement used in North Africa and elsewhere in the world. These agreements, however, take different forms and the way that the state participates in joint companies varies considerably across countries [4].

In some countries, particularly in West Africa, initially IOCs focused on onshore fields and later on the shallow ones offshore. A shift to deep-water exploration began in the early 1990s and this has also turned out to be a great success. In West Africa most of the recent exploration took place in three prolific provinces: the Niger Delta tertiary system (central West Africa–Gulf of Guinea), the Gabon–Congo province (southern West Africa), and the Mauritania–Senegal–Bissau province (northern West Africa) [5]. The region's first deep-water discovery came in 1995 off Equatorial Guinea and its first oil was produced two years later [6]. Deep-water exploration and development activities in West Africa have substantially increased in the early 2000s due to the rise in oil prices which made more foreign investment available [7]. In the future, the relative importance of the deep water off West Africa is expected to increase further as exploration success continues and advances in subsea technology overcome the deep-water challenges. While deep-water developments may be large in terms of size, cost, and risk, they also carry large potential rewards for the oil companies involved.

Since the early 2000s, Africa may well have been one of the most exciting regions in the world for oil exploration. But for many international energy companies, the continent has also emerged as an important gas player. As elsewhere in the world, an emphasis on gas is gaining ground in Africa. With Europe concerned over Russia's political motives in exploiting its energy deposits and big international companies shut out or deterred from investing in some Middle Eastern countries, many investors have turned to Africa in the hope of securing future supplies.

Traditionally, African gas supply to Europe has come from North African countries, notably via pipelines from Algeria and Libya, just across the Mediterranean. But the evolution of a global gas market through the development of liquefied natural gas (LNG) technology and shipping in the past two decades has also opened up the possibility of North American and Asian countries importing gas from Nigeria, Egypt, and Algeria. The continent already has some of the world's oldest and biggest LNG plants in Algeria, Libya, and Nigeria. New ones have been recently built in Egypt and Angola, among others. In addition to impressive rises in gas production and export (both via pipelines and as LNG), consumption has substantially increased. Gas is used in power generation and to inject into old oil fields to enhance production. The rise in gas consumption can also be explained by the desire to free more oil for export.

Several factors are likely to have a strong impact on the speedy and full utilization of Africa's hydrocarbon resources:

1. Domestic corruption and lack of transparency: despite some recent improvements, oil and gas authorities in most African countries still score very low on transparency scales. Despite billions of dollars in revenue from oil exports, many African oil producing countries continue to suffer poverty and disease, largely due to corruption and mismanagement. Part of the solution is broader disclosure of oil revenues [8]. There is a lot to be desired to contain corruption not only in the energy sector, but also in the broader government structure and policy.
2. Political instability: several oil producers have yet to find a satisfactory solution to ethnic strife. Nigeria and to a lesser extent Sudan illustrate how ethnic divisions can threaten the steady flow of oil and gas supplies.
3. The rise of resource nationalism: where Africa's resources are becoming hot property, big energy producing countries are beginning to push for greater control of their own oil and gas industries [9].
4. International competition: the relative advantages that African producers enjoy have attracted European, American, Russian, Chinese, and Indian energy companies. The rivalry between these companies and the ability, or lack, of African governments to play off one against the others is already being felt in several African countries.

In the following sections I examine in some detail energy policy in the continent's six major oil and gas producers: Algeria, Libya, Egypt, Sudan, Nigeria, and Angola. The goal is to highlight the similarities and differences between them and how they perceive and pursue their energy security. This will be followed by an analysis of US and European approaches to Africa's energy deposits.

6.1   Algeria

Algeria holds the fourth largest oil reserves in Africa (after Libya, Nigeria, and Angola) and in 2010 was the fourth largest oil producer (after Nigeria, Angola, and Libya). The country holds the second largest natural gas reserves in Africa (after Nigeria) and is the largest gas producer in the continent, and the world's sixth largest (after Russia, United States, Canada, Iran, and Norway).

Algeria has a relatively well-developed energy infrastructure. Most of the oil comes from the following major fields: Hassi Messaoud, Tin Fouye Tabankort Ordo, Zarzaitine, Haoud Berkaoui/Ben Kahla, and Ait Kheir. Part of this crude oil is processed in the country's four refineries: Skikda, Hassi Messaoud, Algiers, and Arzew. Hassi R'Mel is Algeria's largest gas field, accounting for about a quarter of total natural gas production. The remainder of Algeria's natural gas reserves comes largely from associated fields (they occur alongside crude oil reserves) [10].

All these fields and refineries are operated by the national oil company Sonatrach (Société Nationale pour la Recherche, la Production, le Transport, la Transformation, et la Commercialisation des Hydrocarbures s.p.a.). Founded in December 1963, Sonatrach's diversified activities cover all aspects of production: exploration, extraction, transport, and refining. The company also produces a significant proportion of Algeria's gross national product (GNP) [11].

The founding of Sonatrach and the expansion of its role and activities underscore the ups and downs of Algeria's relations with IOCs. The country was one of the first OPEC members to nationalize its oil industry. The nationalization process took several stages and by 1980 the Algerian government had complete control over the oil and gas sectors. Sonatrach discouraged IOCs from doing business in Algeria by offering them unattractive fiscal terms. The outcome of this policy was a steady fall in oil production and revenues, leaving the government with no option but to reverse its policy. The Algerian law governing foreign investment was amended, offering more attractive terms. As a result, IOCs returned to Algeria and the country's oil and gas infrastructure received the badly needed investments and technology. This led to a rise in production in most of the 1990s and 2000s. Algeria's massive proven oil and gas resources have attracted a large range of international energy companies including British Gas (BG), British Petroleum (BP), Cepsa, Conoco-Phillips, ENI, Gazprom, Repsol, Ruhrgaz, Shell, Statoil, and Total.

This policy of providing incentives to encourage foreign investment in the country's energy sector was partly reversed in the later part of the 2000s. High oil prices for most of the decade meant that the Algerian authorities accumulated substantial oil and gas export revenues and became less dependent on private and foreign investment. Thus, in 2006, new legislation gave Sonatrach rights to take on 51% of any hydrocarbon project [12]. Two years later (2008) foreign investors who obtained tax exemptions were ordered to reinvest in Algeria amounts equivalent to the breaks they had received within four years [13]. These new regulations underscore the rise of resource nationalism, which has been echoed in several other oil and gas producing nations. Another sign of the assertive role of Sonatrach was made public when the company announced a plan to invest $63.5 billion between 2009 and 2013 to update and expand its infrastructure, production, and export volumes [14].

Sonatrach's growing role and the new regulations on foreign investment have not weakened Algeria's strong ties with oil and gas importing markets. Algeria exports oil to the United States, Netherlands, France, Spain, Germany, and the United Kingdom among others. Most of the country's natural gas exports reach Europe via three main pipelines or in the form of LNG. The first pipeline is the Trans-Mediterranean (Transmed, also called Enrico Mattei) pipeline which runs from Algeria via Tunisia and Sicily to mainland Italy. The Maghreb–Europe (MEG, also called Pedro Duran Farell) pipeline connects Algeria to the Spanish and Portuguese gas transmission networks via Morocco. The Medgaz pipeline links Algeria to Spain and France [15]. In addition, the Galsi scheme connecting Algeria to the Italian national gas network is under construction and is expected to be completed by 2012 [16]. Finally, in 1964 Algeria became the world's first producer of LNG and has since remained a major exporter. The vast majority of its LNG exports go to European importers on the other side of the Mediterranean, particularly France, Italy, and Spain. Algeria also exports LNG to the United States.

6.2 Libya

Unlike Persian Gulf producers such as Iran, Iraq, and Saudi Arabia where oil was discovered early in the twentieth century, oil in Libya was discovered late in the 1950s. In a short period of time oil discoveries were brought on-stream, particularly from the Sirte Basin. Thus, in the late 1960s Libya had become the world's fourth largest exporter of crude oil [17]. This rush to raise production in Libya reflected not only the world's growing appetite for oil, but also certain advantages that the Libyan oil sector enjoys. First, Tripoli holds huge proven oil reserves – estimated at 43.7 billion barrels, or 3.5% of the world's total, the largest in Africa. Second, production costs are among the lowest in the world. Third, Libya produces high-quality, low-sulfur "sweet" crude oil. Fourth, the proximity of Libya to Europe is a big advantage in terms of ease and costs of transportation to a large and growing market.

Given these advantages, it is little wonder that US and European oil companies were heavily involved in exploring and producing oil in Libya. The country's oil production reached a peak of 3.32 million b/d in 1970. This high level of production, however, was proven unsustainable. In 2008 Libya produced 1.84 million b/d, about half of its production in 1970. This decline can be explained more by political factors than geological ones. From the outset the post-1969 revolutionary regime has had tense relations with the US government and US oil companies operating in the country. Eventually, in the mid-1980s, these oil companies completely withdrew from Libya and froze all their operations. For most of the 1990s, comprehensive international sanctions were imposed on Libya by the United Nations Security Council. These bilateral sanctions in the 1980s and multilateral ones in the 1990s deprived Libya's oil industry of badly needed spare parts, new equipment, modern technology and management techniques, and produced a hostile political environment. A competent Libyan National Oil Corporation (NOC) and a handful of European oil companies have kept oil production going over the years, although at a level greatly reduced from that of the booming late 1960s.

NOC was established in November 1970 to assume responsibility for the oil sector operations. It carries out exploration and production operations through its own affiliated companies, or in participation with other companies under service contracts or any other kind of petroleum investment agreements. This is in addition to marketing operations of oil and gas, locally and abroad. NOC owns refineries (Ras Lanuf, Az Zawiya, Tobruk, Brega, and Sarir), petrochemical complexes, and gas processing plant [18]. NOC also owns national service companies which carry out oil well drilling, provide all drilling materials and equipment, lay and maintain oil and gas pipelines, build and maintain oil and gas storage tanks, and carry out related technical and economic studies [19].

Diplomatic re-engagement with Libya and the lifting of sanctions in the early 2000s have been followed by serious efforts by IOCs to resume their oil exploration and production operations in the country. Since mid-2004, the entire Libyan hydrocarbon sector has appeared poised for a new, promising, and bright development fueled by substantial foreign investment. Several characteristics can be identified in the outlook for Libya's energy sector. First, because of the two-decade-long bilateral and multilateral sanctions, Libya remains highly unexplored. The absence of IOCs means that the most updated technology in exploration has yet to be used in Libya. In the mid-2000s Libyan officials and international observers confirmed that only about 25% of the country was covered by active exploration licenses [20]. This suggests that Tripoli has an outstanding prospect for large discoveries.

Second, like many other oil producing countries, Libya is heavily dependent on oil. Indeed, oil export revenues account for about 95% of the country's hard currency earnings and around 75% of government expenditure [21]. The collapse of oil prices in 1998–1999 contributed to deteriorating economic conditions and sharpened the sense of vulnerability to the fluctuation of oil prices. Despite efforts to diversify the economy, Tripoli is still heavily dependent on oil revenues. In order to make up for dropping reserve replacement and to increase the country's oil production capacity, the Libyan leadership came to the conclusion that foreign investment was a necessity. NOC plans to double production capacity by 2014. This target is considered feasible given the country's low-cost production and the existence of pipelines and export facilities. Official at NOC estimate that Libya would need about $30 billion in investment to modernize its oil sector and increase its production capacity [22]. Accordingly, since the UN sanctions were suspended in 1999, Libya has launched several bidding rounds to allow international companies to explore for oil and natural gas.

Third, prior to 1973, foreign oil companies worked in Libya under concession arrangements. Since then another arrangement, called an exploration and production-sharing agreements (EPSA), has been in place. Under this arrangement the government, through NOC, retains exclusive ownership of oil fields while signatory oil companies are considered as contractors. Three rounds of EPSA contracts were held before international sanctions were imposed – one in 1974, another in 1980, and a third one in 1988, with some differences regarding recovery of development and production costs [23]. EPSA contracts usually involve an initial exploration period in which companies assume exploration costs and risks and are required to invest specific sums in exploration. If a discovery is made, the EPSA continues in force for a set period (usually from 20 to 30 years) and the output is divided between NOC and the contracting company. Since 2004, Tripoli has proposed a new, more attractive model called EPSA-IV. Under this revised formula, contracts are awarded on the basis of competitive bidding instead of by closed negotiations. International companies carry all exploration and appraisal costs, as well as training costs for Libyan nationals, during a minimum exploration period of five years. Thereafter, capital expenses for development and exploitation and operating expenses are borne by NOC and the investor according to their primary agreement [24]. In August 2004, NOC launched the first licensing round under the EPSA-IV model. Since then several rounds have been launched. The goal is to attract more IOCs to do business in Libya.

Fourth, despite the focus on oil exploration and development, there has been a growing interest in the country's almost untapped natural gas resources. Libya became a gas exporter in 1971 when an LNG plant at Marsa al-Brega – the world's second LNG facility – started up. After early pricing disputes, exports built up to a peak of 3.6 billion cubic meters in 1977, but in 1980 the government nationalized the facilities from Esso (now ExxonMobil) and imposed more price rises. The main buyers of the facility's LNG either canceled their contracts or scaled down their purchases. Thus, great prospects and discoveries of natural gas resources will lay the foundation to make Libya an important natural gas producer and exporter. NOC has already established a partnership with the Italian company Agip in a joint venture called the Western Libya Gas Project. The goal is to produce 10 billion cubic meters a year over a 20-year period. Most of this gas will be exported to Italy via the Green Stream pipeline which connects the two countries through Sicily. In parallel to the rise in gas exports, domestic consumption has increased since the mid-2000s.

Fifth, Libya enjoys a special relationship with Europe. These special ties are based, at least partly, on historical experience and geographical proximity. Several European countries have extensive trade relations with Libya and nearly all Libyan oil is exported to European countries, particularly Italy, Germany, France, and Spain. Furthermore, European oil companies maintained their operations in Libya after their US rivals left in the 1980s. The major foreign oil company in Libya since the departure of the US companies is ENI/Agip of Italy. ENI is the largest and longest established foreign oil company in Libya, extracting more than half of the country's production, and Italy is the largest purchaser of Libyan oil. In December 2008, Libya bought 10% shares in ENI, cementing the close ties between the two sides [25]. French, Spanish, Dutch, and UK companies are also heavily involved. The prospects for continuing close cooperation between Europe and Libya in the future are very strong given the growing European dependence on imported oil and gas supplies and the European Union's policy of diversifying its suppliers. Since sanctions were lifted, top European officials have visited Tripoli and the Libyan leader Muammar Gaddafi has visited several European countries.

Sixth, US oil companies have been involved in Libya's oil industry since the early discoveries in the late 1950s and early 1960s. Some of the largest fields were found in concessions held by independent US companies. Marathon, Amerada Hess, and Conoco formed the Oasis Group, which, with NOC, achieved world-class commercial oil discoveries in Libya's Sirte Basin in 1962 [26]. The outcome of this partnership was a steady and substantial increase in Libya's oil production. However, this mutual profitable partnership was interrupted by political tension between Washington and Tripoli. In January 1986, former US President Ronald Reagan issued an executive order imposing unilateral sanctions against Libya. US companies' assets in the country were put in "suspended animation [27]." In order to protect their concessionary interests, five US firms signed a Standstill Agreement with NOC. Under this arrangement, the US companies retained the original rights and obligations in the fields they operated while NOC became responsible for the development of these fields until the return of the US firms.

In the following two decades the Standstill Agreement survived the extreme political tension between the two nations. Since the mid-1980s, NOC and its subsidiaries have maintained production at these fields, albeit at much lower levels. With the lifting of UN sanctions in 2003 and US sanctions in the following years, US oil companies were allowed to resume their operations in Libya. Shortly after lifting sanctions, top executives from Occidental Petroleum, Conoco-Phillips, Marathon Oil, and Amerada Hess went to Libya and started negotiations to resume their operations. In May 2004, NOC announced its first oil shipment to the United States in over 20 years. Since then the United States has increased its imports of Libyan oil.

Seventh, in addition to warmer diplomatic and commercial relations with Europe and the United States, Tripoli has maintained its traditional good ties with Moscow. In April 2008, Vladimir Putin visited Libya, becoming the first Russian leader to visit the North African state. Putin oversaw the writing off of $4.5 billion of Libya's Soviet-era debt in return for securing multibillion-dollar contracts for state corporations. Gazprom, Russia's natural gas company, signed a memorandum of cooperation to set up a joint venture with NOC [28]. Several months later (October 2008) Libyan leader Muammar Gaddafi reciprocated by visiting Russia, the first such visit since 1985.

6.3   Egypt

In 2010, Egypt's proven oil reserves were estimated at 4.4 billion barrels, 0.3% of the world's total, and sixth largest in Africa (after Libya, Nigeria, Angola, Algeria, and Sudan). Cairo holds larger natural gas reserves of approximately 2.19 trillion cubic meters, 1.2% of the world's total, third in Africa (after Nigeria and Algeria). Egypt's energy output and potential are compounded by the country's large population (most populous in the Arab world) and decades-long heavy government subsidies. This combination means large consumption. In order to overcome these hurdles and raise production, Egypt has sought to attract foreign investment by offering a stable legal system and more generous fiscal terms than in other African countries.

After experimenting with a heavy state-led economic model in the 1950s and 1960s, Cairo introduced reforms in most economic sectors including energy. These reforms in conjunction with high oil prices in the 1970s provided incentives for IOCs. Accordingly, in 1976, Egypt became a net exporter of crude oil for the first time. The oil sector is dominated by the Egyptian General Petroleum Corporation (EGPC). Originally founded in 1956 under the name of the General Corporation of Petroleum Affairs and later in 1976 became known as the Egyptian General Petroleum Corporation, the company supervises all petroleum operations. These include signing agreements with IOCs, exploration, production, refining, domestic consumption, environmental concerns, and training [29].

Egyptian oil production comes from four areas: the Gulf of Suez, the Western Desert, the Eastern Desert, and the Sinai Peninsula [30]. Most of the fields are developed by joint ventures between the EGPC and foreign companies such as ENI, BP, Petronas, Apache, Dana, and Burren Energy. This partnership with foreign companies had slowed down the depletion of the already small oil fields. Still, Egypt's oil production peaked at 910 000 b/d in 1993 and has since been in continual decline. Since the mid-1980s the EGPC and foreign companies have sought to arrest the declining rates using a variety of enhanced oil recovery techniques.

Despite the small and declining oil production, Egypt's oil industry is important in two different areas: transit and refining. Egypt controls two routes for the export of Persian Gulf oil: the Suez Canal and the Sumed pipeline. The former is an important route for oil tankers and the Suez Canal Authority works on several enhancement and enlargement projects. The Sumed pipeline became operational in 1977 and serves as an alternative and supplement to the Suez Canal. It is a joint venture between Egypt, Saudi Arabia, Kuwait, the United Arab Emirates, and Qatar [31]. In addition to these two important routes, Egypt has the largest refining sector on the African continent.

Egypt's limited oil reserves have prompted the authorities to focus on natural gas. The first gas field, Abu Madi, came on-stream in 1975, but the real momentum came a decade later when Shell and the EGPC signed the first Gas Clause in 1988, which became the standard agreement for future gas concessions [32]. Under this arrangement natural gas was to be treated similar to oil in PSAs. In August 2001, the Ministry of Petroleum established the Egyptian Natural Gas Holding Company (EGAS) to implement a broad strategy aimed at increasing production to meet the growing domestic demand and export [33].

Working with major foreign companies such as BG, BP, ENI, and Shell, Egypt has established itself as an important gas exporter both via pipelines and in the form of LNG. In 2008, Egypt became the world's sixth largest LNG producer and a significant supplier to Europe, Arab countries, and other markets [34]. In December 2008, the European Commission and the Egyptian government signed a memorandum of understanding (MOU) to enhance energy cooperation between Cairo and the European Union. This MOU underscores five priorities including the establishment of a work program to gradually converge Egypt's energy markets with the EU's and the development of energy networks such as the Arab gas pipeline, which could transport Egyptian natural gas to Europe. Other areas covered include market reforms, promotion of renewable energy and energy efficiency, and technological and industrial cooperation [35].

Egypt's most expansive export project is the Arab gas pipeline that connects Egypt to Jordan and Syria. Turkey and Syria also signed an agreement to connect the pipeline to the Turkish grid and extend the pipeline into Europe [36]. Another pipeline, the Arish–Ashkelon, became operational in 2008 and began transporting natural gas to Israel. Egypt also has a number of LNG trains that carry its gas for onward shipment to distant markets including the United States [37].

6.4   Sudan

Sudan, the largest country in Africa, has limited oil reserves, namely 6.7 billion barrels, 0.5% of the world's total. In 2010, its production was 490 000 b/d, 0.6% of the world's total. The exploration and development of these limited deposits have been further complicated by political instability due to prolonged civil war between the north and south and ethnic conflict in Darfur in the western part of Sudan. These two conflicts and accusations of massive abuse of human rights by the Sudanese government have forced most Western oil companies out of the country. In response the government in Khartoum has pursued a two-fold strategy. First, it created an attractive investment environment, including removing key price controls, liberalizing the investment code and exchange regime, and reducing trade restrictions [38]. Second, the Sudanese government established partnerships with several Asian oil companies to develop the country's oil deposits and infrastructure. Both China and India see Sudan as an important strategic oil supplier and "snapped up licenses in the wake of the Western companies' departure" [39]. Sudan exports the majority of its oil to Asian markets (China, Japan, Indonesia, India, South Korea, Taiwan, Thailand, and Malaysia).

In 1975 the Sudanese government granted Chevron a concessionary area in the south of the country. Following geological and geophysical surveys, Chevron made two significant discoveries: the Unity oil field and the Heglig field in 1980 and 1982 respectively. However, the civil war and attacks and threats against oil installations and expatriate employees forced Chevron to suspend its operations and leave the country. Chevron sold its concession to a Sudanese oil company, ConCorp, which sold it to a private Canadian company, Arakis Energy Corporation. Unable to raise the necessary funding to construct the pipeline needed to link its oil fields to Port Sudan on the Red Sea, Arakis sold most of its shares to the China National Petroleum Corporation (CNPC), Petronas of Malaysia, and the national Sudanese oil company (Sudapet) and formed a new consortium, the Greater Nile Petroleum Operating Company, with Arakis acting as the operator. In 1998, Arakis sold its share to another Canadian oil company, Talisman Energy, which assumed the role of an operator. Four years later, Talisman sold its interest to India's Oil and Natural Gas Corporation Limited [40].

Despite all these changes, Sudan's oil production has risen steadily since 1999, when CNPC completed a crude oil pipeline from the Unity and Heglig fields to Port Sudan on the Red Sea and started production. The majority of reserves are located in the south in the Muglad and Melut Basins. All oil fields and the entire energy sector are managed by the Sudan National Petroleum Corporation (Sudapet). Founded in 1997, Sudapet serves as the government's arm in negotiating with IOCs, training, transferring technology, exploration, and development [41]. In October 2005, the Sudanese government established the National Petroleum Commission (NPC) to bolster the development of the country's oil resources. The NPC allocates new oil contracts to ensure an equal sharing of oil revenues between the national government in Khartoum and the government of Southern Sudan [42].

Factional fighting in the south and rebel attacks on oil infrastructure have kept oil production and exploration from reaching their full potential. Sanctions have added more hurdles. The United States prohibits US nationals from engaging in any transactions or activities related to the petroleum or petrochemical industries in the entire territory of Sudan. The United States has maintained economic sanctions against Sudan since 1997. In Executive Order 13067 of November 1997, President Clinton imposed trade sanctions and blocked all property and interests on property of Sudan within the United States. In April 2006, President Bush expanded the scope of the Sudan sanctions by issuing Executive Order 13400 to block all property of persons determined to be contributing to the conflict in Darfur. Six months later (October 2006), President Bush issued Executive Order (EO) 13412 calling for support of the government of Southern Sudan and assistance for the peace efforts in Darfur [43].

Khartoum also faces sanctions from the UN and the EU which include arms embargoes, travel bans, and restrictions on financial activities that may impede the peace process between the government in Khartoum and its opponents in south Sudan and in Darfur. These sanctions are documented in two UN Security Council Resolutions: Resolution 1556 of July 2004 and Resolution 1591 of March 2005 [44].

6.5 Angola

Angola holds 13.5 billion barrels of proven reserves, 1.0% of the world's total and third largest in Africa (after Libya and Nigeria). The country's oil production has steadily increased since the early 1990s reaching 1.784 million b/d in 2010, 2.3% of the world's total. This rise in production reflects geological and geopolitical developments. Initially, oil production was largely concentrated in numerous onshore blocks. In recent years major discoveries have been made in shallow-water, deep-water, and ultra-deep-water blocks. Angola's oil production is projected to reach 2.7 million b/d by 2030 [45]. The majority of oil is medium to light crude with a low sulfur content.

Geopolitically, Angola has experienced a prolonged period of transformation following its independence from Portugal in 1975. For 27 years the country was weakened by a civil war between the Popular Movement for the Liberation of Angola (MPLA), led by Jose Eduardo Dos Santos, and the National Union for the Total Independence of Angola (UNITA), led by Jonas Savimbi. Peace seemed imminent in 1992 when Angola held national elections, but fighting picked up again by 1996. Savimbi's death in 2002 ended UNITA's insurgency and strengthened the MPLA's hold on power.

This relative peace, however, should not be overestimated. The US Department of State describes internal security in Angola as fragile and unstable. Borders remain porous and vulnerable to movements of small arms, diamonds, and other possible sources of terrorist financing. Angola's high rate of dollar cash flow made its financial system an attractive site for money laundering, and the government's capacity to detect financial crimes remains limited. Corruption, lack of infrastructure, and insufficient capacity continue to hinder Angola's border control and law enforcement capabilities [46].

On the threshold of independence, a working group was set up in the oil industry both to support the industry and to mobilize Angolans working in the business. The main objective was the strategic preparation of the oil industry after the proclamation of independence. Decree 52 of 1976 instituted the Sociedade Nacional de Combustíveis de Angola (Sonangol) as the state national oil company whose mission was the management of hydrocarbon resource exploration. Despite having the government as the sole shareholder, Sonangol has always been run as a private company and is under strict performance standards to ensure efficiency and productivity. The absence of qualified nationals for the local oil industry forced Sonangol to begin paying special attention to the training and professional development of its employees. In the late 1970s the company sent students to acquire oil industry knowledge and experience in several countries including Algeria and Italy. Upon return, these students became the driving force in Sonangol [47].

Sonangol works with foreign companies through joint ventures and PSAs, while funding its share of production through oil-backed borrowing. Major international oil companies operating in Angola include BP, Chevron, ENI, Total, ExxonMobil, Devon Energy, Maersk, Occidental, Roc Oil, Statoil, and Sinopec. Some of these IOCs are involved in the country's refining production and have expressed interest in developing natural gas deposits [48]. Most of Angola's oil is exported to China and the United States and to a lesser extent to Europe and Latin America.

As a sign of its growing role in global energy markets and its rising volume of production and export, Angola joined OPEC on January 1, 2007. Like other oil producing countries, oil revenues provide a large proportion of the country's national income. It is important to re-emphasize that Angola's energy security and its rising role as an important oil producer and exporter are restrained by domestic violence and corruption. Most onshore exploration and production activities have mainly focused around the Cabinda province. This area is home to separatist movements demanding access to oil revenues and greater participation in the oil industry. While the government has sought to accommodate members of these separatist movements by appointing them to political and security positions, clashes still occur between government forces and rebels in the area. Equally important, allegations of corruption and lack of transparency in public finance persist. In 2010 the World Bank ranked Angola as one of the most difficult places in the world to do business as a result of cronyism and bureaucracy (169th out of 183 economies) [49].

6.6 Nigeria

With approximately 150 million people (2009) Nigeria is the most populous country in Africa. It holds 44.3 billion barrels of proven oil reserves, 3.3% of the world's total (Africa's second largest after Libya) and in 2010 average production was 2.061 million b/d, 2.6% of the world's total. Nigeria's natural gas reserves are equally impressive. It holds 5.28 trillion cubic meters of proven reserves, 2.8% of the world's total and the largest in Africa. In 2010 it produced 24.9 billion cubic meters, 0.8% of the world's total and third largest in Africa after Algeria and Egypt. The country has been an active member of OPEC since 1971. Nigeria's oil production is projected to grow by an average of 1.4% per year from 2008 to 2030, reaching 3.4 million b/d, and its geological structure has "the greatest potential to increase natural gas production" [50].

In short, Nigeria is a major player in global energy policy and has the potential to increase its contribution to international energy security and raise the standard of living of its large population. In order to achieve these goals Nigeria has to aggressively address problems related to corruption, mismanagement, and lack of transparency and to find a way to reduce ethnic and religious tension and violence in the Niger Delta region, where a significant proportion of its oil comes from.

Nigeria achieved its independence from British occupation in 1960. For most of the following two decades the country was under military rule. In 1999 a new constitution was adopted and in most of the 2000s Nigeria has experienced a civilian type of government and a civilian transition of power. These civilian governments have not succeeded in containing the long-standing ethnic and religious violence over oil deposits in the Niger Delta.

This violence is mainly between the Nigerian government and the Movement for the Emancipation of Niger Delta (MEND). MEND launched itself onto the international stage in January 2006 by claiming responsibility for the kidnapping of four foreign oil workers. Since then the group's attacks on oil pipelines and kidnappings have continued and substantially reduced oil output from the Niger Delta. These attacks and kidnaps have forced some IOCs working in Nigeria to reduce their operations, declare force majeure on oil shipments, or completely pull out of the country. MEND's stated goals are to localize control of Nigeria's oil, redistribute oil wealth, and secure reparations from the federal government for pollution caused by the oil industry [51]. Oil exploration and development activities have been blamed for pollution that has damaged air, soil, and water, leading to losses in arable land and decreasing fish stocks. In 2007, President Umaru Yar'Adua tried to reach a compromise with the rebels and to hold a peace conference to end violence in the Niger Delta, but various armed groups pulled out of the talks, accusing the government of a lack of goodwill [52].

In addition to the Niger Delta, most of the oil is found offshore in the Bight of Benin, the Gulf of Guinea, and the Bight of Bonny. The first successful well was drilled in 1956 and two years later (1958) Nigeria started exporting oil [53]. In recent years exploration activities have mostly focused on deep water and ultra-deep water offshore. Most of the output is light, sweet, and low-sulfur crude. Oil is processed at four refineries (Port Harcourt I and II, Warri, and Kaduna) [54]. All exploration, development, refining, and transportation activities are managed jointly by the national oil company, Nigerian National Petroleum Corporation (NNPC), and a number of IOCs including Royal Dutch Shell, ExxonMobil, Chevron, Conoco-Phillips, Total, Agip and Addax Petroleum. Royal Dutch Shell has the longest history of any oil company in Nigeria. Since the mid-2000s oil companies from Russia, China, and India started competing over Nigeria's oil deposits.

In 1977 the government established the NNPC by Decree 33. In addition to exploration activities, the NNPC was given power and operational interests in refining, petrochemicals, transportation, and marketing. In 1988, the Nigerian government divided the NNPC into 12 subsidiary companies in order to better manage the country's oil industry. These 12 units cover the entire spectrum of oil industry operations [55]. As in other oil producing countries, in recent years Nigeria has pushed for increasing participation by local companies.

Nigeria's large natural gas reserves are largely underutilized. Most of the gas is often flared and wasted. The Nigerian government has sought to develop gas deposits both to meet growing domestic demand (particularly for power generation) and for export, mainly in the form of LNG. In 2009 the Gas Master Plan was developed to achieve these goals. Nigeria exports a large volume of its natural gas and oil to the United States. It also exports oil to Europe, Brazil, India, and South Africa, among others.

In addition to security concerns, the full utilization of the country's hydrocarbon wealth is restrained by widespread corruption and lack of transparency. Nigeria ranks very low (121st out of 180 countries) on the Transparency International Corruption Perception Index [56] and 125th out of 183 on the World Bank Ease of Doing Business Index [57]. Addressing these issues would greatly improve the chances for Nigeria to reach its full potential as a major oil and natural gas producer and exporter and enhance its energy security and overall global one.

6.7 United States and Africa

For decades the United States has viewed imported oil as a national security concern [58]. US indigenous oil production peaked in the early 1970s and the nation has since grown more dependent on imported supplies. Despite calls and efforts to reduce the "addiction to oil," petroleum is still the main fuel in the nation's energy mix and in 2010 the United States imported approximately 60% of its oil needs. Within this context, Africa's hydrocarbon resources play a significant role in Washington's energy, commercial, and strategic interests.

As far as the United States is concerned, Africa's oil and gas deposits enjoy several advantages. Geographically, the United States is close to African oil and gas producers in both the northern and western parts of the continent. Algeria, Angola, Egypt, Libya, and Nigeria have easy maritime access to the United States. Environmentally, Africa's oil is generally light, sweet, and low sulfur. This good-quality oil is suitable for US refineries and yields less pollution than heavy oil, common in other producing regions [59]. Strategically, a large proportion of Africa's oil and natural gas deposits are offshore, therefore less vulnerable to domestic violence. Furthermore, oil and LNG exports from Africa do not cross the crowded Straits of Hormuz. This means that tension in the Persian Gulf is less likely to interrupt shipments from Africa. Given these advantages, Africa's oil and natural gas supplies account for a growing share of the US energy mix and imports. Egypt, Algerian, Nigeria, Angola, and recently Libya are important suppliers of oil and gas to the United States.

In addition to these supplies, major US oil companies have taken the lead in developing hydrocarbon deposits in several African countries. Unlike some other producing regions, most African states welcome foreign investments in their oil and gas sectors. As a result, US oil companies have invested billions of dollars in Africa. Protecting these huge investments is a major US objective. Thus, US energy and foreign policy in Africa seeks to promote and reinforce political stability and economic reform in order to protect US investments and to prevent any interruption of oil and gas supplies from the continent.

Throughout the Cold War, US relations with North Africa were defined by the broader struggle with the Soviet Union. For example, once the development of long-range bombers and intercontinental ballistic missiles made US Strategic Air Command bases in Libya and Morocco less necessary to deter a Soviet nuclear strike, those bases were quickly and easily dispensed with. Similarly, Algeria's leadership in the non-aligned movement throughout the Cold War strained its relations with the United States, even though Algeria was not a close ally of Moscow [60]. Since the 1980s other issues have become important and have since shaped US policy not only with North African nations, but also with Sub-Saharan ones.

A prominent US concern is the threat of terrorism on political stability in the region and on global security. In order to coordinate counter-terrorism efforts between Washington and African countries, the United States launched a number of initiatives including the Pan-Sahel Initiative (PSI), the Trans-Sahara Counter-Terrorism Initiative (TSCTI), the Africa Contingency Operations Training and Assistance (ACOTA), and the Middle East Partnership Initiative (MEPI). All these programs share the same goal – reinforcing domestic security and political stability. Stable African governments with growing and transparent economies would serve the interests of their peoples, US energy and strategic objectives, and global energy security and peace.

The PSI was initiated in November 2002 and was designed to protect borders, track the movement of people, combat terrorism, and enhance regional cooperation and stability. The PSI was a state-led effort to assist Mali, Niger, Chad, and Mauritania in detecting and responding to suspicious movements of people and goods across and within their borders through training and cooperation. Its goals support two US national security interests in Africa: waging war on terrorism and enhancing regional peace and security. Despite its successes, the PSI was constrained from its inception by limited funding and a limited focus [61]. The PSI was succeeded by the broader and more comprehensive TSCTI.

The TSCTI started in June 2005 and was planned as a follow-up to the PSI. The overall approach is straightforward: build indigenous capacity and facilitate cooperation among governments in the region that are willing partners (Algeria, Burkina Faso, Chad, Libya, Mali, Mauritania, Morocco, Niger, Nigeria, Senegal, and Tunisia). The rationale behind this initiative is that the Trans-Sahara region is an area of acute vulnerability due to the vast expanses of desert and porous borders. With a long history of being a center through which arms and other illicit trade flow, it became increasingly important as terrorists seek to use these routes for logistical support, recruiting grounds, and safe havens.

The TSCTI helps strengthen regional counter-terrorism capabilities, enhances and institutionalizes cooperation among the region's security forces, promotes democratic governance, and ultimately benefits US bilateral relationships with each of these states. In addition to the US military's leading role, other US government agencies also became active players in the program. The US Agency for International Development (USAID), for example, addresses educational initiatives; and the State Department and the Department of Treasury train indigenous officials on counter money-laundering techniques [62].

Initiated in 2004, the ACOTA program is managed and funded by the State Department. It is an initiative designed to improve African ability to respond quickly to crises by providing selected militaries with the training and equipment required to execute humanitarian or peace support operations. Once trained, forces can be deployed in multinational units to conduct operations under the auspices of the African Union, the UN, or regional security organizations [63].

The MEPI was launched in 2002 to assist efforts to expand political participation, strengthen civil society and the rule of law, empower women and youth, create educational opportunities, and foster economic reform throughout the Middle East and North Africa. In support of these goals, the MEPI works with non-governmental organizations, the private sector, and academic institutions, as well as governments [64].

The activities of most of these initiatives are reflected and coordinated by the US–Africa Command (AFRICOM). In February 2007, President Bush and Defense Secretary Robert Gates announced the creation of AFRICOM. The decision was the culmination of a 10-year thought process within the Department of Defense acknowledging the emerging economic and strategic importance of Africa, and recognizing that peace and stability on the continent impact not only Africans, but the interests of the United States and the international community as well. The creation of AFRICOM was driven by the desire to better focus the US government's resources to support and enhance existing US initiatives that help African nations, the African Union, and the regional economic communities to succeed [65].

A major focus of AFRICOM is the strong connection between security and development in Africa. AFRICOM is more about preventing wars than fighting them. Its main objective is to take the lead in establishing a secure environment. This security will, in turn, set the groundwork for increased political stability and economic growth. As a result, AFRICOM reflects a much more integrated staff structure, one that includes significant management and staff representation by the State Department, USAID, and other US government agencies involved in Africa. The command also cooperates with partner nations and humanitarian organizations, from Africa and elsewhere, to work alongside US staff on common approaches to shared interests [66].

To sum up, since the early 2000s, US oil and natural gas imports from Africa have been on the rise. Their share of the US energy mix is projected to increase. Additionally, US energy companies have invested substantial resources in developing hydrocarbon infrastructures in several African countries. These two dynamics (oil and gas imports and large investments) have shaped short- and long-term US policy in Africa. US initiatives seem to underscore the indispensability and strong connection between political stability and economic development. The thrust of the US strategy is that a stable and prosper Africa would enhance US efforts in the war on terrorism as well as improve energy security. Essentially, Europe shares similar goals and strategy.

6.8   Europe and Africa

Energy security is a major objective of Europe in order to sustain economic development and ensure the well-being of its citizens. A major theme of Europe's energy security is that the more the region is exposed to a high concentration of energy markets, the lower is its energy security. Stated differently, the diversification of energy suppliers is a major component of Europe's energy strategy. Another important theme is interdependence between energy consumers and producers. Generally, European leaders understand that the notion of "energy independence" cannot be applied to Europe, given the continent's limited indigenous resources and the globalization of energy markets. Against this backdrop, Europe's overdependence on Russia's oil and gas supplies has heightened the continent's sense of vulnerability. Meanwhile, Africa, with its large hydrocarbon deposits and promising potential, is geographically close to Europe and the two regions have had close historical and cultural ties for centuries.

European policy in Africa has dramatically changed since the early 1990s. This fundamental change reflected the end of the Cold War. Africa is no longer seen as a mere playground for the East–West conflict [67]. Rather, European investments and technology and Africa's largely underutilized resources lay the foundation for a strong and mutually beneficiary partnership. These new dynamics and perceptions created appropriate conditions for holding the first Africa–EU Summit in Cairo in 2000. Since then top officials from the two sides have held regular meetings and have articulated an Africa–EU strategic partnership, which covers cooperation in several areas of mutual interest. These include peace and security; democratic governance and human rights; trade, regional integration, and infrastructure; millennium development goals; migration, mobility, and employment; science, information society, and space; climate change; and energy [68].

Under the energy partnership, the two sides defined a strategy to address key challenges covering energy security, access to renewable energy sources, and climate change issues. They agreed on the need to bolster African institutions and establish the legal, fiscal, and regulatory environments best able to promote private investments in energy and establish national transparency plans and guidelines for energy companies. Furthermore, they agreed on promoting cooperation in improving energy efficiency and launching a renewable energy program [69]. Finally, they agreed on the urgent need to promote the electrification of Africa and to launch an Electricity Master-Plan for Africa [70].

   The Trans-Sahara gas pipeline illustrates this growing EU–Africa energy partnership. The project stretches a distance of 4300 kilometers across the Sahara Desert from Nigeria to Niger and Algeria. Upon completion it will connect Nigeria's gas reserves to Europe via Algeria's Mediterranean coast. It will supply 20 billion cubic meters a year of gas to Europe by 2016, reducing its dependence on Russia and contributing to the continent's overall energy security. In 2009 the European Union offered to invest $21 billion in the project [71].

   Louis Michel, the European Commissioner for Development and Humanitarian Aid, summed up the underlying drive for EU–Africa partnership: "Helping Africa to make the most of its advantages and overcome its weaknesses, and, in doing so, respond to our own economic and political interests, is the purpose of this Africa–Europe partnership" [72].

6.9   Conclusion: The Way Ahead

Africa is a large and diverse continent. Its massive hydrocarbon deposits, like some other African resources, are largely underexplored and underdeveloped. Individual African states and the continent as a whole face daunting socio-economic and political challenges. The full utilization of Africa's energy deposits to meet its fast-growing consumption and to export to North American, Asian, and European markets is essential to overcome these challenges. Oil and gas revenues already represent a significant proportion of the national income in most African producers.

   The full utilization of these resources, however, is restrained by the shortage of a well-trained labor force, managerial skills, necessary investments, and modern technology. Some African countries have already made significant progress in educating and training their indigenous labor force and acquiring technological and management skills. The continent as a whole, however, still has a long way to go. Partnership with North American, European, Chinese, Indian, and Russian companies has served the interests of all involved parties. These IOCs provide the financial and technological resources that African oil and gas producers need. On the other hand, major energy consumers seek to diversify their energy suppliers. African oil and gas supplies are essential in such diversification and in enhancing global energy security. This brief discussion of six African producers and the US and European approaches underscores the common interests all sides share. Political stability and economic prosperity would benefit African oil and gas producers and foreign consumers. It is a win–win situation.

References

[1] British Petroleum (2010) *BP Statistical Review of World Energy*, British Petroleum, London, pp. 6, 8, 22, 24.
[2] Izundu, U. (2008) Competition reshaping African deal structures. *Oil & Gas Journal*, **106** (20), 20–25, 20.
[3] England, A. (2008) North Africa: At the Center of Attention. Financial Times (January 28).
[4] Fattouh, B. (2008) *North African Oil and Foreign Investment in Changing Market Conditions*, Oxford Institute for Energy Studies, Oxford, p. 3.
[5] McLennan, J. and Williams, S. (2005) Deepwater Africa reaches turning point. *Oil & Gas Journal*, **103** (6), 18–25, 18.
[6] Barkindo, M. and Sandrea, I. (2007) Undiscovered oil potential still large off West Africa. *Oil & Gas Journal*, **105** (2), 30–34, 31.
[7] Barkindo, M. and Sandrea, I. (2007) West Africa second only to Russia in non-OPEC supply contribution. *Oil & Gas Journal*, **105** (3), 43–48, 43.

[8] Izundu, U. (2008) African producers grapple with issues of transparency. *Oil & Gas Journal*, **106** (20), 24.

[9] Mahtani, D. (2008) The New Scramble for Africa's Resources. Financial Times (January 28).

[10] Energy Information Administration, *Country Analysis Briefs: Algeria*. Available at http://www.eia.doe.gov/emeu/cabs/Algeria/full.html (accessed May 12, 2009).

[11] Sonatrach, *Sonatrach: A world leader in the field of energy*. Available at http://www.Sonatrach-dz.com (accessed September 28, 2009).

[12] Mahtani, D. (2008) Gas: Highly Prized Reserves in a Tightening Market. Financial Times (January 28).

[13] Saleh, H. (2008) Algeria Tightens Foreign Investor Rules. Financial Times (August 11).

[14] Hoyos, C. (2009) Algeria's Sonatrach to Boost Investment by 41%. Financial Times (March 18).

[15] Middle East North Africa Financial Network, *Medgaz Pipeline between Algeria, Spain Operational by end of 2009*. Available at http://www.menafn.com/qn_news_story_s.asp?storyID=1093259755 (accessed September 30, 2009).

[16] Nicholls, T. (2007) Galsi gas to flow in 2012, but export targets will be missed. *Petroleum Economist*, **74** (12), 24–26, 24.

[17] Gurney, J. (2000) Opportunities and risk in Libya. *Energy Economist*, (221), 1–12, 4.

[18] Energy Information Administration, *Country Analysis Briefs: Libya*. Available at http://www.eia.doe.gov/emeu/cabs/Libya/full.html (accessed July 28, 2009).

[19] National Oil Corporation – Libya, *NOC in Brief*. Available at http://en.noclibya.com.ly/index.php?opion=com_content&task=view&id=167&Itemid=55 (accessed September 27, 2009).

[20] Quinlan, M. (2004) Libya's oil still tantalizes. *Petroleum Economist*, **71** (3), 32–34, 32.

[21] Energy Information Administration, *OPEC Revenues: Country Details*. Available at http://www.eia.doe.gov (accessed June 16, 2008).

[22] Banerjee, N. (2004) Libya to Open Eight Oil Projects to Bidders. New York Times (May 5).

[23] Bahgat, G. (2004) Libya's energy outlook. *Middle East Economic Survey*, **47** (42), 22–24.

[24] Khodadad, N., DeBeer, S., and Rogers, D. (2004) Swings and roundabouts. *Petroleum Economist*, **71** (7), 22–24, 24.

[25] Dinmore, G. (2008) Rome's Colonial Past Key to Libya's ENI Stake. Financial Times (December 9).

[26] Logan, E. (2004) Libya looks to a bright future post-sanctions. *Energy Economist* (272), 30–33, 31.

[27] Khadduri, W. (2000) Libya set to start signing major oil and gas agreements with IOCs by June 2001. *Middle East Economic Survey*, **43** (48), 18–22.

[28] Moscow Times (2008) $4.5 Billion Debt Deal Crowns Libya Trip. Moscow Times (April 18).

[29] Egyptian General Petroleum Corporation, *Corporate Profile*. Available at http://www.oilegypt.com (accessed October 1, 2009).

[30] Editorial (2004) Egypt's western desert onshore oil and gas production strong. *Oil & Gas Journal*, **102** (42), 34–39, 34.

[31] Arab Petroleum Pipelines Company SUMED, *About SUMED*. Available at http://www.sumed.org (accessed October 1, 2009).

[32] Fattouh, B. (2008) *North African Oil and Foreign Investment in Changing Market Conditions*, Oxford Institute for Energy Studies, Oxford, p. 12.

[33] Egyptian Natural Gas Holding Company, *Introduction*. Available at http://www.egas.com/eg/Corporate_Overview/Introduction.aspx (accessed September 28, 2009).

[34] Mahtani, D. (2009) Gas: Highly Prized Reserves in a Tightening Market. Financial Times (January 28).

[35] Editorial (2008) EU, Egypt sign energy cooperation agreement. *Oil & Gas Journal*, **106** (47), 5.

[36] Daily World EU News, *Arab Gas Pipeline Approved by Turkey-EU*. Available at http://www.turks.us/article.php?story=20080506092427239 (accessed April 15, 2009).

[37] Energy Information Administration, *Country Analysis Briefs: Egypt*. Available at http://www.eia.doe.gov/emeu/cabs/Egypt/full.html (accessed August 13, 2008).

[38] Cohen, A. and Alasa, R., Heritage Foundation, *Africa's Oil and Gas Sector: Implications for U.S. Policy*. Available at http://www.heritage.org/research/africa/bg2052.cfm (accessed September 28, 2009).

[39] Editorial (2003) Sudan could support growth as non-OPEC oil exporter. *Oil & Gas Journal*, **101** (41), 46–48, 46.

[40] Fattouh, B. (2008) *North African Oil and Foreign Investment in Changing Market Conditions*, Oxford Institute for Energy Studies, Oxford, p. 15.

[41] Sudapet, *Sudapet Profile*. Available at http://www.sudapet.sd.aboutus.hph (accessed October 2, 2009).

[42] Energy Information Administration, *Country Analysis Briefs: Sudan*. Available at http://www.eia.doe. gov/emeu/cabs/Sudan/full.html (accessed September 15, 2009).

[43] US Department of the Treasury, *An overview of the Sudanese Sanctions Regulations*. Available at http://www.treasury.gov (accessed October 2, 2009).

[44] United Nations Security Council, *Security Council Committee established pursuant to resolution 1591 (2005) concerning the Sudan*. Available at http://www.un.org/sc/committees/1591/index.shtml (accessed October 2, 2009).

[45] Energy Information Administration (2009) *International Energy Outlook*, United States Government Printing Office, Washington, DC, p. 225.

[46] US Department of State (2009) *Country Reports on Terrorism*, United States Government Printing Office, Washington, DC, p. 16.

[47] Sonangol, *History of Sonangol*. Available at http://www.sonangol.co.ao/wps/portal/ep/sonangol/history (accessed September 28, 2009).

[48] Energy Information Administration, *Country Analysis Briefs: Angola*. Available at http://www.eia.doe. gov/emeu/cabs/Angola/full.html (accessed March 14, 2008).

[49] World Bank, *Doing Business: Angola*. Available at http://www.doingbusiness.org (accessed October 3, 2009).

[50] Energy Information Administration (2009) *International Energy Outlook*, United States Government Printing Office, Washington, DC, p. 40.

[51] Hanson, S. (2007) *MEND: The Niger Delta's Umbrella Militant Group*, Council on Foreign Relations, Washington, DC. Available at http://www.cfr.org/publication/12920/# (accessed March 22, 2007).

[52] Green, M. (2008) Nigeria: Unfulfilled Potential. Financial Times (January 28).

[53] Nigerian National Petroleum Corporation, *History of the Nigerian Petroleum Industry*. Available at http://www.nnpcgroup.com/history (accessed September 27, 2009).

[54] Energy Information Administration, *Country Analysis Briefs: Nigeria*. Available at http://www.eia.doe. gov/emeu/cabs/Nigeria/full.html (accessed August 28, 2009).

[55] National Nigerian Petroleum Corporation, *About NNPC*. Available at http://www.nnpcgroup.com/Corporate-profile/about-NNPC (accessed October 3, 2009).

[56] Transparency International, *Transparency International 2008 Corruption Perceptions Index*. Available at http://www.transparency.org/news_room/in_focus/2008/cpi2008/cpi_2008_table (accessed February 26, 2009).

[57] World Bank, *Rankings on the Ease of Doing Business*. Available at http://www.doingbusiness.org/Documents/DB10_Overivew.pdf (accessed October 3, 2009).

[58] Klare, M. and Wolman, D. (2004) Africa's oil and American national security. *Current History*, **103** (673), 226–231, 231.

[59] Lake, A. and Whitman, C.T. (2005) *More than Humanitarianism: A Strategic U.S. Approach Toward Africa*, Council on Foreign Relations, Washington, DC, p. 28.

[60] Hemmer, C. (2007) U.S. policy towards North Africa: three overarching themes. *Middle East Policy*, **14** (4), 55–67, 56.

[61] US Department of Defense, *Special Forces Support Pan Sahel Initiative in Africa*. Available at http://www.defenselink.mil/news/newsarticle.aspx?id=27112 (accessed October 4, 2009).

[62] Global Security, *Trans-Sahara Counter-terrorism Initiative*. Available at http://www.globalsecurity.org/military/ops/tscti.htm (accessed October 4, 2009).

[63] Africom, *Fact Sheet: Africa Contingency Operations Training and Assistance (ACOTA)*. Available at http://www.africom.mil/printStory.asp?art=1886 (accessed October 4, 2009).

[64] US Department of State, *Middle East Partnership Initiative*. Available at http://mepi.state.gov (accessed October 4, 2009).

[65] United States Africa Command, *About AFRICOM*. Available at http://www.africom.mil/AboutAFRICOM.asp (accessed September 29, 2009).

[66] United States Africa Command, *Questions and Answers About AFRICOM*. Available at http://www.africom.mil/africomFAQs.asp (accessed September 29, 2009).

[67] European Center for Development Policy Management, *The EU-Africa Partnership in Historical Perspective*. Available at http://www.ecdpm.org (accessed December 20, 2006).

[68] European Union, *The EU-Africa Strategic Partnership*. Available at http://europa.eu/rapid/pressReleasesAction.do?reference=MEMO/08/601&format=HTML (accessed October 1, 2008).

[69] Leblond, D. (2008) Africa, EU develop energy cooperation actions. *Oil & Gas Journal*, **106** (38), 40.

[70] European Union, *African Union Commission and European Commission Launch an Ambitious Africa-EU Energy Partnership*. Available at http://europa.eu/rapid/pressReleasesAction.do?reference=IP/08/1298&format=HTML (accessed October 21, 2008).

[71] Africa News Agency, *EU Offers $21 Billion for Trans-Saharan Pipeline*. Available at http://www.afrol.com/articles/30850 (accessed October 26, 2010).

[72] European Commission, *Investing in Africa: A European Commission's Perspective*. Available at http://ec.europa.eu (accessed April 29, 2009).

7

Caspian Sea

The Caspian Sea is 700 miles (1130 km) long and located in north-west Asia. Azerbaijan, Iran, Kazakhstan, Russia, and Turkmenistan share the Caspian Basin. Their policies on the exploration and development of the region's hydrocarbon resources since the collapse of the Soviet Union in late 1991 are the focus of this chapter. The region is important to the United States, Europe, Asia, and other major energy consumers for at least two reasons. First, Caspian producers' potential massive hydrocarbon resources can contribute to the world's oil and natural gas production and satisfy a significant proportion of global consumption. Second, the Caspian Basin can contribute to energy security by reducing dependence on the Middle East and allowing more diversification of energy sources.

The region is not new to the petroleum and natural gas industries. Commercial energy output began in the Caspian Basin in the mid-nineteenth century, making it one of the world's first energy provinces. Some energy analysts claimed that F.N. Semyenov was the first to drill a well on the Apsheron Peninsula, near Baku in Azerbaijan, in 1848 [1]. By 1900, the Baku region produced half of the world's total crude oil. This impressive level of production was the result of combined efforts and investment by the Nobel brothers, the Rothschilds, and the leaders of Royal Dutch Shell, who helped Russia to develop Caspian oil resources [2]. This oil carried considerable strategic weight in both world wars. Short of fuel, the German Army sought unsuccessfully to capture the Baku region. Germany's failure to secure access to the Caspian's oil resources was a major reason for its defeat in 1918 and 1945. Indeed, some of the most brutal battles during World War II were fought north of Azerbaijan.

Since the early 1950s, however, several developments contributed to a substantial reduction of Caspian oil production. Concern over Baku's vulnerability to attack during World War II, along with the discovery of oil in the Volga–Urals region of Russia, and later in western Siberia, led to a switch in the Soviet Union's investment priorities. This new policy resulted in decreased exploration and production in the Caspian for most of the second half of the twentieth century. Since the late 1980s, however, Azerbaijan, Kazakhstan, and Turkmenistan have gradually occupied center stage in the global energy markets. The three countries have succeeded in attracting massive foreign investment in their oil and gas sectors. International oil companies' growing interest in the region is in response to several developments:

Energy Security: An Interdisciplinary Approach, First Edition. Gawdat Bahgat.
© 2011 John Wiley & Sons, Ltd. Published 2011 by John Wiley & Sons, Ltd.

- Production has declined in such great oil provinces as the Alaskan North Slope in the United States and the North Sea in Europe.
- The Caspian region contains some of the largest underexplored and underdeveloped oil and gas reserves in the world.
- Some of the major producers in the Middle East either completely ban foreign investment in their oil sector or impose very strict conditions on international oil companies' participation.
- Finally, since independence, Azerbaijan, Kazakhstan, and to a lesser extent Turkmenistan have welcomed foreign investment in their energy sector. Upon the dissolution of the Soviet Union, their economies were weak. The only way to stop the deterioration of the standard of living was (and still is) to fully utilize their hydrocarbon resources. But the three states lack the necessary financial resources to explore and develop the oil and gas fields. Their leaders concluded that foreign investment was essential for their economic growth.

Thus, shortly after the collapse of the Soviet Union, the Caspian states became open to foreign investment, and the region has re-emerged as a potentially significant player in global energy markets. International oil companies have negotiated and signed several deals worth billions of dollars with the Caspian states, particularly Azerbaijan and Kazakhstan. One of the earliest (and largest) deals was a $20 billion joint venture between Chevron and Kazakhstan, concluded in April 1993. The scheme was to develop the Tengiz oil field, with an estimated 6–9 billion barrels. Similarly, the Azerbaijan International Operating Company (AIOC), an international consortium, signed an $8 billion production-sharing agreement with Baku in September 1994. They agreed to develop three fields – Azeri, Chirag, and the deep-water portions of Gunashli – with total reserves estimated at 3–5 billion barrels.

This large foreign investment in exploration and development operations in the Caspian Sea region indicates that its proven reserves and growing contribution to global oil and gas production and energy security are not in doubt. The story, however, does not stop there. For the region to reach its full potential and to utilize all its hydrocarbon resources in a timely fashion, several geo-economic and geopolitical hurdles need to be addressed and overcome. The rates of investment and the speed of development have been influenced by the success or failure of addressing these domestic, regional, and international obstacles.

In the following sections I provide an assessment of the Caspian Sea littoral states' oil and gas potential. This is followed by a detailed discussion of the unsuccessful efforts to reach an agreement on the legal status of the Basin. Finally, regional and international rivalries will be analyzed and how these rivalries have played out in the race and complexities involved to build pipeline routes.

7.1 Hydrocarbon Resources – An Assessment

In the early 1990s some top officials in Azerbaijan and Kazakhstan described their countries as "another Middle East," "another Saudi Arabia," and "another Kuwait." This euphoria was proven unrealistic. Nevertheless, the region holds large proven reserves and has since made a good contribution to global oil and gas production.

Iran and Russia are believed to have little oil and gas resources in their share of the sea. However, since the late 1990s, both Tehran and Moscow have sought to challenge this notion. In the mid-1990s six exploratory wells were drilled in the Iranian sector of the Caspian, but they did not yield commercially available discoveries. In 1999, Lasmo and the Royal Dutch Shell Group began a seismic survey for oil exploration off Iran's Caspian coast. The preliminary results of this survey showed that there are approximately 10 billion barrels of in-place crude of which 3 billion barrels are recoverable [3]. Tehran signed agreements with the Brazilian company Petrobras and with China's Oilfield Services Ltd to explore deep-water parts of its sector. In 2009, Oil Minister Gholam Hussein Nozari announced that 46 fields had been identified, eight of them ready for exploitation [4]. Iran's section of the Caspian Sea remains largely unexplored.

The Russian oil company LUKOIL began exploration of the north Caspian in 1995. It announced in early 2006 that it had found a large oil prospect at the V. Filanovskogo off-shore field. The company planned to bring six fields in the Russian section of the Caspian Sea on line in the closing years of the 2000s. The six fields contain roughly 6.5 million barrels of hydrocarbons and are expected to reach a maximum output of 140 000 b/d by 2016 [5]. In short, Russia's section, like Iran's, is largely underexplored. The other three littoral states – Azerbaijan, Kazakhstan, and Turkmenistan – hold most of the deposits that have been developed since the early 1990s.

7.1.1   Azerbaijan

As discussed above, oil wells were drilled in the Baku area as early as the mid-nineteenth century. In 1901, Azerbaijan was the world's largest oil producer. The Azeri government claims that offshore oil recovery was originated in its sector of the Caspian Sea in November 1949 [6]. Since independence in 1991 the country's oil and natural gas production has increased several fold. The national oil company, State Oil Company of Azerbaijan Republic (SOCAR), was established in September 1992 with the merger of Azerbaijan's two state oil companies, Azerineft and Azneftkimiya. SOCAR and its many subsidiaries are responsible for the production of oil and natural gas, for operation of the country's refineries, for running most of the pipeline system (a major exception is the Baku–Tbilisi–Ceyhan pipeline), and for managing hydrocarbon imports and exports [7]. Although government officials negotiate exploration and production agreements with foreign companies, SOCAR is party to all of the international consortia developing oil and gas fields in the country.

Azerbaijan's proven oil reserves are estimated at 7 billion barrels (0.6% of the world's total) and 1.20 trillion cubic meters (0.6% of the world's total) in 2008 [8]. More than their Kazakh and Turkmen counterparts, Azeri leaders took the lead in inviting international oil companies (IOCs) to develop the Azeri energy sector shortly after independence. This strategic development was driven by at least two factors. First, the lack of indigenous financial resources and the desperate need for foreign investment; and, second, the desire to ensure the country's economic, as well as political, independence from Moscow. Given the region's hydrocarbon history, the IOCs were eager to start exploration and development operations in Azerbaijan.

In September 1994 a consortium of oil companies led by British Petroleum (BP) signed an $8 billion production-sharing deal with SOCAR. The 30-year contract, or "contract of the century" as it is commonly known, called for production of 700 000 b/d. It provided for the development of the Azeri, Chirag, and Guneshli offshore oil fields in the Caspian Sea. Other investors include the US companies Amoco (17%), Pennzoil (4.8%), Unocal (9.5%), and

Exxon (5%), Russia's LUKOIL (10%), Norway's Statoil (8.5%), Japan's Itochu (7.45%), the UK's Ramco (2%), Turkey's TPAO (6.75%), Saudi Arabia's Delta (1.6%), and Azerbaijan's SOCAR (10%) [9].

Since then SOCAR has signed more than two-dozen agreements with IOCs to develop its oil and natural gas reserves both offshore and onshore. Tables 7.1 and 7.2 provide a brief summary of these agreements:

This active and large foreign investment participation in the Azeri energy sector is a major reason for the country's several-fold increase in both oil and gas production. The increase in oil production came mainly from the international consortium AIOC. The consortium includes BP, Chevron, SOCAR, Inpex, Statoil, ExxonMobil, TPAO, Devon Energy, Itochu, and Delta/Hess. It operates the offshore Azeri, Chirag, and deep-water Guneshli (ACG) mega-structure [10]. It is important to point out that not all IOCs' explorations were successful. There were some major disappointments. For example, in the mid-2000s ExxonMobil and LUKOIL failed to discover commercially viable hydrocarbon reserves in the Zafar–Mashal and Yalama blocks, respectively.

Virtually all of Azerbaijan's natural gas is produced from offshore fields. Two fields are particularly significant. The Bakhar oil and gas field is the country's leading natural gas reservoir, accounting for almost half of the gas output. The production, however, seems to have peaked and output has been declining in recent years. The other important natural gas field is Shah Deniz. A consortium led by BP has been working on developing the structure since the late 1990s. The consortium comprises BP Amoco (25.5%), Statoil (25.5%), SOCAR (10%), Iran's OIEC (10%), LUKOIL (10%), TotalFinaElf (10%), and TPAO of Turkey (9%) [11]. Shah Deniz is considered one of the largest natural gas fields to have been discovered in the last few decades. It is being developed in two phases and production started in early 2007, which transformed Azerbaijan into a net natural gas exporter for the first time.

7.1.2   Kazakhstan

Kazakhstan holds the largest proven oil reserves in the Caspian Sea (39.8 billion barrels, 3.2% of the world's total) and has the potential, or indeed is already on its way, to become a major oil producer. Oil was first found in Kazakhstan in 1899 and was first produced in 1911 [12]. Oil production from Kazakhstan (and Azerbaijan) was the backbone of the Soviet oil industry until fields in western Siberia went on-stream in the 1960. Like Azerbaijan, shortly after independence Kazakhstan invited foreign investment to revive its oil industry. These investors signed agreements with the then national company KazakhOil. In March 2004 the company was renamed Joint Stock Company KazMunaiGaz National Company (KMG). It was created to pursue a comprehensive development of the country's petroleum industry and to ensure rational and efficient operation of hydrocarbons. As in other oil countries, one of KMG's goals is to create jobs for the Kazakhs in their growing oil and gas industry. The company has stakes in virtually all major oil and gas projects in the country [13].

The first and biggest joint venture involving foreign oil companies, TengizChevrOil, was established in 1993. Since then, several joint ventures with foreign companies have followed as Table 7.3 shows.

The majority of growth in Kazakhstan oil production is projected to come from mainly four enormous fields: Karachaganak, Kurmangazy, Tengiz, and Kashagan, particularly the last two.

Table 7.1 Azerbaijan offshore production-sharing agreements, 1991–2009.

Name of PSA	Project partners	Estimated reserves	Projected investment	Project status
Azeri, Chirag, and Deepwater Gunashli (Azerbaijan International Operating Company, AIOC) Signed September 20, 1994; ratified December 1994	BP (34.1%, operator), Chevron-Texaco (10.2%), LUKOIL (10%), SOCAR (10%), Statoil (8.6%), ExxonMobil (8%), TPAO (6.8%), Devon Energy (5.6%), Itochu (3.9%), Amerada Hess (2.7%)	6.5 billion barrels of oil and 140 bcf of gas	$20 billion	Exports began late 1997. AIOC expects 674 000 b/d (33 million tonnes per annum) from the four platforms in total for the full year. Of this, 131 000 b/d is expected from Chirag, 239 500 from Central Azeri, 177 500 from West Azeri, and 126 000 from East Azeri
Shah Deniz Signed June 4, 1996; ratified October 17, 1996	BP (25.5%, operator), Statoil (25.5%), SOCAR (10%), LUKAgip (10%), TotalFinaElf (10%), OIEC of Iran (10.0%) TPAO (9.0%)	2.5 billion barrels of condensate; 22 tcf (630 bcm) of natural gas	Over $3 billion	Consortium plans to produce an average of around 63 000 barrels of oil equivalent per day (or 2.8 billion cubic meters of gas and 0.8 million tonnes of condensate for the entire year) during 2007. Plateau production from Stage 1 will be 8.6 billion cubic meters of gas per annum and approximately 45 000 barrels of condensate per day
Lankaran–Talysh Signed January 13, 1997; effective June 1997	TotalFinaElf (35%, operator), Wintershall (30%), SOCAR (25%), OIEC of Iran (10%)	700 million barrels of oil	$2 billion; $36.6 million invested by 2000	First test well (2001) came up dry
Yalama/D-222 Signed July 4, 1997; ratified November 1997	LUKOIL (80%), SOCAR (20%)	750 million barrels at Yalama field	$2.5–5.5 billion	Drilling to 4400 meters failed to find commercial reserves. LUKOIL does not plan to return to exploration drilling until after 2008. LUKOIL has a commitment with SOCAR to drill the second well
Absheron Signed August 1, 1997; ratified November 1997	SOCAR (50%); Chevron (30%, operator), TotalFinaElf (20%)	858 million barrels of oil; up to 100 tcf of natural gas	$3.5 billion; $10.6 million invested by 2000	Project closed. Chevron abandoned first exploratory well in 2001. In November 2003, Chevron and Total paid $40 million in compensation rather than drill a second well as required under contract

(Continued)

Table 7.1 (*Continued*)

Name of PSA	Project partners	Estimated reserves	Projected investment	Project status
Oguz Signed August 1, 1997; ratified November 1997	ExxonMobil (50%, operator), SOCAR (50%)	290 million barrels of oil and 685 bcf of gas	$2 billion; $5.5 million invested by 2000	Dry well drilled in April 2001. ExxonMobil announced plans to quit the project in April 2002
Nakhchivan Signed August 1, 1997; ratified November 1997	ExxonMobil (50%, operator), SOCAR (50%)	750 million barrels of oil	$2 billion; $22.5 million invested by 2000	ExxonMobil drilled one well but decided not to take the risk of drilling the second exploration and will pay $18 million in compensation to Azerbaijan since no commercial hydrocarbons were found
Kurdashi–Araz–Kirgan Daniz Signed July 7, 1998; ratified July 1998	SOCAR (50%), Agip (25%, operator), Mitsui (15%), TPAO (5%), Repsol (5%)	730 million barrels of oil	$2.5 billion	First test wells drilled, with poor results. Italy's Agip paid compensation to Azerbaijan to be released from the PSA
Inam Signed July 21, 1998; ratified December 1998	SOCAR (50%), BP (25%, operator), Royal Dutch Shell (25%)	2.2 billion barrels of oil	$2 billion; $7.5 million invested by 2000	BP suspended drilling of its first appraisal well in Aug. 2001 due to high pressure. New well is planned to begin drilling when a rig currently digging the Shah Deniz field becomes available
Araz, Alov, and Sharg Signed July 21, 1998; ratified December 1998–	SOCAR (40%), BP (15%, operator), Statoil (15%), ExxonMobil (15%), TPAO (10%), Alberta Energy (5%)	6.6 billion barrels of oil	$10 billion	In exploration phase in 2004. Confrontation with Iranian gunboat in July 2001; exploration suspended, pending resolution of Caspian Sea borders between Azerbaijan and Iran
Atashgah Signed December 25, 1998; ratified June 1999	SOCAR (50%), JAOC consortium (50%). JAOC divided as Japex (22.5%, operator), Inpex (12.5%), Teikoku (7.5%), and Itochu (7.5%)	600 million barrels of oil in Atashgah, Mugan–Deniz, and Yanan Tava fields	$2.3 billion; $35 million invested in 1999	Seismic work being undertaken. Second well at the Yanan–Tava field, part of a concession that also includes Atashgah and Mugan–Deniz, struck gas, but not enough to be commercial. In June 2003, JAOC announced it would leave Azerbaijan

Lerik, Jenab, Savalan, Dalga Signed April 27, 1999	SOCAR (50%), ExxonMobil (30%, operator), unassigned (20%)	1 billion barrels of oil	$3 billion	Exploration D-43, D-44, and D-73 blocks
Zafar–Mashal Signed April 27, 1999; ratified April 2000	SOCAR (50%), ExxonMobil (30%, operator), Conoco (20%)	1–2 billion barrels of oil, 1.8 tcf of gas	$3 billion	Exploration D-9 and D-38 blocks. Reached final drilling point in September 2004; well likely to be shut down due to abnormally high pressure, and ExxonMobil failed to reveal commercial hydrocarbon reserves. ExxonMobil paid $32 million to relinquish its license in 2006
Surakhani Signed August 16, 2005; ratified September 27, 2005	Rafi Oil (75%), SOCAR (25%)	50 million barrels of oil	$400 million	Contract states that oil production at the field should rise 50 % in two years. Rafi Oil will finance SOCAR's stake in the project until it doubles the current rate of extraction. SOCAR will have the right to sever the contract if Rafi Oil does not start exploration within two years. This is initially a 25-year PSA with the possibility of a five-year extension

Table 7.2   Azerbaijan onshore production-sharing agreements, 1991–2009.

Name of PSA	Project partners	Estimated reserves	Projected investment	Project status
Kalamaddin-Mishovdagh (formerly AzPetoil JV) Signed as JV in 1992; converted into PSA in 2000	Nations Petroleum (85%), SOCAR (15%)	200 million barrels of oil	$178 million	Production averaged 5000–7000 b/d during 2007, but Nations Petroleum was reportedly looking to sell its majority stake during 2007. EBRD[a] also a partner
Anshad Petrol Signed as JV in 1993; converted into a PSA in 2000	SOCAR (51%), Attila Dogan (31.5%), Land and General Berhard (17.5%)	219 million barrels at Neftchala, Khilly, Babazanan	—	Drilled four wells 1998–1999. Oil production averaged 77 000 b/d in 2004. Gas production averaged 1.1 mcf/d in 2004
AzGeroil Signed as JV in 1995; converted into a PSA in 2000	SOCAR (51%), Grunewald (49%)	140 million barrels at Ramany, Balkhany, and Sabunchi fields	—	Production averaged 1000 b/d in 1999
Southwest Gobustan Signed June 2, 1998; ratified November 1998	SOCAR (20%), CNPC of China (62.83%), and Arawak Energy of Canada (37.17%)	147 million barrels of oil; up to 7 tcf of natural gas	$700 million	In February 2006 two wells are producing a total of 1 mcf/d, expecting to produce 10 mcf/d
Zykh–Govsany Signed June 5, 2000, ratified June 2001, annulled 2005; re-signed (w/Russneft) November 2006	SOCAR (25%), Russneft (75%),	66–140 million barrels of oil	$150 million (new investment by Russneft)	LUKOIL pulled out of investment due to high environmental costs, but Russneft plans to apply new technology to enhance oil recovery

Field	Partners	Reserves	Investment	Notes
Kursangi–Garabagli Signed December 15, 1998; ratified April 1999	SOCAR (50%), CNPC (30%), Amerada Delta-Hess JV (20%)	182.5 million barrels of oil	$1 billion; proposed $50 million in 2006	Ten additional wells drilled in 2003 to increase production; fields producing 6600 b/d in June 2004. Eight additional wells planned for 2006. Operators reportedly looking to send oil through BTC
Muradkhanli–Jafarli–Zardab Signed July 21, 1998; ratified November 1998	Ramco (50%, operator), SOCAR (50%)	730 million barrels of oil	$1 billion	First test well at Muradkhanli shut down in April 2001. CNPC won a tender to develop the block, although no new PSA has been signed yet
Padar–Kharami Signed April 27, 1999	Nations Petroleum (85%, operator), SOCAR (15%)	580–750 million barrels of oil	$140 million	3–4 exploration wells planned for 2006. EBRD also a partner
Shirvanoil Signed as JV in 1997; converted into a PSA in 2000	SOCAR (49%), Caspian Energy Group (UK) (51%)	650 million barrels of oil at Kyurovdag field	$36 million	Rehabilitating existing wells since 1997. Has produced 11.6 million barrels of oil since 1997
West Absheron (Karadag–Kergez–Umbaki fields) Signed August 10, 1994	BMB (100%)	200 million barrels of oil	$700 million	Project area including the Karadag, Kergez, and Umbaki fileds sold in a contract block to SOCAR in 1999. SOCAR subsidiary Azneft started pre-drilling program at field in 2005

[a]European Bank for Reconstruction and Development.

Source: Energy Information Administration, *Azerbaijan: Production-Sharing Agreements.* Available at http://www.eia.doe.gov/emeu/cabs/Azerbaijan/azerproj.html. Accessed September 5, 2009.

Table 7.3 Kazakhstan: major oil and natural gas projects, May 2008.

Name of field/project	Project partners	Estimated reserves	Projected investment	Project status
Abai	Kazmunaigaz, Statoil	2.8 billion barrels of oil		Kazmunaigaz signed an MOU with Statiol
Aktobe	CNPC Aktobemunaigaz 88% (within Block ADA partners include Korean National Oil Corp. (KNOC), LG International Corp, and Vertom)	1.17 billion barrels of oil	$4.1 billion	JV producing 116 660 bbl/d of oil (2008), 101 bcf per annum of natural gas (2007). Oil production more than doubled between 1997 and 2006. New field Unit discovered in 2005
CPC: **(Tengiz–Novorossiysk pipeline)**	Caspian Pipeline Consortium (CPC): Russia 24%; Kazakhstan 19%; Chevron (US) 15%; LukArco (Russia/US) 12.5%; Rosneft-Shell (Russia–UK/Netherlands) 7.5%; ExxonMobil (US) 7.5%; Oman 7%; Agip/ENI (Italy) 2%; BG (UK) 2%; Kazakhstan Pipeline Ventures LLC 1.75%; Oryx 1.75%	990 mile (1600 km) oil pipeline from Tengiz oil field in Kazakhstan to Russian's Black Sea port of Novorossiysk; Phase I capacity: 565 000 b/d; Phase II capacity: 1.34 million b/d (2015)	$2.6 billion for Phase 1; $4.2 billion total when completed	First tanker loaded in Novorossiysk (October 2001); Target expansion to 1.3 million b/d by 2010
Darkhan	Kazmunaigaz (Kaztransgas), possibly Chinese consortium including CNPC, and Repsol	11 billion barrels of oil		Negotiations still under way with Chinese consortium. Located between the two offshore fields of Kurmangazy and Karazhambas
Egizkara	LG International Corp. 50%, and others	200 million barrels of oil		Exploration beginning in October 2006 with drilling starting in late 2007

Emba	Kazakhoil-Emba (Kazmunaigaz subsidiary) 51%, MOL Rt, Vegyepszer (Hungary) combined 49%	500 million barrels of oil	—	Producing 56 000 b/d of oil (2008); produced 5.58 bcf of natural gas (2007); 37 fields and over 2000 producing wells
Istatai	Undisclosed	1.75 billion barrels of oil		Negotiations with undisclosed partner continuing
Karachaganak	Karachaganak Integrated Organization (KIO): Agip (Italy) 32.5%; BG (UK) 32.5%; Chevron (US) 20%; LUKOIL (Russia) 15%	2.3–6 billion recoverable barrels of oil and gas condensate reserves; 16–46 tcf of recoverable natural gas reserves	$4 billion for Phase Two (completed in 2004)	Producing 252 000 bbl/d of oil (2008), 70% of oil exported through CPC. Produced 421 bcf in 2007
Karakuduk	LUKOIL	Total estimated proved plus probable reserves of approximately 63 million barrels	$190 million through 2000s with $170 million expected between 2006 and 2010	Producing 24 000 bbl/d of oil (2008); 2.1 bcf in 2007
Karazhanbas	Nations Energy	400 million barrels of oil	$250 million since 1997, $120 million in 2005	Producing 44 800 b/d (2005) (80–90 000 b/d planned in next two years); produced 1.8 Mcf/d natural gas (2005)
Kashagan	Agip Kazakhstan North Caspian Operating Company (Agip KCO) (formerly OKIOC): ENI, Total, ExxonMobil, and Shell 16.66%, Conoco-Phillips 8.28%, Kazmunaigaz 16.81%, Inpex 8.28%	9 billion to 13 billion recoverable (up to 38 billion probable)	Originally costed at $29 billion but estimates put final total approaching $50 billion	Production starting no sooner than 2012 (initial production slated for 75 000 b/d, max 1.2 million b/d by 2015)

(Continued)

Table 7.3 (*Continued*)

Name of field/project	Project partners	Estimated reserves	Projected investment	Project status
Kazgermunai	Petrokazakhstan 25%,[a] Kazmunaigaz 50%	100 million barrels of oil	$300 million	Produced 37 300 b/d of oil; 32 mcf/d of natural gas (2005)
Khvalinskoye	Kazakhstan and LUKOIL	400 million barrels of oil; 12.3 tcf of natural gas. Target start date 2014	$3.5 billion for petrochemicals plant	Field is located on the Kazakh–Russian border in the Caspian Sea and in Russia's jurisdiction. Agreement was prepared during 2007, signed in 2008
Kumkol (North)	Turgai Petroleum: PetroKazakhstan 50%,[a] and LUKOIL (Russia)	97–300 million barrels of oil	—	Producing 71 000 bbl/d of oil, 18.3 mcf/d of natural gas (2007). Legal dispute between PKZ and LUKOIL has stopped production in the past
Kumkol (South) and South Kumkol	PetroKazakhstan Kumkol Resource (PKKR), wholly owned by PetroKazakhstan[a]	116 million barrels of oil		Producing 74 000 b/d of oil, 18.1 mcf/d of natural gas (2007). Development of export pipeline infrastructure will allow for production growth
Kurmangazy	AO Kazmunaiteniz Offshore Oil Company (a Kazmunaigaz subsidiary) 50%, Rosneft subsidiary OOO RN-Kazakhstan 25%. Russia's Zarubezhneft has an option on 25% in the project	2.2–8.8 billion barrels of oil	$23 billion allocated over 55 years	Russia and Kazakhstan signed PSA in July 2005; start date of 2010; Rosneft reports first assessment well drilled in 2006 yielded disappointing results, although additional drilling is scheduled
Mangistau	Mangistaumunaigaz (Kazmunaigaz subsidiary)	500 million barrels of oil	—	Producing 112 000 bbl/d of oil, 33.3 mcf/d of natural gas (2008)
North Buzachi	LUKOIL 50%, China National Petroleum Corp. 50%	1 to 1.5 billion barrels of oil	Over $800 million	Producing 33 000 b/d of oil, 4.5 mcf/d of natural gas (2008), Accelerated development plan approved in 2004

Nursultan ("N" Block)	Kazmunaigas operating independently. Conoco-Philips and Shell mentioned as participants	4.65 billion barrels of oil		PSA signed in 2008; $40 million allocated
Satpayev	Kazmunaigaz, Oil and Natural Gas Corp. (ONGC)	1.85 billion barrels of oil		PSA signed in 2007
Tengiz	TengizChevroil (TCO): Chevron (US) 50%; ExxonMobil (US) 25%; Kazmunaigaz 20%; LukArco (Russia) 5%; discovered in1979, agreement signed in 1993	9 billion barrels of oil	$23 billion over 40 years	Producing 350 000 b/d of oil (2008); expected max production of 1 million b/d by 2012; produced 580 mcf/d of natural gas in 2005
Tsentralnoye	Kazmunaigas, Gazprom, LUKOIL	3.8 billion barrels of oil, and 3.24 TCM of gas	N/A	A first deep-water exploration well was drilled at the end of 2007
Tyub–Karagan	LUKOIL 50%, Kazmunaigaz 50%	7 billion barrels of oil		Exploration well in 2006 yielded disappointing results. Second delayed until 2008. PSA signed in 2004 covers Tyub–Karagan and adjacent Atashskaya field in Kazakh sector of Caspian Sea. Plateau production forecast at 144 000 b/d
Uzen	Uzenmunaigaz (Kazmunaigaz subsidiary) 100%	147 million barrels of oil	—	Producing 132 000 b/d of oil (2008), 29.8 bcf of natural gas (January–September 2004), 30% improvement from 2003 from advanced technologies

(*Continued*)

Table 7.3 (*Continued*)

Name of field/project	Project partners	Estimated reserves	Projected investment	Project status
Zhambyl	Kazmunaigaz 73%, Korean National Oil Consortium 27%; KNOC: KNOC 35%, SK Corp. 25%, LG Corp. 20%, Daesung and Samsung 10% each	1.26 billion barrels of oil	—	MoU signed by Kazakh government and South Korean consortium (KNOC)
Zhanazhol	CNPC Aktobemunaigaz which is 85.4% owned by CNPC (94.5% of voting shares). Other Kazakh and foreign entities own 5.07% of ordinary shares (5.5% of voting shares); 9.53% of preferred non-voting shares belong to the enterprise's employees and other parties	—	—	20 year license signed in 1995. Field discovered in 1978
Zhemchuzina (a.k.a. Pearls Block)	Shell 55%, Kazmunaigaz 25%, Oman Oil Company 20%	733 million barrels of oil		A joint operating company was set up at the end of 2006 with Shell financing 100% of the appraisal program

CNPC acquired PetroKazakhstan and its assets in Kazakhstan in October 2005 and sold a 33.3% stake in PetroKazakhstan to Kazmunaigaz in order to secure Kazakh government approval for the deal. Production figures for many of these entities, if producing, can be obtained from Energy Intelligence's *Nefte Compass* publication.

Source: Energy Information Administration, *Kazakhstan: Major Oil and Natural Gas Project*. Available at http://www.eia.doe.gov/emeu/cabs/Kazakhstan/Kazaproj. html. Accessed September 7, 2009.

The Tengiz field is the largest source of oil production in the country. It was discovered in 1979 by Soviet geologists, but has been largely developed since 1993 by the Tengizchevroil joint venture. The Kashagan field is one of the largest oil discoveries in the world in the last few decades. These massive reserves aside, Kashagan is one of the most technically challenging oil projects in the world. The development is in shallow water that ices over in winter, and the oil is under extremely high pressure and contains huge amounts of lethal hydrogen sulfide.

In August 2007, the Kazakh government effectively shut down the project, after the operating company ENI said costs would soar to $136 billion from $57 billion and pushed back the startup date. After months of negotiations, all parties agreed to double KazMunaiGaz's stake to 16.81% with all of the partners reducing their stakes on a pro rata basis. They also agreed that a newly created North Caspian Operation Company, jointly owned by all the consortium members, would take over responsibility for developing the field from ENI with management duties being rotated among the partners [14].

Commenting on this dispute with ENI, President Nursultan Nazarbayev said, "We are consistently strengthening state influence in the strategically important energy sphere. We will continue our work in that direction" [15]. Indeed, this dispute with ENI should be seen as another sign of the rise of the so-called "resource nationalism" or a growing assertiveness by the Kazakh authorities over the country's natural resources. During 2007, the government announced it would review all energy and mineral resources contracts in a bid to generate more revenues and diversify the sources of investment. President Nursultan Nazarbayev signed an amendment into law in October 2007 that allows the government to unilaterally break contracts with oil companies. Another amendment to the country's subsoil law in 2005 extended the government's power to buy back energy assets by limiting the transfer of property rights to strategic assets in Kazakhstan [16].

Kazakhstan holds much smaller proven natural gas reserves (1.82 trillion cubic meters or 1.0% of the world's total) and in 2010 it produced 32.2 billion cubic meters or 1.0% of the world's total [17]. Four characteristics of the Kazakh natural gas industry can be identified:

1. Almost all of the country's gas is "associated" gas found in oil fields. Most of the Kazakh gas is produced from two major oil fields: Tengiz and Karachaganak.
2. Much of the gas is reinjected in oil fields to maintain reservoir pressure and enhance oil output. Gas is also used in power generation but on a small scale.
3. As in some other countries, natural gas was largely unutilized. Indeed, much of the Kazakh gas was being flared and the country was one of the largest gas flarers in the world. Since 2005 the government has ordered oil companies to avoid natural gas flaring.
4. Much of the gas is produced in the western part of the country, while the main consumption centers are in the south. As a result, these population and industrial centers in the south depend on imported gas supplies from neighboring Uzbekistan. Since 2007 Kazakhstan has been exporting a small volume of natural gas.

7.1.3   Turkmenistan

Turkmenistan is different from the other two former Soviet republics of Azerbaijan and Kazakhstan in two fundamental ways. First, the country's proven oil reserves and production are much smaller than those of its two Caspian Sea neighbors. But, on the other hand,

Ashgabat's natural gas proven reserves and production are the largest in the region. Second, as discussed above, both Azerbaijan and Kazakhstan have welcomed foreign investment and created a favorite environment for IOCs since the early 1990s. Turkmenistan was ruled by President Saparmurat Niyazov till his death in December 2006, when he was succeeded by President Gurbanguly Berdymukhammedov. The autocratic regime of Niyazov created a hostile political and business environment. Slow-paced political and economic reforms made the majority of the international energy companies that entered the country withdraw their investments. As one energy analyst put it, "Turkmenistan has distinguished itself among the ex-Soviet Caspian nations as a virtual black hole of foreign investment" [18]. Shortly after ascending to power, President Berdymukhammedov promised a new approach toward foreign investment.

These two characteristics have shaped Turkmenistan's oil and gas production. In 2010 the country's proven reserves were estimated at roughly 600 million barrels [19]. Most of the oil fields are located in the south Caspian Basin in the west of the country. Oil production has grown at a slow pace due to lack of foreign investment. Another important impediment is high domestic consumption. Turkmenistan consumes over nine times as much oil per unit of gross domestic product (GDP), making it the most oil intensive country in the world [20]. Finally, oil production faces another hurdle. Many of the prime oil deposits are located in disputed areas of the Caspian Sea. The lack of a border-delineation agreement between Turkmenistan and its neighbors has further delayed the exploration and development of oil fields.

Despite its considerable natural gas resources, Turkmenistan has faced a significant political challenge. Since the collapse of the Soviet Union, the Russian government and its state gas company Gazprom have seen Turkmen gas as a competitor. The Soviet legacy meant that all the Turkmen gas was exported to Russia with no access to other markets. This geopolitical situation has left Ashgabat with limited options and given Gazprom tremendous leverage. Within this asymmetrical strategic context, Russia has the upper hand and Turkmenistan's gas production has fluctuated based on its relationship with Moscow. After a dramatic collapse in most of the 1990s, production started a steady recovery in the early 2000s, partly due to agreement with Gazprom, under which most of the Turkmen gas is exported to Russia [21].

Like its oil consumption, Turkmenistan's gas consumption is one of the highest in the world on a per capita basis. Natural gas is provided free of charge to residential consumers and is heavily subsidized for industrial use. Much of the gas comes from two major fields: Dauletabad and Shatlyk. Both of them have been producing since the Soviet era and, therefore, have recently exhibited signs of natural depletion. However, in the mid-2000s a super-giant gas field, South Yolotan–Osman, was discovered. After being audited by the UK firm Gaffney, Cline and Associates (GCA), the field was confirmed among the world's five biggest [22]. The South Yolotan–Osman field faces significant development challenges. These include substantial sulfur and carbon dioxide contents and higher-than-average temperatures and pressure [23].

Turkmenistan plans to develop the field in series of phases. It is not clear what role international energy companies will be allowed to play. Given the geological challenges, it is doubtful that the Turkmen authorities would be able to develop South Yolotan–Osman on their own without tremendous foreign investment and technology. The Turkmen government announced that it would offer only service contracts – not the PSAs energy companies crave, which allow them to book reserves and receive a share of future output [24].

This brief survey of the oil and gas industry in the Caspian Sea suggests that the region has the geological potential to contribute to global production and the diversification of hydrocarbon sources. As has been discussed, oil and gas production from Azerbaijan, Kazakhstan, and Turkmenistan has already increased several fold since the early 1990s. For the region to reach its full potential in a timely fashion, geopolitical challenges need to be adequately addressed.

7.2   The Legal Status of the Caspian Sea

The Caspian Sea is the largest, completely enclosed body of salt water in the world and constitutes a particularly fragile ecosystem. For much of the twentieth century it was the exclusive domain of Iran and the Soviet Union, with the latter enjoying naval dominance. With the disintegration of the Soviet Union, the geopolitical situation in the region changed significantly. Instead of two, there are now five riparian states, each filing differing legal claims. This uncertainty surrounding the legal regime that will eventually govern hydrocarbon development and exports from the Caspian Sea is one of several risk factors that investors have had to consider in doing business in the region.

Part of the problem regarding the legal status is the lack of consensus on the definition of the Caspian: is it a "sea" or a "lake?" If the Caspian is a sea, the United Nations' Convention on the Law of the Sea (UNCLOS) would be applicable. In 1982, 135 countries signed the UNCLOS, which states that every nation bordering a sea or ocean may claim 12 miles (19 km) from shore as its territorial waters and an Exclusive Economic Zone for 200 miles (320 km) beyond that. Everything farther out is considered the common property of the world's nations [25]. If the Caspian is a lake, then customary international law concerning border lakes would apply. An international lake is a lake that is surrounded by the territory of various states. Use of the waters of border lakes is regulated by the international agreements of border states, which determine the lines of state borders, right of navigation, and terms of use of waters for non-navigational purposes [26]. In other words, the Caspian and its resources would be developed jointly – a division referred to as the condominium approach.

Unfortunately, the Caspian does not appear to fall into either category. It is therefore necessary to take into consideration the evolution of the legal system that has governed the relationship between the littoral states. One of the earliest treaties on the demarcation and cession of certain territories was the Treaty of Resht (1729) concluded between the Russian and the Persian Empires, which provided for freedom of commerce and navigation. It was followed by the Golestan Treaty (1813) and the Turkomanchai Treaty (1828). The most recent ones are the Treaty of Friendship (1921) and the Treaty of Commerce and Navigation (1940). These two treaties indicated that transport and fishing were free in the Caspian for all Iranian and Soviet ships. The two treaties did not differentiate between warships and other types of ships (i.e., passenger and transport), did not refer to environmental issues, and said nothing about seabed resources. Since the disintegration of the Soviet Union, the five Caspian states have sought unsuccessfully to reach an agreement on the legal status of the Caspian. Their stances can be summarized as follows.

7.2.1   *Azerbaijan*

Azerbaijan has called for the Law of the Sea to be applied and has advocated the establishment of maritime boundaries into national sectors based along median lines. Boundaries would follow those established and recognized under the Soviet Union to delineate republic sectors for oil exploration and development. In line with this policy, Azerbaijan signed agreements with Kazakhstan and Russia to divide the northern part of the Caspian.

7.2.2   Kazakhstan

Kazakhstan has supported Azerbaijan's view for the establishment of national sectors, but has stated that cooperation on the environment, fishing, and navigation would be beneficial. Kazakhstan also signed a bilateral agreement with Russia dividing the Caspian along median lines between the two countries.

7.2.3   Russia

The Russian position has varied over time. Initially Russia argued that neither the Law of the Sea nor its precedents applied because the Caspian is an enclosed sea, and that regional treaties signed in 1921 and 1940 between Iran and the former Soviet Union are valid. In December 1996, Russia called for joint navigation rights, joint management of fisheries and environmental protection, and the establishment of an interstate committee of all boundary states to license exploration in a joint-use zone in the center of the Caspian beyond an exclusive national zone of 45 nautical miles (83 km), and a joint corporation of these states to exploit these resources. In the following years Russia again changed its stance and signed bilateral agreements with Azerbaijan and Kazakhstan.

In another development, Russia suggested that the airspace above the Caspian Sea, the surface of the sea, and the water of the sea should have open access and be administered jointly. Meanwhile, the sea floor should be divided roughly along median lines between the littoral states to permit the development of mineral resources. These median lines would not be drawn according to strict rules from the shores of the Caspian and its islands, but would be open to negotiation between the five littoral states to take into account other issues such as equity and history [27].

7.2.4   Turkmenistan

This country's position has evolved over time. Initially it supported Russia's proposal for a 45 nautical mile zone. In 1996, Turkmenistan signed a protocol with Iran and Russia to develop a joint-stock company to develop the energy resources in the national zones of the three countries. However, Turkmenistan has changed its position and called for a division of the Caspian based on the Soviet-era policy until the five littoral states agree upon a new legal system.

7.2.5   Iran

Iran's position has been the least flexible, insisting on one of three alternatives: (a) initially Iran insisted that the 1921 and 1940 treaties were valid and wanted all the littoral states to approve any offshore oil developments until the legal status of the Caspian is agreed upon by all of them; (b) Iran indicated a willingness to divide the Caspian into national sectors, provided there is equal division of the sea, so that both the sea floor and surface would be divided into five equal part of 20% each; (c) Iran accepted the so-called condominium approach, where there is no division of the Caspian into national sectors and any development would be jointly undertaken by all of the littoral states [28].

Given these contradictory positions, the five littoral states have failed to reach an agreement on how to divide the Caspian Sea between them despite multilateral negotiations that started in the early 1990s. This failure to reach a five-party agreement paved the way for bilateral negotiations and unilateral initiatives by the Caspian states to develop the hydrocarbon deposits within what they perceive as their own sections. These unilateral actions have complicated the full utilization of the region's resources. For example, Turkmenistan and Azerbaijan have long disputed the ownership of oil and gas fields such as Serdar, Omar, and Osman [29]. After long, unsuccessful negotiations, Turkmenistan decided to take the dispute with Azerbaijan to the International Court of Arbitration [30].

To sum up, this legal uncertainty has hindered but not prevented oil and gas development. All the five littoral states have decided not to wait and have actively sought to develop oil and natural gas in their respective sections in cooperation with international energy companies. Nonetheless, an agreement on the legal status of the Caspian Sea would be useful, but such an agreement does not appear to be imminent.

7.3   Geopolitical Rivalry and Pipeline Diplomacy

Oil and natural gas are not only economic commodities, but also strategic ones. The world economy runs on energy. This why decisions on prices, production, shipment, and trade of fuels are driven by political considerations as much as by economic interests. Stated differently, geo-policy plays an equal role to supply and demand laws in shaping energy markets. The Caspian region holds considerable hydrocarbon reserves. The low level of domestic consumption means that much of the Caspian production is exported to global energy markets. In addition, the Caspian region is strategically located between two large and growing energy consuming regions: China and Europe. The Caspian producers are also sandwiched between two global powers: China and Russia. Finally, all these neighbors, as well as the United States, have made substantial investments in the Caspian energy sector. US, European, Russian, and Chinese companies (both private and state owned) have taken the lead in developing Caspian oil and gas fields and transporting much of the output to markets around the globe. In short, geopolitical and geo-economic dynamics have made the rivalry between regional and international powers inevitable.

These geopolitical and geo-economic rivalries have been most apparent in the race to build oil and natural gas pipelines. Azerbaijan, Kazakhstan, and Turkmenistan are landlocked, with no access to the high seas. For their hydrocarbon resources to reach global markets, they have to pass through one or more transit countries. Since the early 1990s, existing pipelines have been upgraded and expanded, new ones have been built, and others are being negotiated. The stakes are high and each regional and global power seeks to secure its strategic and economic interests.

7.3.1   Iran

Iran has a multi-dimensional and centuries-long relation with its neighbors in the Central Asia/Caspian Sea region. Culturally, the two sides share a linguistic and religious heritage [31]. Sharing borders with Azerbaijan and Turkmenistan means that any domestic instability in these two countries and their foreign policy orientations have a direct impact on the Islamic Republic's national security. Economically, Tehran sees these former Soviet republics as important markets for its non-oil exports [32]. Given these broad interests, Tehran has sought to institutionalize cooperation with these neighbors. In 1985, Iran, Turkey, and Pakistan created the Economic Cooperation Organization (ECO) as an intergovernmental regional body to promote economic, technical, and cultural cooperation. At the 1992 Tehran ECO Summit, the six Muslim former Soviet republics (Azerbaijan, Kazakhstan, Kyrgyz, Tajikistan, Turkmenistan, and Uzbekistan) and Afghanistan joined the three original members [33]. The enlargement of the ECO was an Iranian initiative.

Strategic and geological dynamics have shaped Iran's relations with the other Caspian littoral states. Pipeline routes from Azerbaijan, Kazakhstan, and Turkmenistan via Iran to the lucrative energy markets in Asia are among the shortest and most convenient options [34]. US economic sanctions against Iran (imposed since 1979) have largely prevented the Caspian states and international energy companies from pursuing such an option.

On the other hand, almost all of Iran's oil and gas fields are located in the southern part of the country around the Persian Gulf. Meanwhile, significant population and industrial centers are concentrated in the north. This geological/geographical landscape has made swap deals attractive. Iran imports oil and gas from its Caspian neighbors in the north and sells roughly equal amounts for them from its ports on the Persian Gulf in the south. These swap deals include the following:

- Due to lingering tension between Azerbaijan and Armenia, Azerbaijan provides natural gas to its geographically separate Nakhchivan enclave via Iran. Azerbaijan sends the gas through the Baku–Astara pipeline, and Iran then delivers the gas via a 30 mile (48 km) pipeline to the enclave.
- Iran is considering building a north–south gas pipeline that will be used for swap deals from its northern neighbors to the Oman Sea. The pipeline will stretch from Sarakhs on the Turkmen border to the southern port of Jask.
- Since the early 1990s, Iran has concluded and implemented several oil swap deals with Azerbaijan, Kazakhstan, and Russia. Iran has also built an oil pipeline that links its Caspian port of Neka with refineries in Tehran and Tabriz. Since the mid-2000s, Iran has upgraded Neka to allow swap capacity to increase [5].
- In 2008, Iran conducted feasibility studies for a cross-country pipeline to transfer oil from the Caspian Sea port of Neka to the Persian Gulf port of Jask and then to world markets [35]. The pipeline will have the capacity to carry 1 million barrels of oil per day [36].

In addition to these swap deals, in 1997 Turkmenistan and Iran completed the $190 million Korpezhe–Kurt Kui pipeline linking the two countries, thereby becoming the first natural gas export pipeline from Central Asia to bypass Russia. According to the terms of the 25-year contract between the two countries, 35% of Turkmen supplies are allocated as payment for Iran's contribution to building the pipeline [37].

7.3.2 Russia

Compared to the other regional and global powers competing over hydrocarbon sources in the Caspian Sea, Russia enjoys several advantages and a few drawbacks. The Soviet legacy, which expanded over seven decades, left Russia with strong economic and cultural ties with the former Soviet republics. In many developing countries, European languages, particularly English and to a lesser extent French, are widely used in business and by social and political elites. In the Central Asia/Caspian Sea states, it is Russian. Two decades after the disintegration of the Soviet Union, many students from these states choose to get their higher education from Russian universities. Furthermore, Russian ethnic communities are all over the region. Economically, the energy infrastructure built during the Soviet era has survived the collapse of the Soviet Union and is fully operational. True, some of these pipelines need updating and expansion, but this is much cheaper than building new ones. On the other side, the United States and Europe enjoy superior technology to Russia, particularly in offshore exploration and development operations. Finally, Chinese companies have much larger financial resources than their Russian competitors.

In short, Russian leaders still perceive the Central Asia/Caspian Sea region as their backyard where Russia should maintain a special relationship. Russian leaders are sensitive to what they see as attempts by the US, European, and Chinese governments and companies to penetrate the area and threaten Moscow's national interests. Specifically, Russia's energy goals in the Central Asia/Caspian Sea region have been three-fold. First, to compel these states to use the export infrastructure Russia already has in place. Russia is eager to maintain its dominance over the delivery routes of oil and gas to the West from the region [38]. Second, to promote Russian oil and gas companies and help them obtain the maximum shares possible in available projects. Third, to use a variety of instruments to block projects that do not promote Russia's perceived interests [39].

The extensive pipeline network that connects Russia on one side and Azerbaijan, Kazakhstan, and Turkmenistan on the other includes the Central Asia Center, Baku–Novorossiysk, and the Caspian Pipeline Consortium (CPC), among others.

The Central Asia Center pipeline, built in 1974, has two branches. The western branch delivers Turkmen natural gas from near the Caspian Sea region to the north, while the eastern branch pipes natural gas from eastern Turkmenistan and southern Uzbekistan in a north-westerly direction across Uzbekistan. The pipeline branches meet in western Kazakhstan, where they run further directly north and enter the Russian natural gas pipeline system. Turkmenistan has been the chief exporter of natural gas via the Central Asia Center pipeline. Gazprom transits Central Asian gas to the Russian and export markets as well as acting as operator for Turkmen gas transit across Uzbekistan and Kazakhstan [40].

The Baku–Novorossiysk pipeline, also known as the northern route, opened in 1997. The pipeline runs 868 miles (1398 km) from Baku via Chechnya to the Russian Black Sea port of Novorossiysk. Initial exports through the pipeline were limited to approximately 40 000 b/d; however, due to pumping limitations, disputes over transit tariffs, and the conflict in Chechnya, up to 70 000 b/d of oil was forced to bypass Chechnya by rail from Dagestan to Stavropol. The conflict and instability in Chechnya prompted Russian pipeline operator Transneft to construct a 120 000 b/d Chechnya pipeline bypass. This bypass, which was completed in 2000, includes an 11 mile (18 km) spur to Russia's Caspian Sea port of Makhachkala. The pipeline and spur have enabled additional exports from Azerbaijan, Kazakhstan, and Turkmenistan to flow through the pipeline from Baku and Makhachkala [41].

The 980 mile (1580 km) long CPC connects Kazakhstan's Caspian Sea oil deposits with Russia's Black Sea port of Novorossiysk. The governments of Russia, Kazakhstan, and Oman developed the CPC project in conjunction with a consortium of international oil companies.

The CPC system is one of the largest operating investment projects with foreign participation on the territory of the former Soviet Union. The pipeline is an extension of transit infrastructure surrounding the Caspian Sea. Newly constructed components of the line run from the Russian town of Komsomolskaya directly west to Novorossiysk. The CPC is supplied with Kazakh oil through the Soviet-era links surrounding the Caspian Sea, which the consortium members have refurbished. The CPC was inaugurated in October 2001 [42].

Finally, in May 2007, the Presidents of Russia, Kazakhstan, and Turkmenistan signed a widely reported declaration on the construction of the Caspian Coastal Pipeline (CCP) [43]. The declaration was supplemented in December of the same year by a trilateral agreement. The CCP is designed to bring gas from western Turkmenistan and from Kazakhstan northwards to join the Central Asia Center lines in Kazakhstan. Tukmengaz, KazMunaiGaz, and Gazprom agreed to build the pipeline [44].

These pipeline schemes underscore Russia's strategy of seeking to restore and reinforce its control of oil and gas deposits in the former Soviet republics. Two decades after the disintegration of the Soviet Union, Moscow has succeeded in maintaining considerable leverage. However, the leaders of Azerbaijan, Kazakhstan, and Turkmenistan have sought to diversify their energy export routes and destinations. China, Europe, and the United States have been eager to establish diplomatic and economic cooperation with these Caspian Sea states. This cooperation has weakened Moscow's control over the region's hydrocarbon resources.

7.3.3   China

China is a major player in the Central Asia/Caspian Sea energy landscape. The Chinese economy has been the world's fastest growing economy in the last few decades. China's indigenous energy sources could not keep pace with its growing demand. The large gap between consumption and production has been filled by foreign supplies. China is contiguous with landlocked Central Asia. In addition to this geographical proximity, China enjoys another advantage – substantial cash reserves.

Like other consumers, China seeks to diversify its energy sources. Chinese officials have pushed for the development of less vulnerable, land-based oil and gas pipelines that would direct Central Asian energy resources toward their country [45]. Against this background, Beijing has pursued a two-fold strategy: acquiring oil and gas fields; and building pipelines.

In recent years Chinese companies, supported by the authorities in Beijing, have used their financial muscle to acquire oil and gas fields and infrastructure all over the world. In 2005 the state-owned China National Petroleum Company (CNPC) bought PetroKazakhstan for $4.2 billion, then China's biggest foreign acquisition. A year later, the CNPC and Kazakhstan's Kazmunaigas completed the second stage in an oil pipeline from Atasu, in north-western Kazakhstan, to Alashankou, in China's north-western Xinjiang region. The first stage of the project was completed in 2003 and the third and final stage was completed in 2009. The quantity of crude oil supplied to China through this route still represents only a small percentage of China's oil demand [16].

In 2007 the CNPC was granted exploration rights for the Bagtiyarlik territory in Turkmenistan and construction of a gas pipeline from Turkmenistan to China via Uzbekistan and Kazakhstan started [46]. China plans to import 30 billion cubic meters a year of Turkmen gas for 30 years through this pipeline. In June 2009 the two sides agreed to increase the volume to 40 billion cubic meters [47].

7.3.4   Europe and the United States

US policy in the Central Asia/Caspian Sea region has evolved over time. Unlike Iran, Russia, China, and Europe, US energy trade relations with Azerbaijan, Kazakhstan, and Turkmenistan are limited. Still, US policy has reflected a significant interest in shaping energy rivalry in the region. Equally important, Washington has broad strategic interests. Initially, US policy focused on preventing the proliferation of nuclear weapons [48]. It is to be remembered that Kazakhstan inherited part of the Soviet nuclear arsenal. Located next to Afghanistan, Central Asian states have become particularly important in the war on terrorism since September 2001. Finally, bordering Iran and Russia, the US diplomatic, economic, and military presence is meant, at least partly, to contain Tehran's and Moscow's influence in the Central Asia/Caspian Sea region.

US interests in Azerbaijan's, Kazakhstan's, and Turkmenistan's oil and natural gas deposits are two-fold. First, along with other international oil companies, US companies are taking the lead in developing the Caspian hydrocarbon deposits, particularly in Kazakhstan. Second, generally, US strategy has sought to exclude, or at least weaken, the Iranian and Russian role in the Caspian energy sector and to promote pipeline routes via pro-Western transit countries such as Turkey (a NATO member) and Georgia.

Europe is more dependent on oil and natural gas imports from the Central Asia/Caspian Sea region than the United States. The focus of European policy has been more on energy and commercial interests and less on geopolitical and strategic goals. Since the mid-1990s, Europe has sought to institutionalize its relationship with the region. In 1995 the EU started negotiating a program called Interstate Oil and Gas Transport to Europe (INOGATE). Several East European and former Soviet republics participated in these negotiations [49]. All parties signed the INOGATE Umbrella Agreement, which came into force in February 2001. The Agreement sets out an institutional and legal system designed to facilitate the development of interstate oil and gas transportation systems and to attract the investments necessary for their construction and operation [50].

In addition to interstate cooperation, European companies have taken an active role in developing the Caspian oil and gas deposits and in the negotiation and construction of pipelines connecting the region to Europe. One of the first such schemes is the Baku–Supsa/Western Route Export Pipeline. In March 1996, then Georgian President Eduard Shevardnadze and Azerbaijani President Heydar Aliyev signed a 30-year agreement to pump a portion of the AIOC's early oil via Georgia to its Black Sea port of Supsa. The route became operational in April 1999 and has since been updated and expanded [42].

The Baku–Tbilisi–Ceyhan (BTC) pipeline exports oil from Azerbaijan and Kazakhstan via Georgia to the Turkish Mediterranean port of Ceyhan. At a cost of almost $4 billion, the 1040 mile (1675 km) pipeline allows oil to bypass the crowded Bosporus and Dardanelles Straits. A BP-led consortium operates the pipeline. Construction was completed in May 2005 and the first tanker deliveries began a year later (June 2006) [5].

In March 200,1 BP, operator of the Shah Deniz natural gas field, announced plans to build a pipeline from Baku to Erzurum, Turkey. After lengthy negotiations Georgia and Azerbaijan signed a transit agreement under which Georgia receives 5% of the natural gas in the pipeline, as well as preferential rights to purchase additional gas from the pipeline in exchange for transit rights. This pipeline, known also as the South Caucasus pipeline, is Azerbaijan's main conduit for natural gas exports. It runs parallel to the BTC pipeline for most of its route before connecting to the Turkish gas pipeline network. Deliveries of gas began in December 2006. BP (technical operator for construction and operation) holds a 25.5% stake in the project, Norway's Statoil (responsible for business development and administration) holds a 25.5% share, and SOCAR, Russia's LUKOIL, Turkey's TPAO, France's Total, and UAE's NICO hold around 10% each [10].

These pipelines highlight the important role that the corridor from Azerbaijan through Georgia plays as a major artery for oil and gas transportation to international markets. They also underscore Europe's vulnerability to political instability in the South Caucasus. The outbreak of hostilities between Russia and Georgia in August 2008 did not damage any of the pipelines. The war was not fought over energy and did not result in any lasting disruption to energy transit flows. Oil flows via the BTC stopped before the outbreak of the conflict, due to an explosion at a compressor station in Turkey for which Kurdish separatists claimed responsibility [51]. Nevertheless, the conflict raised awareness of political and security risks. Tension between Moscow and Tbilisi is still alive, especially after the Russian recognition of the independence of South Ossetia and Abkhazia.

The net effect of the BTC explosion and the Russia–Georgia conflict was that, for a period from mid to late August, the only operational route across the South Caucasus was the Baku–Novorossiysk pipeline (or Northern Route), which does not cross Georgia but goes directly from Azerbaijan through the Russian territory to the Black Sea port of Novorossiysk. During this period SOCAR put in place a short-term swap arrangement with Iran and delivered oil to the Iranian Caspian port of Neka. This redirection of a portion of SOCAR oil export suggested that Iran could gain at the expense of routes through the South Caucasus [52].

7.4 Conclusion: The Way Forward

Four conclusions can be drawn from this discussion of Caspian Sea energy. First, since the demise of the Soviet Union, Azerbaijan, Kazakhstan, and Turkmenistan have been among other former Soviet republics introducing economic and political reform. They still have a long way to go in terms of transparency, accountability, and good governance. There is a great deal of uncertainty and ambiguity regarding how much oil and gas reserves these countries hold. Indeed, given the huge deals they have undertaken and negotiated, there is concern that some of them, particularly Turkmenistan, have overcommitted themselves and promised more than they can deliver.

Second, the rivalry between Tehran, Moscow, Beijing, Washington, and Brussels has strengthened the bargaining power of the Central Asian and Caucasus states. It seems that some of the regional leaders are trying to play off one of these external powers against the others. The perceived national interests of Central Asian/Caspian Sea states are not identical to those of external powers.

Third, the construction of pipeline routes costs billions of dollars and takes years to accomplish. The decisions to build these routes are driven by both commercial interests and strategic considerations. The decision about which one to pursue will be better made if it is driven more by an economic cost–benefit analysis and less by geopolitical rivalry.

Finally, it is true that if Central Asian/Caspian Sea oil and gas go to one consumer, it would be at the expense of other consumers. But it is also true that full utilization of the region's hydrocarbon resources would contribute to overall energy security. Instead of dividing the region into spheres of influence, Europe, Russia, China, and the United States would benefit more by promoting political stability and economic prosperity. Energy should not be seen as a zero-sum game.

References

[1] Campbell, C. (1997) *The Coming Oil Crisis*, Multi-Science, Brentwood, Essex, p. 32.

[2] Forsythe, R. (1996) *The Politics of Oil in the Caucasus and Central Asia*, Oxford University Press, London, p. 4.

[3] Financial Times (2000) Caspian Reserves Assessed, Financial Times (September 19).

[4] Fars News, *Iran Discovers 46 Oil Fields in Caspian Sea*. Available at http://english.farsnews.net/printable.php?nn=8805020617 (accessed July 24, 2009).

[5] Energy Information Administration, *Country Analysis Briefs: Caspian Sea*. Available at http://www.eia.doe.gov/emeu/cabs/caspian/full.html (accessed January 3, 2007).

[6] State Oil Company of Azerbaijan Republic, *Brief History of Oil and Gas Recovery in Azerbaijan*. Available at http://www.socar.az/oilhistory=en.html (accessed September 5, 2009).

[7] State Oil Company of Azerbaijan Republic, *About*. Available at http://www.socar.az/about-en.html (accessed September 5, 2009).

[8] British Petroleum (2010) *BP Statistical Review of World Energy*, British Petroleum, London, pp. 6, 22.

[9] Bolukbasi, S. (1998) The controversy over the Caspian Sea mineral resources: conflicting perceptions, clashing interests. *Europe-Asia Studies*, **50** (3), 397–414, 398.

[10] Energy Information Administration, *Country Analysis Briefs: Azerbaijan*. Available at http://www.eia.doe.gov (accessed December 6, 2007).

[11] Shammas, P. and Nagata, K. (2000) Profiles of the petroleum sectors in Caspian region countries and the potential for a New Caspian to Middle East gulf export line through Iran. *Energy Exploration and Exploitation*, **18** (5), 473–568, 494.

[12] Shammas, P. and Nagata, K. (2000) Profiles of the petroleum sectors in Caspian Region countries and the potential for a New Caspian to Middle East gulf export line through Iran. *Energy Exploration and Exploitation*, **18** (5), 473–568, 510.

[13] Olcott, M.B. (2007) *KazMunaiGaz: Kazakhstan's National Oil and Gas Company*, James A. Baker III Institute for Public Policy, Rice University, Houston, TX, p. 7.

[14] Chazan, G. (2008) Western Oil Companies Resolve Kazakh Dispute. Wall Street Journal (November 1).

[15] Nurshayeva, R. (2008) Kazakh State Seeks Bigger Energy Role. Moscow Times (February 7).

[16] Energy Information Administration, *Country Analysis Briefs: Kazakhstan*. Available at http://www.eia.doe.gov/emeu/cabs/kazakhstan/full.html (accessed January 31, 2008).

[17] British Petroleum (2010) *BP Statistical Review of World Energy*, British Petroleum, London, pp. 22, 24.

[18] Neff, A. (2005) Caspian nations pursuing oil exports at greatly varying paces. *Oil & Gas Journal*, **103** (22), 34–39, 36.

[19] Editorial (2008) Worldwide look at reserves and production. *Oil & Gas Journal*, **106** (48), 22–23, 22.

[20] Energy Information Administration, *Country Analysis Briefs: Central Asia*. Available at http://www.eia.doe.gov/cabs/Centasia/Full.html (accessed February 7, 2006).

[21] Watson, N.J. (2007) Turkmenistan: east versus west. *Petroleum Economist*, **74** (10), 24–28, 26.

[22] Radio Free Europe, *Turkmenistan Says Gas Field in World's Top Five*. Available at http://www.rferl.org/articleprintview/1329670.html (accessed October 14, 2008).

[23] Farey, B., Oil and Gas Eurasia, *Yolotan-Osman Field a "Super-Giant."* Available at http://www.oilandgaseurasia.com/news/p/0/news/4068 (accessed May 3, 2009).

[24] Chazan, G. (2008) Turkmenistan Dashes Gas Hopes – Nation Holds Vast New Field as Its Own, Sending Foreign Firms to Drill Offshore. Wall Street Journal (December 19).

[25] Jonas, T. (2001) Parting the sea: Caspian littoral states seek boundary disputes' resolution. *Oil & Gas Journal*, **99** (22), 66–69, 66.

[26] Janusz, B. (2005) *The Caspian Sea Legal Status and Regime Problems*, Chatham House, London, p. 4.

[27] Energy Information Administration, *Caspian Sea Region Legal Issues*. Available at http://www.eia.doe.gov/emeu/cabs/casplaw.html (accessed July 8, 2000).

[28] Payvand, *Iran's Changing Perspectives and Policies on the Caspian Sea: Interview with Abbas Maleki*. Available at http://www.payvand.com/news/01/mar/1112.html (accessed March 22, 2001).

[29] Pannier, B., Radio Free Europe, *Flare-Up in Turkmen-Azerbaijani Dispute Latest Nabucco Challenge*. Available at http://www.rferl.org/articloprintview/1786632.html (accessed July 27, 2009).

[30] Moscow Times (2009) Turkmens to Sue Azeris over Caspian Gas Fields. Moscow Times (July 29).

[31] Mohsenin, M. (1995) Iran's relations with Central Asia and the Caucasus. *Iranian Journal of International Affairs*, **8** (3), 834–853, 848.

[32] Herzig, E. (2004) Regionalism: Iran and Central Asia. *International Affairs*, **80** (3), 503–517, 508.

[33] Economic Cooperation Organization, *Member Countries*. Available at http://www.ecosecretariat.org/MainMenu/member_countries.htm (accessed September 10, 2009).

[34] Stauffer, T.R. (1997) *The Iranian Connection: The Geo-economic of Exporting Central Asian Energy Via Iran*, International Research Center for Energy and Economic Development, Boulder, CO, p. 1.

[35] Mehr News, *Iran Opposes Caspian Pipeline*. Available at http://www.mehrnews.com/en/NewsPrint.aspx?NewsID=822130 (accessed January 24, 2009).

[36] Payvand, *Iran Plans Oil pipeline from Caspian to Persian Gulf*. Available at http://www.payvand.com/news/08/jun/1030.html (accessed June 4, 2008).

[37] Energy Information Administration, *Country Analysis Briefs: Iran*. Available at http://www.eia.doe.gov/emeu/cabs/iran/full.html (accessed February 4, 2009).

[38] Kashfi, M. (2008) Did Caspian summit share the sea or Iran's oil riches? *Oil & Gas Journal*, **106** (4), 20–22, 21.

[39] Allison, R. (2004) Strategic reassertion in Russia's Central Asia policy. *International Affairs*, **80** (2), 277–293, 290.

[40] Energy Information Administration, *Country Analysis Briefs: Central Asia*. Available at http://www.eia.doe.gov/cabs/Centasia/Full.html (accessed February 7, 2006).

[41] Energy Information Administration, *Caucasus Region*. Available at http://www.eia.doe.gov/cabs/caucasus.html (accessed March 29, 2002).

[42] Caspian Pipeline Consortium, *General Information*. Available at http://www.cpc.ru/portal/alias!press/lang!en-us/tabID!3357/DesktopDefault.aspx (accessed May 3, 2009).

[43] Radio Free Europe, *Kazakhstan Commits to Russian-led Europe Gas Link*. Available at http://www.rferl.org/articleprintview/1731061.html (accessed May 13, 2009).

[44] Delany, M. (2007) LUKoil Wins Turkmen Energy Deal. Moscow Times (June 14).

[45] Weitz, R. (2006) Averting a new great game in Central Asia. *Washington Quarterly*, **29** (3), 155–167, 160.

[46] Editorial (2007) Work starts on Turkmenistan-China gas line. *Oil & Gas Journal*, **105** (34), p. 10.

[47] Moscow Times (2009) Turkmens Get $4 Billion Loan. Moscow Times (June 26).

[48] Macfarlane, N. (2004) The United States and regionalism in Central Asia. *International Affairs*, **80** (3), 447–461, 450.

[49] INOGATE, *INOGATE in Brief*. Available at http://www.inogate.org/html/brief/brief5.htm (accessed March 31, 2005).

[50] INOGATE, *INOGATE Umbrella Agreement*. Available at http://www.inogate.org/html/brief/brief4.htm (accessed March 31, 2005).

[51] Fernandez, Y., Payvand, *BTC project crisis and the Caspian Basin energy game*. Available at http://www.payvand.com/news/08/sep/1030.html (accessed September 3, 2008).

[52] Gould, T. (2008) *Perspectives on Caspian Oil and Gas Development*, International Energy Agency, Paris, p. 49.

8

Russia

Russia is a major player in the global energy markets. It is the world's largest natural gas producer and exporter and is the world's second largest oil producer (after Saudi Arabia). It is also the dominant gas and oil exporter to Europe and has substantially increased its hydrocarbon exports to Asian markets in recent years. Given historical ties and geographical proximity, Russia enjoys close energy ties with most of the Caspian Sea/Central Asian states. Furthermore, Russian oil and gas companies are actively and aggressively pursuing partnerships with other national and international energy companies to explore and develop hydrocarbon deposits in Africa, the Middle East, Latin America, and elsewhere. Finally, oil and gas revenues provide a large share of Russia's national income and the overall gross national product. In short, the crucial role that Russia plays on the global energy scene and Moscow's heavy dependence on oil and gas revenues cannot be overstated.

The Russian energy outlook, however, faces serious geological, geopolitical, and geo-economic challenges. Unlike other major oil producers, Russia's proven oil reserves are limited. While Russia falls into the category of top producing and exporting countries, it ranks much lower in proven oil reserves (the world's seventh after Saudi Arabia, Iran, Iraq, Kuwait, United Arab Emirates, and Venezuela). This inconsistency between the level of proven oil reserves and the volume of production suggests that the nation's oil fields are being depleted at a high rate. At current levels of production, proven reserves are projected to last only approximately 22 years (by comparison, the ratio of production to reserves in the Middle East is more than 80 years).

Since the early 2000s, Russia's government-controlled and private oil and gas companies have been involved in exploration and development deals all over the world. In addition to traditional economic competition from rival companies, several partners view the rising Russian role with a great deal of suspicion. Many countries, particularly in Europe, accuse Moscow of using its energy leverage to advance its political and strategic interests. Indeed, it can be argued that Russia's energy policy is driven by both economic interests and geopolitical considerations. The frequent interruptions of gas supplies to Ukraine underscore this suggestion.

Finally, geo-economic restraints have had a significant impact on oil and gas outputs. Since the late 2000s, Russia's hydrocarbon production has been stagnant. One reason for

Energy Security: An Interdisciplinary Approach, First Edition. Gawdat Bahgat.
© 2011 John Wiley & Sons, Ltd. Published 2011 by John Wiley & Sons, Ltd.

this is the unpredictability of the Russian government's approach toward private and foreign investments. Unlike most energy producers in Africa, Latin American, and the Middle East, where a national company largely manipulates the oil sector, in Russia, several government-controlled companies compete with each other and with other Russian private companies and international corporations. The rules of these competitions are ill-defined and constantly changing. This investment environment is a major reason for the slowdown in oil and gas production.

In recent years Russia has sought to define itself as an "energy superpower" [1]. This idea has become doctrinal for the Russian leadership. There is a conviction that the oil and gas wealth will last a very long time and hydrocarbon resources could be made the foundation of the Russian economy and international power. In the following sections I examine the accuracy of these assumptions. This chapter addresses the following topics: Russia's oil and gas potential; the roles that national and international, government-controlled, and private companies play; and energy relations with major producers (i.e., the Middle East and OPEC) and major consumers (i.e., Europe and China). The analysis highlights some of the main challenges that Russia's oil and gas industry faces both domestically and internationally. The study suggests that for the foreseeable future Russia is likely to remain a major player in the global energy markets. Its role, however, should not be exaggerated.

8.1   Oil Sector

Energy historians do not agree on where oil was originally discovered: either in Titusville, Pennsylvania in the United States in 1859, or Bibi-Aybat near Baku, then part of the Russian Empire, in 1846. This disagreement aside, Russia's oil industry started in the mid-nineteenth century and has since played a crucial role in the nation's economy and both domestic and foreign policies. The Rothschild family and the Nobel brothers played a major role in the development of the oil industry in Baku. Between 1890 and 1900 oil production tripled and the Russian Empire accounted for over 40% of global production in 1900. Shell Transport and Trading, which later became part of Royal Dutch/Shell, began life by ferrying oil produced by the Rothschild family to Western Europe.

In the second half of the nineteenth century, Russia began to explore and develop oil fields in its northern Caucasus and Central Asian territories. The rapid development of oil production was paralleled by the construction of various plants to refine and process crude oil. The 1917 Russian Revolution further underscored the importance of oil to the new regime in Moscow. Shortly after securing Azerbaijan, the Red Army delivered oil and oil products to Russia. The development of the oil industry was very important for the new Soviet leaders. The expansion of the oil industry and its operations continued under the Soviet Union. The North Caucasus/Central Asia region, particularly the Baku area, remained the center of the Soviet oil industry until World War II. Shortly after the war, the Soviet authorities shifted their focus to the Volga–Urals region, which was closer to major economic and population centers and where the geology was favorable. These old and new oil fields provided major sources of income to the Soviet economy and state and made the Soviet Union a major oil producer and exporter for most of the decades preceding its collapse [2].

The Soviet Union's oil production reached its peak of 12.5 million b/d in 1988. However, the political and economic turmoil that accompanied the dissolution of the Soviet Union dealt a heavy blow to the overall economy including the oil industry. Production dropped by about 50%, reaching a low level of approximately 6 million b/d by the mid-1990s. In this period there was a sharp reduction in drilling and little or no investment in new wells or in new technology to increase recovery from depleted wells [3]. Several factors contributed to the resurgence of the oil industry by the end of the decade. These include political stability, the devaluation of the ruble, and the rise in world oil prices. The international reach of Russian companies expanded along with production.

This decade (1998–2008) of rising production and expansion in the overall oil sector was abruptly interrupted in 2008 when oil production stagnated. Despite rising prices for most of the decade, Russia's oil production fell in 2008. Industry watchers and Russian officials generally blamed the country's production slowdown on a combination of weather and tight electricity supplies in some parts of the country. Other reasons include the expansion of state control over the oil sector. In the long term the concern is that the decline in production is a sign of aging Siberian fields. For the last few decades most of the nation's oil production has come from giant oil fields located in Western Siberia, between the Ural Mountains and the Central Siberian Plateau. Eastern Siberia is one area where little exploration has taken place. Its reserves, however, are expected to increase significantly with further exploration. Nevertheless, these new fields are not expected to come into full production for several years.

In recent years the Russian government has taken steps to promote oil exploration and development in Eastern Siberia. In late December 2009, Prime Minister Vladimir Putin inaugurated a new oil export terminal at Kozmino on the Pacific Ocean in Russia's far east [4]. The terminal has since been used to export oil from new fields in East Siberia. The terminal is being served by the East Siberia Pacific Ocean (ESPO) oil pipeline that runs across East Siberia to China and the Pacific region. In addition to contributing to the development of Eastern Siberia oil reservoirs, the ESPO pipeline consolidates a Russo-China energy partnership and helps Moscow to diversify its oil exports and reduce its dependence on the European market. Some Kremlin-friendly oil companies (i.e., Rosneft, the state oil company, and TNK-BP, the Russian–UK oil major) were given tax breaks as incentives to encourage them to develop East Siberian reserves. In addition to building new terminals and pipelines, there are plans to upgrade existing port facilities and build new tankers to ship oil and liquefied natural gas (LNG).

In addition to the ESPO pipeline, Russia depends on a number of pipelines and facilities to export its oil. The Baltic Pipeline System (BPS) carries crude oil from Russia's West Siberian and Timan–Pechora provinces westward to the port of Primorsk on the Russian Gulf of Finland. The BPS gives Russia a direct outlet to Northern European markets, allowing the country to reduce its dependence on transit routes through Estonia, Latvia, and Lithuania. The Druzhba pipeline was completed in 1974 and was originally designed to load Middle Eastern oil at Omisalj, then pipe it northward to Yugoslavia and on to Hungary. However, given Russia's booming production, the pipeline's operators and transit states have since considered reversing its flow, thus giving Russia a new export outlet on the Adriatic Sea. Another important outlet is the Murmansk area, which enjoys two advantages. First, unlike many Russian ports, the Murmansk port is ice-free most of the year. Second, it is deep enough to allow huge tankers to be loaded [5].

Given the declining indigenous oil production, Russian companies have aggressively pursued oil deals overseas in Africa, Asia, Latin America, and the Middle East. These partnerships, however, have not succeeded in offsetting the decline in domestic production. For the foreseeable future, it seems that Russia has developed most of the "easy oil." Most of the new fields are located in geologically and environmentally challenging areas.

8.2   Natural Gas

Russia is by far the world's natural gas superpower. It holds approximately 23.7% of the world's proven reserves and is the largest producer and exporter [6]. As in the oil industry, the Rothschild family and the Nobel brothers played a prominent role in the exploration and development of natural gas in Russia. At the end of the nineteenth century, gas was largely used to light major Russian cities and was essentially produced and consumed locally. Russia lagged behind the United States and other countries in laying long-distance gas pipelines. This method of transporting natural gas was utilized shortly after the end of World War II in response to a surge in gas consumption and the need to satisfy a fast-growing demand.

The development of the gas industry, along with the oil industry, was a significant part of overall broad strategy to modernize the Soviet economy in the two decades following World War II. Soviet officials realized the great hydrocarbon potential of Siberia, and consequently the region attracted most of the attention and investment. Gas development, however, was proceeding more slowly than oil development, partly because the infrastructure requirements for gas are more complicated than those for oil and partly because gas was not seen as essential as oil to such industries such as petrochemicals and transportation.

The global oil shock of 1973 gave the Soviet gas industry a much needed boost. Political instability in the Middle East led to a substantial surge in oil and gas prices. Developing the natural gas industry in Russia (and elsewhere) became more profitable in two senses: to take advantage of higher gas prices and to replace oil and increase the volume of oil available for export. Soviet leaders used their expanding natural gas exports to serve both geopolitical and economic interests. A large volume of Russian gas was exported to fellow communist countries in Eastern Europe at low prices to consolidate common ideological and political orientations. On the other hand, Moscow sold a large volume of its gas to West European countries, at much higher prices than those for East Europeans, to earn much needed hard currencies. This dual gas export policy had served the Soviet Union both strategically and financially.

The Soviet invasion of Afghanistan in 1979, coupled with Ronald Reagan's assumption of power in the United States and Margaret Thatcher's in the United Kingdom, had substantially weakened the relative détente of the 1970s between the Soviet Union and the West and reignited the Cold War between the two sides. Economic sanctions were imposed on the Soviet Union and Soviet plans to expand oil and gas exports to Europe came to a halt.

In the late 1980s and the early 1990s communism in Eastern Europe was defeated and the Soviet Union dissolved. This political turmoil had several negative economic impacts. First, the East European, former Soviet republic, and Russian economies substantially shrank and, consequently, their demand for natural gas (and other sources of energy) declined. During this period Russia's gas production declined at a lower rate than consumption. This meant that there was more gas available for export. Second, the ideological war between the West and the Soviet Union was over. European–Russian relations, including oil and gas exports, became driven largely by commercial interests. Third, Russian leaders had to face a new geo-economic landscape: their growing oil and gas exports to Western Europe have to transit countries such as Ukraine and Belarus – former Soviet republics. Since the early 1990s all involved parties have sought to find a satisfactory formula to ensure the non-interruption of these supplies. Finally,

Russian leaders have decided gradually to charge former Soviet republics market prices for oil and gas supplies. Initially, Russia accepted lower prices in order to maintain a level of unity between these former allies.

Russia enjoys two important advantages with regard to gas exports. First, unlike its limited proven oil reserves, it has the world's largest natural gas proven reserves. These reserves can support high volumes of production for a long time. Second, Russia is located next to two large and growing consuming markets – Europe and Asia. The two regions are already dependent on Russian supplies and their dependence is likely to deepen further.

Despite these advantages, Russia's gas production has stagnated in recent years. At least four reasons contributed to this stagnation. First, most of the gas fields located close to industrial, commercial, and population centers have been producing for a long time and are showing signs of aging. In other words, the "easy gas" is close to being depleted. Other reserves are geographically distant from markets and are located in climatically and geologically challenging environments [7]. Second, the new and potential gas discoveries require huge capital investment and highly sophisticated technology. International energy companies' participation can accelerate the development of these reservoirs. The investment environment, however, keeps changing and is highly unpredictable. Third, the gas sector is overwhelmingly dominated by the state monopoly Gazprom. Private companies have no interest in producing natural gas. As a result, gas flaring is a common practice and Russia is losing valuable assets. In the last few years the Russian government has sought to reduce the level of gas flaring. Fourth, domestic gas prices are only a fraction of the price charged to European customers. Low prices have impacted the gas industry's ability to finance capital spending and have hurt incentives to increase efficiency. With about 14% of world gas consumption, Russia is the world's second largest gas consumer (after the United States). Large domestic consumption means that little is left for export abroad.

In an attempt to enhance its credentials as a major gas exporter, the Russian government has sought to export part of its gas as LNG. Russia is lagging far behind other gas producing countries when it comes to LNG production and shipping. Since the early 1990s the costs of LNG have substantially decreased and its share in the overall gas trade has risen. This trend is projected to continue.

One of the main projects to export LNG is on Sakhalin Island, a former penal colony located off the east coast of Russia and to the north of Japan. The area holds almost 12 billion barrels of oil and approximately 90 trillion cubic feet of gas [8]. In February 2009, Dmitry Medvedev, the Russian president, inaugurated the LNG project on the island. The infrastructure includes three offshore platforms, an onshore processing facility, 300 km of offshore pipelines, 1600 km of onshore pipelines, an oil export facility, and the LNG plant. Shareholders in Sakhalin Energy Investment Co., the operator of the Sakhalin-2 project, are Gazprom with 50% plus 1 share, Royal Dutch Shell PLC with 27.5% minus 1 share, Mitsui & Co. Ltd 12.5%, and Mitsubishi Corp. 10% [9]. The company agreed to sell 65% of the gas to Japan, 20% to North America, and 15% to South Korea [10]. The project represents the first significant outflow of Russian energy to non-European markets.

8.3   The Energy Strategy-2030

Over the years the Russian government has issued a number of policy statements and strategies to articulate its energy policy. In November 2009 it announced that the energy strategy approved in 2003 was invalid and to be replaced by a new one that highlights the main investment needs and production and export goals up to 2030 [11].

The strategy states that Russia plans to invest up to $625 billion over the next two decades to raise oil production by about 10% and a further $590 billion to add at least 33% to its gas output. The oil and gas investment, part of a $2 trillion plus plan to develop the Russian energy sector by 2030, also envisages Asian markets taking a much larger share of Russia's exports as the country develops resource fields in Siberia and the Far East.

8.3.1 Oil Sector

The strategy projects crude oil output to average between 530 and 535 million tons in 2030, up from 400 million tons in 2008. To meet this target, Russia must replace depleted West Siberian deposits with expensive new developments further east. The Energy Ministry said that East Siberian fields, which contributed only 3% of Russia's oil production in 2008, will grow their share to 18% by 2030. The Energy Strategy-2030 envisages crude oil exports largely flat within a range of 222–248 million tons by 2030, compared to 243 million tons exported in 2008. Asia-Pacific markets, led by China, Japan, and South Korea, will raise their share of Russian crude oil and refined product exports to 22–25% by 2030 from the current 6%. The Ministry said projected oil investment of between $609 and 625 billion by 2030 would comprise $491–501 billion on production and exploration, $47–50 billion on refining, and $71–74 billion on transport.

8.3.2 Gas Sector

Investment in the natural gas sector is projected at between $565 and 590 billion. Transport – including ambitious pipeline projects – contributes $277–289 billion of this sum. Russia plans to boost natural gas production to between 885 and 940 billion cubic meters (bcm) by 2030 from 664 bcm in 2008. Exports are seen rising to between 349 and 368 bcm in 2030, up 45–53% on 2008. Asian markets are expected to boost their share of Russian gas exports to 19–20%, from practically zero in 2008. Domestic demand for natural gas is forecast to rise between 32 and 40% to a range of 605–641 bcm, from 457 bcm in 2008. The new gas regions would increase their share of Russian natural gas output to 38–39% from only 2% in 2008. The Arctic region of Yamal should contribute 23–24% of the Russian total by 2030 [12].

8.4 The Arctic Hydrocarbons

The Energy Strategy-2030 underscores Russian leaders' growing interest in exploring and developing the Arctic hydrocarbon resources. The size of the Arctic Shelf is approximately 4.5 million square kilometers. The Arctic Ocean is subdivided into several bodies of water, including the Barents, Kara, Laptev, East Siberia, and Chukchi Seas and their adjacent water-ways. Various sources have offered diverse forecasts for the potential of Arctic hydrocarbon reserves. Most analysts agree that the region holds substantial oil and gas deposits. They also agree that Russia is likely to play a dominant role in the development of these deposits [13].

The western part of Arctic Russia is considered to be one of the nation's most important future oil and gas provinces, containing about 8.2 billion tons of hydrocarbons. Significant oil and gas reserves have already been discovered in the Barents, Pechora, and Kara Seas and in the Timan–Pechora Basin. The Barents Sea includes the Shtokman gas field and Prirazlomnoye oil field. Gazprom controls both of them. The Kara Sea Basins include the Russanov and Leningrad gas fields. The Timan–Pechora Basin is the only part of Barents Russia currently producing oil and gas. Minor oil and gas deposits have been discovered in the onshore territories near the Bering Sea, indicating that there may be more hydrocarbons on the adjacent seabed. However, due to the severe climate, this area has not been properly explored. In the coming decades, oil and gas production from these areas is expected to grow as production declines in traditional Russian hydrocarbon regions, such as the Volga and Urals [14]. Russian experts project that gas production in the region will reach around 800 million cubic meters of natural gas per day (more than half of the rate of gas production in Russia in 2007) [15]. Similarly, potential Arctic oil reserves could prove highly valuable to the Russian oil sector.

The full utilization of the Arctic hydrocarbon potential is conditioned on overcoming fundamental geopolitical, geo-economic, and climatic hurdles. The legal status of the region is uncertain. In addition to Russia, the United States, Canada, Norway, and Denmark claim sovereign territory within the Arctic Circle. The United Nations Convention on the Law of the Sea (UNCLOS) states that countries are entitled to an exclusive economic zone of 200 miles (320 km) beyond their coastlines. The five powers have different interpretations of the UNCLOS.

Russia claims that the Arctic Ocean seabed is an extension of the Siberian continental platform. In order to assert its claims, the Russian government sent a scientific team aboard the mini-submarine *Mir* to explore the ocean floor below the North Pole in August 2007 [16]. These Russian activities provoked angry reactions from the other powers that share the Arctic.

Since 1982 the United States Senate has failed to ratify the UNCLOS, adding more confusion to Washington's legal argument. In order to counter Russia's claims, the US Coast Guard icebreaker *Healy* was dispatched to the Bering Sea in August 2007. The Canadian Prime Minister Steven Harper toured the Northwest Territories and Nunavut in the days following the Russian expedition and repeatedly stressed the need to use military power to protect Canada's Arctic interests. The Danish government sent a multinational team to the North Pole and instructed the Danish scientists to gather evidence that an underwater feature known as the Lomonosov Ridge is an underwater extension of Greenland, rather than Russia. The research has been used to support Denmark's territorial claims in the Arctic. Finally, due to close cooperation between Norway and Russia over offshore hydrocarbon development, Oslo's reaction has been muted.

This controversy over the legal status of the Arctic was replayed in 2009 with the issuing of a new Russian national security strategy that identified the question of Arctic ownership as a source of potential military conflict within a decade. Dmitri Rogozin, the Russian Ambassador to NATO, warned the military alliance not to meddle in the Arctic, saying that there was "nothing for them to do there" [17].

How the controversy over the ownership of the Arctic and the utilization of the hydrocarbon resources will be solved is not clear. However, there is serious doubt that Russia is currently capable of exclusively developing the Arctic oil and gas deposits, given the region's severe climate and vulnerable habitats. Lack of relevant experience and technologies and absence of essential industrial equipment and vital infrastructure are certain to restrain Moscow's plans. On the other hand, international energy companies such as ExxonMobil, BP, Royal Dutch

Shell, Statoil, and Norsk Hydro have demonstrated experience and records in developing offshore resources. Cooperation between all these players seems necessary to develop the Arctic hydrocarbon potential. Similarly, Russia's relations with its largest import market, Europe, and the other main oil and gas producing region, the Middle East and OPEC, can be characterized as a combination of cooperation and rivalry.

8.5 Russia–EU Energy Partnership

An energy partnership between Russia and Europe is almost inevitable. Russia is the world's largest natural gas producer and exporter and the second largest oil producer. The EU, with a population of nearly half a billion and one of the highest standards of living in the world, is a major energy consumer. Geographical proximity further cements these hydrocarbon ties. Little wonder that the EU imports a large proportion of its gas and oil needs from Russia and that the revenues Russia receives from Europe represent a major source of government income and overall gross national product. Finally, European energy companies play a significant role in oil and gas exploration and development in Russia.

Recent Russian–European energy cooperation goes back to 1968, when the Soviet Union started selling natural gas to Austria. Five years later (1973), Germany started buying Soviet gas. In the ensuing decades more European countries were added to the list and Russia emerged as the major oil and gas supplier to the EU as a bloc and to several individual European countries.

The two sides sought to institutionalize their emerging energy cooperation by negotiating and signing the Energy Charter Treaty (ECT). The roots of the ECT go back to a political initiative launched in Europe in the early 1990s, at a time when the end of the Cold War offered an unprecedented opportunity to overcome the previous economic divisions on the European continent [18]. The ECT and the Energy Charter Protocol on Energy Efficiency and Related Environmental Aspects were signed in December 1994 and entered into legal force in April 1998. By 2010 the ECT had been signed or acceded to by 51 states plus the European Communities. The ECT is a legally binding multilateral instrument dealing specifically with intergovernmental cooperation in the energy sector [19]. The ECT is designed to promote energy security through the operation of open and competitive energy markets, while respecting the principles of sustainable development and sovereignty over energy resources.

Specifically, the ECT's provisions focus on five broad areas: the protection and promotion of foreign energy investments, based on the extension of national treatment, or most favored-nation treatment; free trade in energy materials, products, and energy-related equipment, based on WTO rules; freedom of energy transit through pipelines and grids; reducing the negative environmental impact of the energy cycle through improving energy efficiency; and mechanisms for the resolution of state-to-state or investor-to-state disputes [20].

The EU spent years trying to get Russia to abide by the provisions of the ECT, which compelled Russia to open up the development of its hydrocarbon reserves and the running of its pipelines to foreign commercial involvement. Moscow, on the other side, signed the ECT and applied its rules on a provisional basis, but never ratified the Treaty. For years Russian officials had complained that the ECT was outdated and favored consumers. In August 2009 Prime Minister Vladimir Putin signed an order withdrawing from the ECT [21].

In another attempt to cement Russian–European energy cooperation, the two sides launched an Energy Dialogue on the occasion of the Sixth EU–Russia Summit (Paris, October 30, 2000). It was agreed to institute an Energy Dialogue between the EU and Russia in order to enable progress to be made in the definition and arrangements for an EU–Russia energy partnership. The overall objective of this partnership is to enhance the energy security of the European continent by binding Russia and the EU into a closer relationship in which all issues of mutual concern in the energy sector can be addressed while, at the same time, ensuring that the policies of opening up and integrating energy markets are pursued.

The Energy Dialogue aims at improving the investment opportunities in Russia's energy sector in order to upgrade and expand energy production and transportation infrastructure as well as improve their environmental impact, to encourage the ongoing opening up of energy markets, to facilitate market penetration of more environmentally-friendly technologies and energy resources, and to promote energy efficiency and energy savings [22]. The Energy Dialogue has permitted a good and frank debate at different levels between the EU and Russia and has allowed broad participation and involvement of the various Russian governmental bodies, the European Commission, EU Member States, and international financial institutions such as the European Bank for Reconstruction and Development, as well as a wide variety of EU and Russian energy companies.

Of course, the Energy Dialogue does not exist in a political vacuum. Rather, it simultaneously reflects and contributes to a broader economic, security, and strategic relationship between the two sides. For example, a more economically and politically stable Russia is less likely to show signs of compromise and accommodation with the EU on a variety of issues including pipeline routes and stability in transit countries such as Ukraine and Georgia. Thus, despite heavy mutual dependence, both Brussels and Moscow are pursuing separate strategies to improve their energy security and the overall perceived national interests.

One major reason for frequent disputes between Russia and some of the former Soviet republics (FSR) is Moscow's sensitivity to the political orientations of these former allies. Strategically, some Russian leaders do not wish to see Western influence in their "near abroad." Economically, Russian officials resent being beholden to these FSR, mainly Ukraine and to a lesser extent Belarus, for access to pipelines they once built and controlled. Against this backdrop, in the 1990s Moscow allowed a number of FSR to buy gas at hugely discounted prices, hoping to buy their loyalty. Apparently this policy did not work and Russia started demanding market prices close or similar to the ones West European consumers pay.

As early as 1990, Moscow cut energy supplies to the Baltic States in a futile attempt to stifle their independence movement. A similar episode took place in 1992 in retaliation for Baltic demands that Russia remove its remaining military forces from the region. In 1993 and 1994 Russian punished Ukraine, the conduit for about 80% of Russia's gas exports to Europe, by reducing gas supplies to force Kiev to pay for previous supplies and to pressure it into ceding more control to Russia over the Black Sea Fleet and over Ukraine's energy infrastructure. In addition, Russia resented the "Orange Revolution" that brought President Viktor Yushchenko to power in Kiev and his avid push for Ukraine to join NATO and the EU. A similar technique was applied to Belarus in 2004. In December 2005 and December 2006, Russia again cut or threatened to cut gas supplies to Ukraine and Belarus respectively to demand higher prices. In January 2009, Russia again cut off gas deliveries to Ukraine. Little wonder that a recent report by the International Energy Agency (IEA) stated that the flow of Russian gas through Ukraine may be subject to disruption "at almost any time" [23]. In order to face this challenge, the European Commission proposed new regulations in July 2009. These require all Member States to have a competent authority that would be responsible for monitoring gas supply developments, assessing risks to supplies, establishing preventive action plans, and setting up emergency plans. The regulations also obliged Member States to collaborate closely in a crisis, including through a strengthened Gas Coordination Group and through shared access to reliable supply information and data [24].

Like his counterpart in Ukraine, President Mikheil Saakashvili of Georgia promoted economic and political reform at home and sought close relations with the West and membership of NATO and the EU. His domestic and foreign policy orientations further complicated relations with Russia. Tension between Tbilisi and Moscow was further escalated in August 2008 when Russian troops attacked Georgia in support of the breakaway Abkhazia and South Ossetia regions.

For many years, Georgia has been considered by the EU and the United States as one of the main building blocks in the formation of alternative energy routes which bypass the territory of Russia [25]. Major pipelines that carry Caspian oil and gas to Europe via Georgia had been built since the late 1990s and others are in the planning process. The Baku–Tbilisi–Ceyhan (BTC) oil pipeline and Baku–Tbilisi–Erzurum (BTE) gas pipeline are the most prominent. Russian air strikes did not hit any of the international oil and gas pipelines crossing the country or any oil ports, but they forced BP, which is an operator of both the BTC and BTE, to stop oil and gas shipments through Georgia as a precautionary measure [26]. In the aftermath of this military operation Russia recognized Abkhazia and South Ossetia as independent states and signed defense pacts with them. These defense agreements allow Russia to establish and maintain military bases in the two regions for the next 50 years [27]. These uncertain security conditions in Georgia raise serious doubts about the country's ability to maintain its role as a major corridor between the Caspian Sea/Central Asia and Europe.

Some Europeans perceive their reliance on Russian energy, in particular gas, as a threat. In order to mitigate this perceived threat, the EU and several individual European countries have taken several steps to reduce their dependence on Russia. Most prominently, Europe is investing in alternative energy sources, particularly renewable fuels and nuclear power. Equally important, Europe is establishing energy partnerships with other major oil and gas producers in Africa, Caspian Sea/Central Asia, and the Middle East. Finally, Europe is seeking to diversify pipeline routes away from Russia. A major part of this strategy is the Nabucco pipeline project. It would bring Central Asian and Middle Eastern gas to Europe without passing through Russian territory. It would run from eastern Turkey through Bulgaria, Romania, and Hungary, ending in Austria. In May 2009, Azerbaijan, Egypt, Georgia, and Turkey signed an agreement committing themselves to the project. Two months later (July 2009), the five transit countries (Austria, Bulgaria, Hungary, Romania, and Turkey) agreed a deal allowing work on the pipeline to start.

In addition, the EU is encouraging the construction of new intra-EU interconnecting pipelines. Several schemes have already been built, are under construction, and are being planned. These include routes connecting Hungary and Romania, Bulgaria with Romania and Greece, and Greece and Italy.

On the other hand, Russia is pursuing a two-fold strategy that seeks to further consolidate the EU's dependence on its hydrocarbon supplies and simultaneously open up new markets, mainly in Asia, to its oil and gas exports. In recent years the Russian government has invested great financial and political capital in promoting two pipelines – the Nord Stream and South

Stream. The Nord Stream (also called the North European Gas Pipeline) will pass under the Baltic Sea starting from Vyborg in Russia to Greifswald in Germany. It will transport gas to Germany where it can be shipped to Denmark, the Netherlands, Belgium, the UK, France, and other countries. The shareholders are Gazprom (51%), two German companies, and one Dutch company: Wintershall Holding AG (20%), E.ON Ruhrgas AG (20%), and NV Nederlandse Gasunie (9%) [28]. The Nord Stream scheme underscores Russia's strategy of avoiding transit countries and building direct pipelines to Europe.

South Stream is a joint venture between Gazprom and ENI, the giant Italian oil company. The pipeline will run from Beregovaya in Russia, underneath the Black Sea, to Bulgaria. From there the pipeline would branch off in two directions: one toward the north-west, crossing Serbia and Hungary and ending in Austria; the other directed to the south-west through Greece and Albania, linking to the Italian network.

In May 2009 Gazprom and ENI agreed to double the capacity of the South Stream pipeline from 31 to 63 bcm. The agreement was signed in the presence of Prime Minister Vladimir Putin and his Italian counterpart, Silvio Berlusconi. The agreement also defined how Gazprom and ENI would divide the gas between them. At the same time, Gazprom and national gas companies from Bulgaria, Serbia, and Greece signed deals to create joint ventures in these countries to perform feasibility studies and construction for the project [29].

Two other major pipelines carry Russia's oil and gas to Europe: the Druzhba pipeline and Blue Stream. The Druzhba pipeline, also known as the Friendship pipeline, is one of the oldest pipelines supplying Russian oil to Europe. It was built in the early 1960s to supply oil to the former Soviet bloc and to Western Europe. It carries Russian and Kazakh oil to points in Ukraine, Hungary, Poland, Germany, and other destinations in Central and Eastern Europe.

Blue Stream connects the Russian system to Turkey underneath the Black Sea. It is a joint venture between Gazprom and ENI. The pipeline became operational in December 2002. Part of this Russian gas is re-exported to Europe via the Turkey–Greece interconnector (inaugurated in November 2007), and another link connecting Turkey to Greece and ending in Italy is planned.

It is also important to point out that Moscow has skillfully exploited divisions among EU Member States by striking bilateral deals that undermine Brussels' efforts to forge a common energy policy. The gas trade divides the EU almost as much as it unites it. The EU's new Member States depend on Russia's gas to a far greater degree than Western Europe does. Thus, big Western customers such as Germany, Italy, and France are in a position to strike bilateral deals with Moscow, while Eastern states, particularly the most vulnerable ones such as Bulgaria, the Baltic States, Slovakia, and Hungary, plead for EU-wide solidarity [30].

Finally, European efforts to develop alternative fuels and to forge partnerships with oil and gas suppliers from Africa, Central Asia, and the Middle East have raised concern among Russian officials about the security of demand for their energy supplies to Europe. Accordingly, Russia has negotiated oil and gas deals with other consumers, particularly in the fast-growing and energy-hungry Asian market. In December 2009 a new oil export terminal at Kozmino, near the port city of Nakhodka on the Pacific Ocean, was inaugurated. The terminal has since been used to export oil from fields in East Siberia to China and other Asian markets. At the same time Russia launched the East Siberia–Pacific Ocean (ESPO) pipeline which runs from Irkutsk Oblast to Skovorodino near the Chinese border [31]. The terminal and the pipeline open the way for East Siberian oil to the Asia-Pacific region and contribute to the diversification of Russia's export markets.

This discussion of the uneasy energy partnership between Russia and the EU suggests a number of conclusions. First, both Moscow and Brussels depend on each other. Russia's oil and gas supplies are crucial to maintain Europe's economic prosperity and high standard of living while the revenues Russia receives in return provide a major proportion of the nation's national income. Second, this mutual dependence or interdependence is good economically and strategically for both sides. It raises the stakes that each side has in the other's prosperity. It can serve, and indeed has served, as the core for broader European integration. Third, Russia's stagnant oil and gas production, its unstable legal system, and changing attitude toward private and foreign investment mean that Europe has more reasons to worry about Russia's ability, rather than willingness, to deliver sufficient quantities of oil and gas to the EU in the future. Fourth, Russia's geographical proximity to Europe and the long and extensive historical and strategic ties between the two sides mean that Russia will always be an important player in Europe's energy outlook. At the same time, the EU's aggressive efforts to establish partnerships with energy producers in Africa, Caspian Sea/Central Asia, and the Middle East suggest that Russia's share in the European oil and gas imports is likely to decrease [32].

8.6   Russia, the Middle East, and OPEC

Soviet policy in the Middle East was largely driven by a combination of ideological orientation, Cold War geopolitical considerations, and perceived Soviet national interests. The rise of military leaders with leftist orientations in key Arab countries like Egypt, Syria, Iraq, Algeria, and Libya provided a golden opportunity for Moscow to establish itself in the region and to counter US and European influence. The Soviet Union had very little contact with Iran and the rich Arab states on the Persian Gulf. The opportunity to mend fences came on the eve of the Gulf War (1990–1991), when Mikhail Gorbachev, the last Soviet president, supported the anti-Iraq coalition and, in return, secured major loans from Saudi Arabia and Kuwait [33].

Interestingly, Russian policy in the Middle East since the early 1990s has not been a complete departure from that of the Soviet Union in the preceding decades. Russian officials still seek to counter US and European influence and present their country as a superpower and an alternative to the West. A major difference, however, is that perceived Russia's national interests, rather than Cold War ideological considerations, have taken a prominent role in driving Russia's policy in the region [34].

The Shah of Iran was a close ally of the United States and was highly suspicious of what he perceived as Soviet imperialism. The hostility between the Islamic Republic in Tehran and the United States since 1979 has provided Moscow with a great opportunity to forge a closer relation with Iran. Iran needs the backing of global powers. Russia, and to some extent China, fulfill this role. Cooperation between Moscow and Tehran includes a variety of important issues such as arms sales, nuclear technology, the Caspian Sea, and energy. Iran buys a substantial proportion of its weapons from Russia. Russia is building the nuclear reactor in Bushehr and is protecting Iran from severe economic sanctions promoted by the United States and European powers based on allegations that Iran is trying to build nuclear weapons. Meanwhile, Tehran and Moscow do not agree on how to divide the Caspian Sea, but they do agree on containing the US role in the region. Finally, Russian firms are taking advantage of Western companies' hesitancy to do business with Iran. In December 2009 the French oil giant Total dropped out of a multi-billion-dollar gas investment to develop Phase II of the South Pars gas field on the grounds that it was too risky politically to invest in the Islamic Republic. Shortly after Total's withdrawal, Iran negotiated and signed a deal with Gazprom to replace the French company [35]. Similarly, Iran and Russia agreed swap arrangements, under which Gazprom delivers Turkmen gas purchased by Russia to northern Iran in exchange for gas deliveries from southern Iran to Persian Gulf countries [36].

Moscow has maintained close relations with Baghdad since 1958 when the monarchy was overthrown and a republican system was established. Despite Moscow's opposition to the Iraqi invasion of Kuwait and support for the international coalition, relations with Saddam Hussein were not as bad as his relations with Western powers. Indeed, the former Iraqi leader sought to divide the international coalition by offering lucrative oil deals to Russian (and Chinese) oil companies. Despite the prominent US role in Iraq since the 2003 war, Russian companies managed to negotiate and sign oil deals with the post-Hussein government in Baghdad. In 1997 Saddam Hussein granted the Russian oil company LUKOIL rights to develop the West Qurna 2 field, believed to contain massive reserves. The Iraqi leader rescinded the contract shortly before his regime was toppled. In December 2009, the Iraqi government signed a new contract with LUKOIL, along with Statoil of Norway, to develop the field [37].

The close relations between the United States and the Arab states on the Persian Gulf left Russia with limited room to maneuver. Nevertheless, Moscow has not given up on approaching these oil-rich countries. Russian companies won deals to develop Saudi Arabia's natural gas fields. In 2007 President Vladimir Putin visited Saudi Arabia, Qatar, and the United Arab Emirates, the first visit ever by a Russian leader.

Despite being the world's second largest oil producer and a major exporter, Russia has chosen not to join OPEC. The Russian case is not unique; some other major producers, such as Norway and Mexico, have chosen not to join OPEC for different reasons. However, most of these non-OPEC oil producers and exporters coordinate their oil policies with OPEC and attend its meetings as observers, without voting powers.

Russia and most members of OPEC share some similarities and differences. First, most OPEC members established national oil companies which are completely owned by the state and enjoy exclusive control over the oil industry. The Russian case is a little different. The Russian government owns the majority of stocks in a number of major energy companies, but other stocks and other companies are owned by private sector or foreign corporations. Second, the export policy for the majority of OPEC members is driven by both geopolitical and commercial interests. Thus several OPEC members consciously maintain an idle or spare capacity. Saudi Arabia in particular keeps a large idle capacity as an insurance policy against any unpredicted political, economic, or environmental upheavals. Russia's export policy, on the other hand, seems to be largely driven by commercial interests, seeking to sell as much oil as it can. It does not keep any idle capacity. Third, like most OPEC members, Russia is heavily dependent on oil and gas revenues. The Russian economy, however, is a little more diversified than most OPEC economies. For example, arms sales are a major source of public revenues in Russia. Furthermore, as one of the largest countries in the world, Russia holds a variety of resources and the nation has somewhat well-developed industrial, commercial, and agricultural sectors. Fourth, given its relatively developed economy and large population, Russia consumes a large proportion of its oil and gas production, unlike several OPEC members, such as Kuwait, Qatar, and the United Arab Emirates, where domestic consumption is limited and a large proportion of production is exported.

Against this background, Russia's cooperation with OPEC is limited and takes place only when this cooperation serves Moscow's interests. Indeed, the relation between Russia and OPEC can be characterized as a combination of cooperation and competition depending on oil prices. When oil prices are low, the two sides try to work together to limit production, so prices can move higher. Occasionally, Russia did agree to reduce its production, in coordination with OPEC, in order to address a glutted oil market. But when prices are high, Russia does not restrain itself by OPEC quota. Rather, as one observer notes, Russia "seeks to make hay while the sun shines" [38]. To put it differently, OPEC has continually reduced its production and market share in order to bolster prices. This means that the greater the increase in Russian oil exports, the lower OPEC's production, in order to keep the price at a reasonable level [39].

Given this complex relation between Russia and OPEC, the former has been accused of being a free-rider, taking advantage of OPEC's efforts to manage global oil prices without committing itself to any production quota system. Russian officials deny this accusation. In OPEC meetings held in 2008 and 2009, the Russian Deputy Prime Minister Igor Schin asked OPEC members to upgrade Russia's status in relation to the organization from observer to that of permanent observer and invited them to hold one of their meetings in Russia [40]. The Deputy Prime Minister also proposed several ideas to consolidate cooperation between the two sides, including the creation of new trading centers for oil, harmonized taxation of oil producing companies, and preferential long-term contracts for oil supplies [41].

8.7   Energy Sector Organization

Russia's energy sector is different and more complicated than in most other oil and gas producing nations. Since the dissolution of the Soviet Union in 1991, fundamental changes have taken place in the ownership, management, and structure of the sector. These changes reflect the dominant political and economic conditions in Moscow and how Russian leaders seek to employ the hydrocarbon wealth to pursue their broad strategies. The rush to privatize most of the nation's oil and gas companies was the underlying feature of the energy sector in most of the 1990s, while "renationalization" or "de-privatization" was the dominant one in the 2000s.

The dissolution of the Soviet Union and the rebirth of Russia were accompanied by a great deal of political and economic chaos. The Russian government suffered a severe financial crisis and was critically out of cash. Then, President Boris Yeltsin decided that one way to overcome this shortage was to sell the state's hydrocarbon assets to the private sector. Thus, several oil producing enterprises and refineries were transformed into open-stock companies. Shares in these companies were auctioned to a group of Russian commercial banks for cash at a very low price. The rush to sell these valuable enterprises created a new super-rich social class, the so-called oil oligarchy. Mikhail Khodorkovsky, former CEO of Yukos (one of the biggest Russian oil companies in the late 1990s and early 2000s), was a prominent figure in this class.

On the positive side, this privatization restrained government corruption and brought efficiency to the oil sector. In addition, privatizing public enterprises encouraged international oil companies to invest in Russia's oil and gas industry. On the negative side, many Russians inside and outside the government resented that the state was deprived of these revenues and only a small group of oil barons were accumulating massive wealth. Two other developments added to this popular and official resentment. First, Vladimir Putin was elected president in 2000. His election signaled a new era in Russian policy with much higher political stability than under his predecessor. Second, the rise in oil and gas prices in the early part of the 2000s highlighted the Russian state's losses and the oil oligarchs' gains.

It is important to point out that the state did not lose all the control it previously had over the oil sector and the privatization was only partial. By holding all the voting stocks of Transneft, a monopoly operating and managing pipelines, the government was, and still is, able to retain control over oil transportation and export. Almost all the nation's oil is transported via pipelines controlled by Transneft.

The more stable Russia and more confident Putin have drastically changed the energy sector's landscape. Under Putin's leadership, the privatization policy came to an end or was tremendously slowed down and the state reasserted its role and power in the oil and gas sectors. Khodorkovsky was arrested and sent to prison and his company Yukos was dismantled and taken over by Rosneft, a state-owned company. The private sector still plays a junior role in Russia's energy industry, mainly in coordination with the state and the state-controlled oil and gas companies. The three most prominent state-controlled companies are Gazprom, Rosneft, and Transneft.

Gazprom is by far the world's largest natural gas company. It is a fully diversified energy company managing the exploration, production, sale, and distribution of gas for both domestic and foreign markets; the production and sale of crude oil and gas condensates; and hydrocarbon refining operations. Since the late 1990s, Gazprom has further diversified and expanded its portfolio to include electricity, petrochemicals, communications, and banking.

Initially, Rosneft was established in 1993 as a state enterprise on the basis of assets previously held by Rosneftegaz, the successor to the Soviet Ministry of Oil and Gas. In 1995, a Russian government decree transformed Rosneft into an open joint-stock company. Like the rest of the energy sector, the company performed poorly in most of the 1990s. In order to strengthen Rosneft, the Russian government appointed a new management team led by Sergey Bogdanchikov in 1998. Under his leadership, which lasted for more than a decade, Rosneft expanded its operations domestically and internationally and established itself as the largest oil company in Russia and one of the largest in the world [42].

Transneft was established in 1992 as the successor to the Soviet Ministry of Oil Industry Main Production Department for Oil Transportation and Supplies (Glavtransneft), with 100% of shares belonging to the state. The company is Russia's largest oil pipeline company, managing pipelines that extend from Siberia to the Baltic [43].

The dominant role that Gazprom, Rosneft, and Transneft play in Russia's energy sector and Putin's policy of renationalization or de-privatization have further complicated the foreign investment environment. Many Russians have grown suspicious of the role that international energy companies play in their hydrocarbon sector and believe that they can develop their own resources. In short, since the early 2000s Russia has become less welcoming of foreign investment in its energy sector. Russian officials retracted many of the obligations they previously made and IOCs were forced to completely withdraw or sell some of their stocks to national companies. The formula for production-sharing agreements illustrates the change in Russia's stance on foreign investment.

A production-sharing agreement (PSA) is an internationally binding commercial contract between an investor and a state. A PSA defines the conditions for the exploration and development of natural resources from a specific area over a designated period of time. According to the terms of a standard oil and gas PSA, the state retains ownership of the hydrocarbons and the investors bear responsibility for extracting the resource. The investors typically receive the majority of early revenue from the project as compensation for the cost of exploration and development. Once the project reaches the cost recovery stage, subsequent revenue is shared between the investors and the state according to a pre-negotiated formula [44].

PSAs were originally devised to protect weak states from the IOCs. However, they later became useful tools to protect foreign energy companies from political risks prevailing in unstable countries. A key attraction of the PSA for foreign investors lies in the fact that it replaces energy-specific taxes, and eliminates many uncertainties about future tax as the division of profits between the company and the state becomes the subject of a contract. By negotiating and agreeing on the terms and conditions of exploration and development for the life of the project, PSAs are designed to protect foreign companies from risks such as arbitrary changes in tax legislation, or unpredictable regulations.

In the early 1990s, President Boris Yeltsin and other Russian politicians and business leaders sought to provide incentives for IOCs to invest in the nation's energy sector. In December 1993, Yeltsin issued a presidential decree establishing the basic regulatory framework for PSAs. In order to stimulate foreign investment in geographically isolated, technologically complex, and environmentally challenging hydrocarbon projects, the Russian government signed three PSAs with major IOCs in 1994 and 1995: Sakhalin-1 in 1995; Sakhalin-2 in 1994; and the Kharyaga project in 1995. Foreign companies such as ExxonMobil, Royal Dutch Shell, and Total led the way in developing these oil and gas projects. Although the three projects vary in size, cost, and production, they rank among the largest foreign investments in Russia. In 1995, the Duma (Parliament) passed legislation granting PSAs the status of legally binding contracts and establishing the basic provisions of the agreements in accordance with international standards.

The policy of renationalization or de-privatization altered Russia's stance on PSAs. Russian oil companies perceived them as giving foreign firms a competitive advantage. After intense lobbying efforts by domestic producers, the PSA structure was relegated to a small list of fields approved by the Duma. In 2003, Putin signed legislation that greatly reduced the number of oil and gas fields eligible for development under PSAs and adjusted the federal tax code to make future PSAs less attractive to foreign investors. A year later (2004), a government commission on the implementation of the PSA annulled the results of the 1993 competition for the right to develop the fields of Sakhalin-3. (ExxonMobil and Chevron Texaco, which planned to operate under PSA terms, had won the tender.)

By 2010 only the three grandfathered PSAs are in operation and the prospects for new ones are dim. IOCs willing to invest in Russia are likely to have no choice but to serve as junior partners or contractors to Gazprom, Rosneft, and other Russian companies.

8.8 Conclusion: The Way Forward

Since the early discoveries of oil and natural gas in Russia/Soviet Union, the nation has played a significant role in global energy markets and hydrocarbon revenues have provided a major proportion of its national income. These facts are not likely to change in the foreseeable future. At least three dynamics are likely to shape Russia's energy outlook in the coming decades. First, despite large volumes of oil and gas production, the nation's energy future will be restrained by the continuing rise in domestic consumption which leaves diminishing supplies for export. Russia holds much larger proven natural gas reserves than oil reserves. Consequently, Saudi Arabia and other Persian Gulf producers will continue to occupy the driver's seat in global oil markets, while Russia will continue to dominate the natural gas industry.

Second, the level of production in Russia will continue to be influenced by the nation's economic and political conditions. Despite great progress in recent decades, Russia lacks and needs Western technology in critical areas such as offshore drilling and LNG. The availability of Western technology is particularly important given that most of the new discoveries are located in geologically and environmentally challenging areas. Russia also needs foreign investment. An open and accommodating investment environment is likely to accelerate the development of Russia's hydrocarbon deposits.

Third, Russia is strategically located between two large and growing energy consuming markets: Europe and Asia. Russia already has extensive energy ties with the EU as a bloc and with individual European countries. Despite occasional disagreements and rising suspicion, these ties are likely to endure. Meanwhile, Russia's energy relations with Asian markets (particularly China, India, Japan, and South Korea) are likely to grow. The two sides need each other. Both Europe and Asia are trying to diversify their energy suppliers, but given Russia's geographical proximity and massive reserves, Moscow will continue to feature in their energy security

References

[1] Inozemtsev, V. (2009) *The Resource Curse and Russia's Economic Crisis*, Chatham House, London. Available at http://www.chathamhouse.org.uk (accessed March 10, 2009).

[2] Victor, N.M. (2008) *Gazprom: Gas Giant under Strain*, Stanford University Press, Stanford, CA. Available at http://pesd.stanford.edu (accessed January 22, 2008).

[3] Hill, F. and Fee, F. (2002) Fueling the future: the prospects for Russian oil and gas. *Demokratizatsiya*, **10** (3), 1–25, 4.

[4] Isabel, G. (2009) Russian Oil Export Terminal Looks to Asia. Financial Times (December 27).

[5] Energy Information Administration, *Country Analysis Briefs: Russia*. Available at http://www.eia.doe.gov/emeu/cabs/Russia/full.html (accessed May 27, 2008).

[6] British Petroleum (2010) *BP Statistical Review of World Energy*, British Petroleum, London, p. 22.

[7] Barysch, K. (2008) *Pipelines, Politics, and Power: the Future of EU-Russia Energy Relations*, Center for European Reform, London, p. 27.

[8] Energy Information Administration, *Country Analysis Briefs: Sakhalin Island*. Available at http://www.eia.doe.gov/emeu/cabs/Sakhalin/Full.html (accessed April 23, 2007).

[9] Editorial (2009) Sakhalin energy exports first LNG cargo to Japan. *Oil & Gas Journal*, **107** (13), 12.

[10] Medetsky, A. (2009) Medvedev Opens LNG Plant on Sakhalin. Moscow Times (February 19).

[11] Daily News Bulletin, *Russian Government Approves Energy Strategy for Period Until 2030*. Available at http://www.istockanalyst.com/article/viewiStockNews/articleid/3669592 (accessed November 29, 2009).

[12] Robin Paxton, Reuters, *Russia Unveils $2 Trillion Energy Growth Plan to 2030*. Available at http://uk.reuters.com/articlePrint?articleId=UKGEE5AP25J20091126 (accessed November 29, 2009).

[13] Clark, M. (2007) Arctic: a tough nut to crack. *Petroleum Economist*, **74** (2), 32–34, 32.

[14] Yenikeyeff, S.M. and Krysiek, T.F. (2007) *The Battle for the Next Energy Frontier: The Russian Polar Expedition and the Future of Arctic Hydrocarbons*, Oxford Institute for Energy Studies, Oxford, p. 2.

[15] Idiatullin, S. (2007) Udar Nizhe Polyusa. Kommersant Vlast (August 13).

[16] BBC, *A Russian Mission to Explore the Ocean Floor below the North Pole Is Back on Track after Engineers Fixed an Engine Problem Aboard One of the Vessels*. Available at http://news.bbc.co.uk/2/hi/europe/6916662.stm (accessed July 26, 2007).

[17] Halpin, T. (2009) Russia Warns of War within a Decade over Arctic Oil and Gas Riches. The Times (May 14).

[18] Energy Charter Organization, *About the Charter*. Available at http://www.encharter.org/index.php?id=7 (accessed May 21, 2009).

[19] Energy Charter Secretariat, *What is the Energy Charter? An Introductory Guide*. Available at http://www.encharter.org (accessed September 24, 2002).

[20] Energy Charter Organization, *1994 Treaty*. Available at http://www.encharter.org (accessed May 21, 2009).

[21] Moscow Times (2009) Putin Rejects Energy Charter Treaty. Moscow Times (August 7).

[22] European Union, *EU-Russia Energy Dialogue*. Available at http://europa.eu/rapid/PressReleasesAction.do?reference=Memo/09/121&format=HTML&Language=en (accessed March 19, 2009).

[23] International Energy Agency, *Natural Gas Market Review 2009*. Available on line at http://www.iea.org (accessed June 29, 2009).

[24] European Commission, *The Commission Adopts New Rules to Prevent and Deal with Gas Supply Crises*. Available on line at http://ec.europa.eu (accessed July 16, 2009).

[25] Yenikeyeff, S.M. (2008) *The Georgia-Russia Standoff and the Future of Caspian and Central Asian Energy Supplies*, Oxford Institute for Energy Studies, Oxford, p. 1.

[26] Medetsky, A. (2008) War Casts Cloud over Pipeline Route. Moscow Times (August 14).

[27] Radio Free Europe, *Moscow Signs Defense Pacts with Breakaway Georgian Regions*. Available at http://www.rferl.org (accessed September 15, 2009).

[28] Nord Stream, *The New Gas Supply Route for Europe*. Available at http://www.Nord-Stream.com (accessed May 22, 2009).

[29] Medetsky, A. (2009) Gas Pipeline Fight Escalates Sharply. Moscow Times (May 18).

[30] Escritt, T., Olearchyk, R., Petrov, N., and Wagstyl, S. (2009) A Slight Thaw. Financial Times (December 4).

[31] Radio Free Europe, *Russia's Putin Launches New Pacific Oil Terminal*. Available at http://www.rferl.org/articleprintview/1915199.html (accessed December 29, 2009).

[32] Volkov, D., Gulen, G., Foss, M., and Makaryan, R. (2009) Gazprom pipeline gas remains key to Europe. *Oil & Gas Journal*, **107** (16), 53–56, 53.

[33] Dannreuther, R. (1993) Russia, Central Asia and the Persian Gulf. *Survival*, **35** (4), 92–112, 107.

[34] Herrmann, R.K. (1994) Russian policy in the Middle East: strategic change and tactical contradictions. *Middle East Journal*, **48** (3), 455–474, 472.

[35] Fars News Agency, *Iran, Russia Ink Energy Deal*. Available at http://www.farsnews.com/English/printable.php?nn=8704241172 (accessed July 14, 2008).

[36] Fars News Agency, *Iran, Russia Sign MoU on Oil Swap*. Available at http://english.farsnews.com/printable.php?nn=8712260779 (accessed March 16, 2009).

[37] Kanter, J. (2009) Oil Field Project in Iraq Won by Lukoil and Statoil. New York Times (December 30).

[38] Boue, J.C. (2004) *Will Russia Play Ball with OPEC?* Oxford Institute for Energy Studies, Oxford, p. 5.

[39] Chalabi, F. (2004) Russian oil and OPEC price policies. *Middle East Economic Survey*, **47** (12), 3.

[40] Moscow Times (2008) Sechin Sends OPEC Weak Signal. Moscow Times (December 18).

[41] Watkins, E. (2009) Russia declines OPEC membership; offers alternative. *Oil & Gas Journal*, **107** (12), 24–25, 25.

[42] Rosneft, *History*. Available at http://www.rosneft.com/about/history (accessed January 11, 2010).

[43] Transneft, *Transneft Russian Pipeline*. Available at http://www.transneft.ru/company (accessed January 11, 2010).

[44] Krysiek, T.F. (2007) *Agreements from Another Era: Production Sharing Agreements in Putin's Russia, 2000-2007*, Oxford Institute for Energy Studies, Oxford, p. 2.

9

OPEC and Gas OPEC

For a long time, energy policy was perceived as a zero-sum conflict where the interests of producers and consumers were mutually exclusive. Each side pursued a strategy to maximize its interests at the expense of the other side. Stated differently, consuming countries were interested in low oil and gas prices while producing nations sought to raise prices. The two sides realized that individual states or companies would have little leverage and creating a collective entity would make it easier to reach their respective goals. Against this backdrop, the Organization of Petroleum Exporting Countries (OPEC) was established in 1960 and the International Energy Agency (IEA) was founded in 1974. The former represents the producers' interests and the latter promotes the consumers' objectives.

The two organizations were created in the midst of price crises. In the late 1950s and early 1960s, producing nations believed that IOCs were paying them very little for their precious product. Driven by this perception and the desire to receive a "fair" price for crude oil and enhance their negotiation leverage, major oil producers founded OPEC. Following the 1973 Yom Kippur War between Israel and Arab countries, most OPEC members imposed an oil embargo on the United States and some other countries to punish them for their support of Israel. More importantly, they incrementally cut off their production. These steps led to the so-called first oil price shock in 1973–1974. The soaring oil prices drove consuming nations, led by the United States, to create the IEA. The goal was to articulate a strategy on how to ensure consumers' energy security.

Under these circumstances, it was almost inevitable that the two organizations adopted conflicting strategies. Their opposing policies failed to assure global energy markets and, indeed, contributed to the wide fluctuation of oil prices and the overall instability of international economy. Little wonder that a growing consensus emerged calling for cooperation between producers, consumers, and other major energy players (i.e., IOCs). Increasingly, more producers and consumers have ceased to perceive their respective interests as mutually exclusive and have identified growing areas of common interest. In order to promote and consolidate cooperation, several dialogues, partnerships, and organizations were created. The International Energy Forum (IEF) is a prominent example of these efforts.

In this chapter I examine the history, objectives, and structure of OPEC. This is followed by a detailed analysis of the ongoing efforts to create a similar organization for natural gas

Energy Security: An Interdisciplinary Approach, First Edition. Gawdat Bahgat.
© 2011 John Wiley & Sons, Ltd. Published 2011 by John Wiley & Sons, Ltd.

producers, the so-called Gas OPEC. The next chapter focuses on the IEA and provides an assessment of the organization's policies to enhance its Member States' energy security. The IEF's efforts to bring together producers, consumers, and major energy companies and to institutionalize energy cooperation are the subject of the last chapter. The goal is to highlight some of the main changes in the global energy landscape and how each side's perception of energy security has evolved.

9.1   OPEC: History and Evolution

For most of the twentieth century the global oil markets were dominated by a few major IOCs, the so-called Seven Sisters: Standard Oil Co. of New Jersey (later Exxon), Standard Oil Co. of New York (originally Socony, later Mobil), Standard Oil Co. of California (Socal, later Chevron), Royal Dutch Shell, Texaco, BP, and Gulf [1]. The oil producing countries of OPEC did not participate in production or pricing of crude oil, but simply received a stream of income through royalties and income taxes as part of the concession system. OPEC countries were too weak to challenge the multinational Seven Sisters' domination of the industry.

The first move towards the establishment of OPEC took place in 1949, when Venezuela approached Iran, Iraq, Kuwait, and Saudi Arabia and suggested that they exchange views and explore avenues for regular and closer communication between them. The need for closer cooperation became more apparent when, in 1959, the Seven Sisters unilaterally reduced the price of oil. This prompted the convening of the First Arab Petroleum Congress, held in Cairo. The Congress adopted a resolution calling on oil companies to consult with the governments of the producing countries before unilaterally taking any decision on oil prices. It also set up a general agreement on the establishment of an oil consultation commission [2].

In 1960 the Seven Sisters reconfirmed their domination by further reducing oil prices. In response, delegates from five major oil producing nations – Iran, Iraq, Kuwait, Saudi Arabia, and Venezuela – met in Baghdad and announced on September 16, 1960 the foundation of OPEC. The goal was, and still is, to protect the interests of these major producing nations. Accordingly, in their first resolution the five OPEC members emphasized that the companies should maintain price stability and that prices should not be subjected to fluctuation. They called for companies not to undertake any change in the posted price without consultation with the host country. They pledged to establish a price system that would secure stability in the market by using various means, including the regulation of production, with a view to protecting the interests of both consumers and producers.

These ambitious goals, however, proved hard to achieve given the little leverage that producing nations had and the dominant role of the Seven Sisters. By the early 1970s, some major dynamics of the global oil industry fundamentally changed. First, some IOCs, such as ENI of Italy and Occidental of the United States, started operations in the Middle East and elsewhere and offered more attractive financial terms than those offered by the Seven Sisters. This gradual process eroded the near monopoly imposed by the multinational companies. Second, economic prosperity in the United States, Western Europe, and Japan accelerated and was fueled by growing demand for oil. These major consumers, however, lacked sufficient indigenous supplies. Thus, the growing appetite for oil was met, mainly, by production from OPEC countries. Stated differently, the world grew more dependent on OPEC supplies. The combination of these two developments left OPEC producers in a stronger bargaining position than in the early 1960s.

Building on this newly acquired bargaining power, OPEC producers sought to increase oil prices. Their demands were rejected by the multinational oil companies and negotiations between the two sides collapsed. The 1973 Arab–Israeli War provided the geopolitical and geo-economic opportunity to fundamentally alter the balance of power between OPEC members and IOCs. In addition to imposing an oil embargo on the United States and a few other countries, and incrementally cutting production, some OPEC governments stopped granting new concessions and started to claim equity participation in the existing concessions, with a few of them opting for full nationalization. Asserting their power, OPEC members decided in October 1973 to unilaterally raise oil prices independently of the multinational oil companies' participation. These developments paved the way for structural changes in the world oil industry.

In the aftermath of the first oil shock, OPEC members consolidated their control over production and prices. However, their lack of the necessary technological and financial infrastructures left them dependent on multinational oil companies. Thus, rhetoric aside, OPEC members continued to sell big proportions of their production to the old concessionaires. Political developments in the Persian Gulf including the 1979 Iranian Revolution followed by the eight-year-long war between Iran and Iraq had a significant impact on OPEC and the broad global oil industry. The interruption of oil supplies from Iran and Iraq triggered widespread chaos leading to soaring oil prices in what became known as the second oil price shock.

The continuing push for higher prices underscored a division within OPEC between two competing strategies. The first strategy was advocated by OPEC members with considerable proven reserves, small populations, and high per capita incomes (i.e., Kuwait, Qatar, Saudi Arabia, and the UAE). These countries sought to moderate prices in order to maintain demand over the long run. The other strategy was pursued by members with larger populations, lower oil exports per capita, and lower per capita income (i.e., Algeria, Indonesia, Iran, and Nigeria). This second group demanded restraint on OPEC production and higher prices [3].

This disagreement between the so-called hawks and doves laid the ground for an awkward situation whereby a two-tiered pricing system prevailed. Saudi leaders perceived high oil prices as harming oil producers in the long run by encouraging investment in high-cost areas outside OPEC and switching to alternative fuels. Accordingly, Riyadh refused to raise prices beyond a certain level.

Against this backdrop, OPEC adopted a quota system in March 1983 which set a production ceiling. By controlling the volume of global production, OPEC sought to influence prices. Within this framework OPEC adjusted its production upward and downward based on the level of production from non-OPEC countries. Saudi Arabia, the major oil producer and exporter, played the role of "swing producer" within OPEC.

Political turmoil and the lack of consensus among OPEC members for a unified strategy prompted IOCs to increase their investments in areas outside OPEC, most notably the North Sea and the Gulf of Mexico. The increasing supplies from outside OPEC coupled with the fall in demand as a result of high prices led to a drastic fall of OPEC's share of the global oil market. This intense rivalry between OPEC and non-OPEC producers and within OPEC proved unsustainable. In the mid-1980s, Saudi Arabia decided to drop its system of selling oil at fixed prices and instead adopted a market-oriented pricing system. Consequently, Saudi Arabia's production started to rise quickly and by 1986 global markets became saturated. This led to a severe collapse in oil prices.

The very low prices in the mid-1980s hurt the interests not only of OPEC producers, but also those of other producers such as the North Sea, the United States, and the Soviet Union as well as the overall global economy. This broad chaos and the emergence of many suppliers and many consumers led to the development of "a complex structure of oil markets which consist of spot, physical forward, futures, options, and other derivative markets" [4]. This structure is based on formula pricing where the price of a certain variety of crude oil is set as a differential to a certain benchmark or reference price. These include Brent Blend, West Texas Intermediate (WTI), Dubai, and Nigerian Forcados, among others. One of the major characters of the oil market in the later part of the 1980s and most of the 1990s was the stability of the long-term oil price at a relatively low level.

From 2000 up to 2008, oil prices soared and, as a result, most oil exporting countries in OPEC and non-OPEC members accumulated substantial revenues. The imbalance between supply and demand was the driving force behind the soaring oil prices. Unlike the supply-interruption oil shocks of 1973–1974 and 1979–1980, the 2000s' surge was a demand-driven one, fueled by strong Asian consumption. Furthermore, the surge reflected not only increasing demand and decreasing supply, but also broader macroeconomic and geopolitical changes such as rising exploration and production costs, the falling value of the US dollar, the re-emergence of "resource nationalism," inadequate refining capacity, and an aging labor force [5].

After reaching a peak of $147 per barrel in 2008, prices significantly dropped and then started a slow process of recovery. OPEC members and non-OPEC producers reacted in different ways to the rise in oil prices. There was a common assumption that in the face of high and rising oil prices, OPEC will respond by increasing supply to moderate prices and stabilize the market. Such an action would help maintain healthy growth in global oil demand and limit the entry of substitutes such as tar sands and ethanol. This view was influenced by OPEC's decision to introduce a price band in 2000, which involved production adjustments when the price moved above $28 for 20 consecutive trading days or when the price moved below $22 for 10 consecutive trading days.

OPEC members failed to put a ceiling on the price and, indeed, most members took advantage of rising prices by increasing their production and exporting as much as they could to maximize their profit. This attitude suggests that OPEC's role is not to prevent oil prices from rising above certain levels or to set a price ceiling. Rather, a key objective of the organization is to avoid oil prices from falling below a certain level deemed unacceptable by its members.

On the other hand, non-OPEC producers' response to the 2000 price boom was weak. They were not able to raise their production to take advantage of rising prices. This suggests that non-OPEC production has peaked or is close to reaching this stage. It is increasingly becoming more costly and technologically and environmentally challenging to maintain or increase production from non-OPEC producers. As one analyst argues, there seems to be an asymmetric response to oil prices: "A sharp rise in oil price induces a modest investment response in non-OPEC countries, while a decline in the oil price generates a sharp fall in investment and a period of underinvestment in the oil sector" [6]. This argument is in line with the IEA's projection, which predicts that as conventional oil production in countries not belonging to OPEC peaks, "most of the increase in output would need to come from OPEC countries, which hold the bulk of remaining recoverable conventional oil resources" [7].

Three conclusions can be drawn from this brief review of the fluctuation of oil prices in the last few decades and the role OPEC played in this process. First, as with any commodity, the role of oil prices is to signal relative scarcity or abundance which in turn causes all energy players (i.e., consumers, producers, national oil companies, and IOCs among others) to adjust to the allocation of resources [8]. Second, rhetoric aside, OPEC does not fix oil prices and does not have a direct impact on their rise and fall. Rather, given the OPEC members' substantial proven reserves and large volumes of production and exports, the organization plays a significant indirect role in influencing price formation [9]. OPEC signals its preferred price and alters its production volume up or down. These signals are perceived by other energy players and, in turn, they respond to these signals. Within this context, it is important to point out that despite impressive improvements, the availability of accurate data on production, consumption, reserves, and other vital information is not perfect. As a result, "market psychology" plays an important role in shaping the movement of oil prices. Third, historically, exporters and importers have had divergent interests, with the former favoring higher prices and the latter favoring lower ones. These perceived opposite interests have recently changed in favor of an emerging realization that too high or too low prices do not serve anyone's interests. Too high prices encourage conservation and alternative energy, and destabilize the overall global economy. Too low prices reduce investment in new exploration and the development of oil deposits and contribute to economic and political turmoil in producing countries.

9.2   OPEC: Objectives, Membership, and Organization [10]

Article 2 of OPEC's Statute clearly spells out the main objectives of the organization. These include coordinating the members' petroleum policies and safeguarding their interests, individually and collectively. Another related goal is to stabilize prices and eliminate fluctuations. Most importantly, OPEC seeks to secure a steady income for its members, sufficient and reliable supplies to consuming countries, and fair return on investments by either NOCs or IOCs.

Chapter 2 of the Statute distinguishes between three types of membership: (a) founder members: these are the five nations (Iran, Iraq, Kuwait, Saudi Arabia, and Venezuela) which were represented at the first meeting in Baghdad in 1960 and signed the original agreement of the establishment of the organization; (b) full members: those countries whose applications for membership had been accepted; (c) associate members: those countries accepted by a majority of three-quarters, including all founder members. Unlike the first two categories, associate members do not have the right to vote, though they can attend meetings and participate in deliberations. All members are major oil producers and exporters and are expected to adhere to the organization's principal goals. Members also have the right to withdraw from OPEC. Since its foundation in 1960, OPEC's membership has expanded and there have been a few changes in the last several decades (Table 9.1).

Ecuador suspended its membership in December 1992 and reactivated it in October 2007. Gabon, which became a full member in 1975, terminated its membership with effect from January 1995. Indonesia, which became a full member in 1962, suspended its membership in December 2008.

In their second meeting OPEC members decided to locate the Secretariat in Geneva, Switzerland. A few years later (1965) they decided to move the headquarters to Vienna, Austria. After successfully negotiating with the Austrian government, a Host Agreement was signed [11].

OPEC as an organization is divided into three main organs: the Conference; the Board of Governors; and the Secretariat. The Conference holds the main authority in the organization.

Table 9.1   Membership in OPEC.

Member	Year of accession
Iran	1960
Iraq	1960
Kuwait	1960
Saudi Arabia	1960
Venezuela	1960
Qatar	1961
Libya	1962
UAE	1967
Algeria	1969
Nigeria	1971
Ecuador	1973
Angola	2007

It has the responsibility of formulating the general policy of the organization; deciding upon applications for membership; confirming the appointment of members of the Board of Governors; approving the budget as submitted by the Board of Governors; and appointing the Secretary General. It consists of heads of delegations – normally the ministers of petroleum, oil, energy, or equivalent portfolio of Member States. The Conference holds two ordinary meetings a year (usually in March and September) and as many extraordinary meetings as required. Delegations review the status of the international oil market and the forecasts for the future in order to agree upon appropriate actions. With the exception of procedural matters, all decisions require the unanimous agreement of all full members. Non-members such as Egypt, Oman, and Russia attend as observers, but cannot vote.

The Board of Governors can be compared to the board of directors of a commercial company. It is composed of one Governor nominated by each Member State and confirmed by the Conference for two years. The Board meets at least twice a year. It implements the decisions taken by the Conference and directs the management of the organization; and draws up the budget of the organization and submits it to the Conference for approval.

The Secretariat carries out the executive functions of the organization, in accordance with the provisions of the Statute and under the direction of the Board. It consists of the Secretary General and such staff as may be required. The Secretary General is the chief officer of the Secretariat. The Conference appoints the Secretary General for a period of three years, which can be renewed for another three years. English is the official language of the Secretariat. The Secretariat is responsible for the implementation of all resolutions of the Conference, as well as decisions of the Board of Governors. It conducts research, the findings of which constitute key inputs to decision making. It also disseminates news and information to the world at large.

In addition to these three main organs – the Conference, Board of Governors, and Secretariat – other divisions contribute to formulating and implementing OPEC policy. The Division of Research monitors, forecasts, and analyzes developments in the energy and petrochemical industries; analyzes economic and financial issues related to international financial and monetary matters and to the international petroleum industry; and maintains and expands data services to support the research activities of the Secretariat and those of Member States.

The Finance and Human Resources Department is responsible for all financial matters and financial control functions of the Secretariat as well as ensuring the financial integrity of the organization. It is also responsible for developing and applying effective human resources management policies to enable the Secretariat as a whole to carry out its functions efficiently. The Public Relations and Information Department is responsible for presenting OPEC's objectives, decisions, and actions in the most desirable perspective; disseminating news of general interest regarding the organization and the Member States on energy and related matters; and carrying out a central information program and identifying suitable areas for the promotion of the organization's aims, objectives, and image. Its activities include press relations, publications, speech writing, managing the web site, and news monitoring.

The Economic Commission is a specialized body operating within the framework of the Secretariat, with a view to assisting OPEC in promoting stability in international oil prices at equitable levels. The Legal Office has the responsibility for providing legal advice to the Secretary General and supervising the Secretariat's legal and contractual affairs. It also evaluates legal issues of concern to OPEC. The Internal Audit Office has the responsibility for independently ascertaining whether the ongoing processes for controlling financial and administrative operations in the Secretariat are adequately designed and functioning in an effective manner. It also investigates irregularities and assesses weaknesses in the accountancy system or budgetary control.

Another important specialized organ is the OPEC Fund for International Development (OFID). In 1975, the Finance Ministers of the Member States proposed the creation of a new multilateral financial facility to channel OPEC aid to developing countries. Known initially as the "OPEC Special Fund," this institution was one of several bilateral and multilateral development institutions set up by OPEC and Arab countries.

The OFID, which was originally intended to be a temporary facility, started operations in August 1976 with an initial endowment of $800 million and within a little over a year its resources had doubled. It directly extends loans to developing countries and channels donations from its Member States to other development institutions including the International Monetary Fund's Trust Fund and the International Fund for Agricultural Development. The OFID became a fully fledged, permanent international development agency in May 1980 [12].

The main objective of the Fund is to reinforce financial cooperation between OPEC Member States and other developing countries by providing financial support to assist the latter countries on appropriate terms in their economic and social development efforts. The Fund is authorized to engage in all necessary activities to achieve this central objective. These include providing concessional loans for balance of payments support; granting concessional loans for the implementation of development projects and programs; making contributions to eligible international agencies; and financing technical assistance activities [13].

9.3   OPEC Summits [14]

In addition to the formal structure (i.e., the Conference, Board of Governors, Secretariat, and other divisions and departments), OPEC heads of state and government of member countries occasionally meet to deliberate and articulate broad strategic guidelines. Since its foundation in September 1960, OPEC has held three summits – Algiers, 1975; Caracas, 2000, and Riyadh, 2007. For each of these summits, the main purpose was to step back from the day-to-day activities of the international oil market and examine issues at the national leadership level, pertaining to the fundamental principles, objectives, and procedures of OPEC. The summits also examined contemporary issues confronting the world, particularly the global economy and the divisions between rich and poor countries and how OPEC, individually and collectively, can help bridge this gap.

The first summit was held in 1975 in the aftermath of the first oil shock and in the midst of intense confrontation between producing and consuming countries. Not surprisingly, the summit's deliberations and resolutions reflected this international confrontation. OPEC leaders rejected allegations attributing to the price of petroleum the responsibility for the instability in the world economy. They claimed that adjustment in the price of oil did not contribute to the high rates of inflation within the economies of developed countries. OPEC leaders reminded the rest of the world that they have contributed through multilateral and bilateral channels to the development efforts and balance of payments adjustments of other developing countries, as well as industrialized nations. They reaffirmed solidarity with other developing countries in their struggle to overcome underdevelopment.

Furthermore, OPEC leaders asserted that the conservation of petroleum resources is a fundamental requirement for the well-being of future generations and urged the adoption of policies aimed at optimizing the use of this essential and non-renewable resource. On the other hand, OPEC leaders reaffirmed their readiness to ensure that supplies will meet the essential requirements of the developed countries, provided that the consuming countries do not use artificial barriers to distort the normal operation of the laws of supply and demand.

By the time the second summit was held in Caracas in September 2000, the international economic system and the global energy markets had experienced some fundamental changes; in particular, environmental issues had attracted significant attention and OPEC's share in global supply was falling in favor of supplies by other producers such as Russia and the Caspian Sea. Finally, oil prices were stable at a relatively low level for most of the 1990s. The Caracas Summit reflected OPEC leaders' concerns over these issues, among others.

OPEC heads of state and government confirmed their commitment to provide adequate, timely, and secure supplies of oil to consumers at fair and stable prices and emphasized the strong link between the security of supply and the security and transparency of demand. They called for a fair share for OPEC in the world oil supply and for growing cooperation on a regular basis between OPEC and other oil exporting countries. On the other hand, they demanded the opening of effective channels of dialogue between oil producers and consumers. OPEC leaders also asserted their association with the universal concern for the well-being of the global environment, and their readiness to continue to participate effectively in the global environmental debate and negotiations, including the UN Framework Convention on Climate Change and the Kyoto Protocol, to ensure a balanced and comprehensive outcome.

OPEC leaders called on consuming nations to adopt fair and equitable treatment of oil by ensuring that their environmental, fiscal, energy, and trade policies do not discriminate against oil. They also expressed their concern that taxation on petroleum products forms the largest component of the final price to the consumers in the major consuming countries, and called upon them to reconsider their policies with the aim of alleviating this tax burden for the benefit of the consumers, for just and equitable terms of trade between developing and developed countries, and for the sustainable growth of the world economy.

The third summit, held in Riyadh in November 2007, came amid rising global concern over climate change and deep global dependence on fossil fuels fueled by soaring oil prices in the previous seven years. OPEC leaders sought to address these concerns. Thus, the summit focused on three major themes: the stability of global energy markets; energy for sustainable development; and energy and the environment.

OPEC leaders pledged to efficiently manage and prolong the exploitation of their exhaustible petroleum resources in order to promote the sustainable development and the welfare of future generations. They re-emphasized the connection between demand security and supply security and recognized that with globalization the economies of the world and energy markets are integrated and interdependent. They urged all parties to find ways and means to enhance the efficiency of financial petroleum markets with the aim of reducing short- and long-term price volatility. They reiterated the need to continue the process of coordination and consultation with other petroleum exporting countries and the necessity to strengthen and broaden the dialogue between energy producers and consumers. They repeated their call on consuming governments to adopt transparent, non-discriminatory, and predictable trade, fiscal, environmental, and energy policies and promote free access to markets and financial resources.

OPEC leaders associated their countries with all global efforts aimed at bridging the development gap and making energy accessible to the world's poor while protecting the environment. They stated that eradicating poverty should be the first and overriding global priority guiding local, regional, and international efforts.

As major oil producers and exporters are heavily dependent on oil revenues, OPEC members have adopted a cautious stance on the climate change controversy. They reiterated that the process of production and consumption of energy resources poses different local, regional, and global environmental challenges. Meanwhile, they stressed that human ingenuity and technological development have long played pivotal roles in addressing such challenges and providing the world with clean, affordable, and competitive petroleum resources for global prosperity. OPEC leaders underscored that they share the international community's concern that climate change is a long-term challenge, and recognize the interconnection between addressing such concerns on the one hand, and ensuring secure and stable petroleum supplies to support global economic growth and development on the other hand. Finally, they stressed the importance of cleaner and more efficient petroleum technologies and demanded that all policies and measures developed to address climate change concerns be both balanced and comprehensive.

9.4 OPEC: Long-Term Strategy [15]

The main guiding texts for OPEC are the OPEC Statute, approved in January 1961; the Summit Declarations of 1975, 2000, and 2007; and the Long-Term Strategy, adopted on the Organization's 45th Anniversary, September 2005. This Strategy, prepared over a period of two and a half years, provides a broad vision and framework for the organization's future.

The Strategy identifies the uncertainties surrounding future demand for OPEC oil as a key challenge. Factors such as future world economic growth, consuming countries' policies, technology development, and future non-OPEC production levels contribute to these uncertainties regarding demand for OPEC oil. The Strategy explores these uncertainties in three scenarios that depict contrasting futures of the global energy scene. These scenarios are dynamics-as-usual, protracted market tightness, and a prolonged soft market.

There are substantial uncertainties over future economic growth arising from the complex interplay of domestic and global determinants of that growth, including such diverse factors as demographics, advances in technology, capital availability, and trends in commodity prices, domestic policies, and global trade developments, regimes, environmental policies, and financial regulations. Another area of uncertainty stems from consuming countries' policies. OPEC claims that taxation of energy products is often seen not only as a means of raising revenue, but also as a means of controlling demand in addressing environmental and energy security issues. The Strategy alleges that consuming countries' policies demonstrate significant discrimination against oil, involving not only higher tax rates, but also subsidies for competing fuels.

A third area of uncertainty with significant impact on oil demand is technological development. For example, in the transportation sector, conventional internal combustion engines could continue to achieve significant fuel economy improvements, while hybrid vehicles may witness an important growth. The introduction of non-oil-fueled vehicles and the use of alternative fuels, such as biofuels, are drivers that could affect oil demand growth patterns in the transportation sector. A fourth area of concern is the development in non-OPEC supply. A number of factors, such as oil prices, upstream legal and fiscal regimes, and investments in non-OPEC countries, technological advances, and exploration successes, will shape future scenarios regarding non-OPEC supply.

A fifth area of uncertainty is related to environmental concerns. The profile of incremental global demand is overwhelmingly for light and clean products, while incremental supply comprises significant volumes of sour, medium, and heavy crude grades. The combination of this with the move to ever-stricter product quality and environmental regulations represents a challenge for the downstream industry. Future refining capability needs to be considered in terms of both the adequacy of secondary processes – for example, to upgrade heavy streams and to meet tight targets for sulfur – and crude distillation capacity.

The combination of these uncertainties signifies a heavy burden of risk in making the appropriate investment decisions. Accordingly, the Strategy calls for several measures to meet these concerns. First, the promotion of the development of technologies that address climate change concerns such as carbon dioxide capture and storage technology. Second, OPEC members should continue to play an active role in the climate change negotiations. OPEC supports the principle of common, but differentiated, responsibilities. It believes that the international community should fulfill its obligations to strive to minimize the adverse effects of policies and measures on developing countries, in particular fossil-fuel exporting developing countries (i.e., OPEC members). This could involve assistance in relation to economic diversification and transfer of technology. Third, dialogue among producers and between producers and consumers should be widened and deepened to cover all issues of mutual concern. These include security of demand and supply, market stability, upstream and downstream investment, and technology. Finally, the Strategy emphasizes OPEC's commitment to support oil market stability. It recognizes that extreme price levels, either too high or too low, are damaging for both producers and consumers, and points to the necessity of being proactive under all market conditions.

To conclude, when OPEC was founded in 1960 the oil industry and the global economy were very different from what they are more than five decades later. During this time the oil industry has experienced several upheavals, OPEC policy contributing to some of them. Over the years many observers have predicted OPEC's demise. However, it has survived and become an important driving force in the global energy markets. Its members' interests are not identical, though they have managed to coordinate their policies and find common ground most of the time. Any assessment of OPEC's role in managing global oil prices would be highly controversial. The push for higher prices, which characterized the first decades of OPEC's life, has waned in favor of an emerging consensus that stable prices at a "reasonable" level would serve both the producers' and consumers' interests.

9.5 Gas OPEC

In recent years, global concern over energy security has intensified, fueled by the surge in oil prices up to 2008. In response, many consuming countries have sought to diversify their energy mix by investing in nuclear power, biofuels, and renewable sources. Despite this renewed interest in these alternative sources of energy, the IEA projects that fossil fuels (coal, natural gas, and oil) will continue to dominate the global energy mix. From 2005 to 2030 oil's share is projected to slightly decline from 35 to 32%, the share of natural gas will increase modestly from 21 to 22%, while that of coal will rise from 25 to 28% [16]. In other words, fossil fuels will continue to provide more than 80% of the world's demand for energy.

Within this context, natural gas has been widely considered the fuel of choice for many consumers. It is abundant and less polluting than coal or oil. Little wonder that natural gas consumption and trade have expanded in the last few decades. Consumers in the United States, Europe, and Asia have grown more dependent on natural gas. Domestic production in these regions, however, cannot keep pace with rising demand. This large and growing gap between demand and domestic production has been increasingly filled by imports from foreign suppliers. The deepening dependence on foreign sources has heightened concerns about security of supply. Stated differently, natural gas consumers are increasingly concerned about potential movements by major gas producers to influence gas markets and prices, similar to the role of OPEC in oil markets. Will a "Gas OPEC" emerge?

On the supply side, natural gas producers have sought to coordinate their policies in order to strengthen their position in negotiating prices and terms of trade with consumers. In a meeting held in Tehran in May 2001, some major gas producing nations created an organization called the Gas Exporting Countries Forum (GECF) to facilitate such cooperation. Since then, some hawkish members, such as Iran and Venezuela, have sought to transform the GECF into Gas OPEC. Others (i.e., Algeria and Qatar) believe it would take some time for such a transformation to take place. Meanwhile, Russia, the major gas producer and exporter, has taken an ambivalent stand, sending conflicting signals.

In this section, I examine the prospects for the evolution of the GECF into Gas OPEC. I argue that gas producers are likely to continue and expand their cooperation. Such cooperation, however, is unlikely to involve attempting to control output or influencing prices like OPEC does. This low probability that the GECF would evolve into Gas OPEC is based on major differences between the oil and natural gas markets. In addition, Russia is not likely to play a leading role in forming a Gas OPEC, similar to the Saudi leading role in OPEC.

9.6 GECF and OPEC

As discussed above, when OPEC was founded the oil trade was characterized by bilateral long-term contracts and oil producers were not satisfied with the low financial returns they were getting for their crude. At the outset, OPEC looked weak and did not have the powerful leverage it came to exert on oil output and prices in the following decades. The 1973 Arab–Israeli War and the subsequent use of oil as a "political weapon" was a turning point in the history of OPEC and the oil industry. Two factors contributed to strengthening OPEC's bargaining position. First, in the early 1970s, US oil production peaked and Washington began its steady reliance on foreign supplies. Second, oil was almost the primary fuel in the transportation, residential, and industrial sectors, providing almost half of global energy requirements [17]. In the early 1980s, OPEC initiated a production-quota system. Such a move has enabled OPEC to adjust its overall production in response to signals from global oil markets.

Despite these efforts to shape production level and influence prices, OPEC members do not perceive it as a cartel. Still, OPEC's huge share of the world's proven oil reserves and production ensures that it exerts a powerful leverage over oil prices and markets.

On the other hand, the GECF is an informally structured organization. Its members meet at the ministerial level once a year to discuss topics of mutual interest. One of the key themes of the GECF meetings has been the existence of long-term contracts between gas exporters and importers. The GECF has been keen to see these long-term contracts maintained in order to assist with the underwriting of large capital projects and to provide a stable income to its members. Another theme has been the pricing formulas that link gas prices to those of oil and how to "de-link" the pricing of these two strategic commodities [18].

Little progress, if any, has been achieved on both themes. This can be explained by the fact that the GECF, unlike OPEC, has recently started building a bureaucratic infrastructure, chosen a formal headquarters, and has loose membership rules. Since its inception in Tehran 2001, the GECF has held meetings in Algiers, Doha, Cairo, and Moscow. Membership has risen to 15 from the inaugural 11. These 15 members control a substantial share of the world's natural gas reserves and production.

It is important to point out that major natural gas producers such as Australia (1.6% reserves and 1.4% production), Canada (0.9% reserves and 5.4% production), and the Netherlands (0.6% reserves and 2.1% production) are not members, while Norway (1.1% reserves and 3.5% production) attends only as an observer. Meanwhile, Iran is barely a natural gas exporter and Venezuela does not export. In short, membership of the GECF is less coherent than in OPEC.

The figures in Tables 9.2 and 9.3 show that while the Middle East is the unparalleled driving force in the global oil market, Russia is the dominant power in the gas market. In other words, natural gas deposits and production are less concentrated than those of oil. In addition, seven nations (Algeria, Iran, Libya, Nigeria, Qatar, UAE, and Venezuela) are members of both OPEC and the GECF. Despite this overlap in membership and interest between the two organizations, major differences between gas markets and oil markets suggest that the GECF is unlikely to evolve into a Gas OPEC, at least in the near future.

9.7   Oil vs. Gas

Members of OPEC and the GECF share similar goals – to coordinate their policies to obtain what they perceive as a "fair" price for their products. However, geological and commercial differences between oil and gas and their respected markets have laid the groundwork for OPEC to become a powerful force in shaping oil prices and markets. These same geological

Table 9.2 OPEC's share of global proven reserves and production, 2010.

Member	Share of reserves (%)	Share of production (%)
Algeria	0.90	2.00
Angola	1.00	2.30
Ecuador	0.10	0.70
Iran	10.30	5.30
Iraq	8.60	3.20
Kuwait	7.60	3.20
Libya	3.30	2.00
Nigeria	2.80	2.60
Qatar	2.00	1.50
Saudi Arabia	19.80	12.00
UAE	7.30	3.20
Venezuela	12.90	3.30
Total	76.60	41.30

Source: British Petroleum, *BP Statistical Review of World Energy*, London, June 2010, pp. 6 and 8.

and commercial characteristics pose challenges that the GECF has to overcome in order to play a similar role in influencing natural gas prices and markets. Specifically, the two commodities and markets differ in how they are traded, priced, and in terms of substitutability.

First, oil is easier and cheaper to transport than gas. Every day, tankers carrying millions of tons of crude oil and petroleum products sail from producing regions to consuming markets.

Table 9.3 GECF's share of global proven reserves and production, 2010.

Member	Share of reserves (%)	Share of production (%)
Algeria	2.40	2.70
Bolivia	0.40	0.40
Brunei	0.20	0.40
Egypt	1.20	2.10
Indonesia	1.70	2.40
Iran	15.80	4.40
Libya	2.80	0.50
Malaysia	1.30	2.10
Nigeria	2.80	0.80
Oman	0.50	0.80
Qatar	13.50	3.00
Russia	23.70	17.60
Trinidad & Tobago	0.20	1.40
UAE	3.40	1.60
Venezuela	3.00	0.90
Total	72.9	41.1

Source: British Petroleum, *BP Statistical Review of World Energy*, London, June 2010, pp. 22 and 24.

On the other hand, given the high costs of transporting gas, the fuel is largely transported via pipelines, with a small but growing proportion in the form of LNG. This difference in shipping expenses is a major reason why oil has developed into a global market while natural gas is still a regional market. In other words, oil consumers buy the fuel from any producer regardless of the distance. Meanwhile, natural gas consumers import most of their needs from nearby producers. Thus most of the imported gas in the United States comes from Canada; Europe is heavily dependent on gas supplies from the North Sea, Russia, and North Africa; while China, India, Japan, and South Korea rely on neighboring producers, particularly Australia, Brunei, Indonesia, and Malaysia.

Second, the quality and prices of crude oil are shaped largely by two characteristics: viscosity (thickness or density) and sulfur content. Crude oil with a lower density is referred to as light crude while that with a higher density is called heavy crude. Similarly, the one with a low sulfur content is known as sweet oil and the one with a high sulfur content as sour crude [19]. The three major benchmark crudes, namely West Texas Intermediate (traded mainly in the United States), Brent (traded mainly in Europe), and Dubai–Oman (traded mainly in the Far East), reflect the level of density and sulfur content.

Gas prices on the other hand are mostly pegged to oil prices. Furthermore, given the high costs of trading natural gas, prices are usually locked in long-term bilateral contracts (20 years or longer) with a Take-or-Pay (ToP) clause. This means that buyers are obligated to pay for the gas whether they take it or not. This rigidity of gas pricing leaves little room for the GECF to try to influence gas production and prices.

Third, volatility of oil prices and consumers' concern over alleged manipulation by producers have encouraged intensified efforts to reduce dependence on oil and rely on alternative sources such as nuclear power and renewable energy. Since the early 1970s, oil has been largely replaced by these alternative sources of energy in the residential, commercial, and manufacturing sectors. In the transportation sector, however, oil still maintains its primacy. Despite impressive technological innovations, aircraft and most cars still run on petroleum products. This oil primacy in the transportation sector is a fundamental advantage for OPEC. High oil prices might force consumers to fly and drive less, but they still have to. Largely, there is no substitute for oil in the transportation sector.

On the other hand, natural gas is used mainly to generate electricity. It faces competition from coal, hydroelectric, and nuclear power. Gas is readily substitutable by these other fuels. A higher elasticity of demand is translated into reduced market power of the GECF [20].

To sum up, the high cost of transporting gas, the lack of a spot global natural gas market, and competition from other fuels have all placed limits on the evolution of the GECF into a Gas OPEC. In the last few years, however, the rapidly growing importance of LNG has introduced new dynamics into the natural gas industry and markets. Theoretically, LNG offers the potential to transform natural gas markets from regional to global ones.

LNG is gas cooled to liquid form, so it can be shipped by cargoes to various markets and is far more flexible than gas shipped by pipeline. Trade in LNG started just a few decades ago when Algeria shipped it to the UK in 1964. This was followed by shipments to France and other shipments from Alaska to Japan. In a short time, Japan had evolved into a major import market of LNG, particularly from Asian exporters such as Brunei and Indonesia. Ships were dedicated to particular importers, but a few speculative vessels were built in anticipation of a growing market [21]. In recent years, technological improvements have led to reduced costs for liquefying, transporting, and regasifying LNG. This increased liquidity in the natural gas industry has promoted the emergence of a spot LNG market similar to that which has emerged in the oil market over the last few decades.

Despite this growing flexibility in natural gas markets, the vast majority of LNG is still traded through long-term contracts due to substantial capital commitment. In short, LNG provides promising potential to alter the dynamics of gas markets. Such changes, however, will take some time to materialize. This leaves Russia, which in 2011 still exports all its natural gas by pipeline, with tremendous leverage.

9.7.1   Russia

Russia enjoys several advantages. It holds the world's largest natural gas reserves and is the world's largest producer and exporter. In addition, it has the advantage of lying between two of the world's largest energy consuming regions: Europe to the west and East Asia to its east. Gazprom, Russia's state-run gas monopoly and the largest gas company in the world, controls the bulk of the country's production. Given these advantages, no effective gas organization can work without the active participation of Russia.

Moscow has participated in GECF meetings and hosted one in November 2008. However, Russian officials have been sending conflicting signals on their country's stance on the potential evolution of the GECF into a Gas OPEC. Since 2002, Vladimir Putin (who served as president from 2000 to 2008 and then as prime minister) has expressed different opinions on Gas OPEC. He said, "We do not intend to set up a cartel, but I think it is right to coordinate our activities"' [22]. The Russian leader also stated, "We do not reject the idea of creating a gas cartel. But this initiative requires more study" [23]. He also called the creation of a Gas OPEC "an interesting idea. We will think about it" [24]. Putin summed up Russia's strategic stance on energy as: "In our opinion, energy security lies in the following principles, For producers of energy resources, it is first and foremost a guarantee of sovereignty over their national stock of energy resources, as well as a responsibility to ensure regular, uninterrupted supplies to their consumers." The Russian leader added, "But there is also a responsibility that must rest with consumer countries. It is their guaranteed ability to buy these resources in required amounts and on predictable conditions" [25]. Russia's Industry and Energy Minister Viktor Kristenko argued that the nature of global gas markets makes it "impossible for exporters to influence prices in the next fifteen to twenty years" [26]. Finally, Gazprom Deputy Chairman Aleksandr Medvedev leveled a threat against the West, saying that if Russia did not get its way in energy negotiations with Europe, it would create "an alliance of gas suppliers more influential than OPEC" [27].

These conflicting statements suggest that Russian leaders are still formulating their stance on a Gas OPEC. Moscow's position on OPEC, however, indicates that Russian leaders prefer to adopt an independent energy policy without limiting their choices by joining a collective grouping. Despite being the world's second largest oil producer and exporter, Russia has never joined OPEC. Indeed, Russia has been able to benefit more from being outside OPEC and responding to oil supply and demand circumstances on its own rather than as part of a group.

In line with this unilateral approach toward energy issues, Russia has been working to strengthen ties with other major gas producers without committing itself to a Gas OPEC. Russian leaders visited and held talks with their counterparts from Algeria, Iran, Kazakhstan, Libya, Nigeria, Qatar, and Turkmenistan, among others, and signed cooperation agreements on gas production. These agreements have further strengthened Russia's leverage as the world's leading gas producer and exporter.

9.7.2   Iran, Qatar, and Algeria

Iran holds the world's second largest natural gas reserves, after Russia, and seeks to become a major gas exporter to Europe and Asia. Despite its potential and ambition, Iran currently is not a major gas exporter. Its status on the world gas stage does not reflect its massive reserves. Several factors have restrained Iran's influence. Iran has one of the largest populations in the Middle East (over 70 million in 2010). This large population consumes a substantial proportion of the country's gas production. The Iranian authorities also tried to substitute oil for gas in order to free up more crude for export. Iran's oil fields are the oldest in the Middle East and some of them are close to maturity. A large volume of gas is reinjected in the oil fields to increase pressure and production. Finally, for the last few years Tehran has been involved in a confrontation with major powers, led by the United States, over its nuclear program. Since the 1979 Islamic Revolution, Iran has been under different types of US economic sanctions, and in the last few years the UN Security Council has imposed waves of sanctions. These sanctions have compelled a number of international companies to freeze their investments in Iran and have complicated Tehran's efforts to finance and develop its massive gas deposits.

Against this background, Tehran has taken the lead in establishing the GECF and calling for its transformation into a Gas OPEC. Such a formal grouping can further improve relations with Russia and put pressure on Western powers. Several Iranian officials have expressed strong support for a Gas OPEC. Heshmatollah Falahatpisheh, a member of the Parliamentary National Security and Foreign Policy Committee, said, "When the world is becoming more dependent on energy, Gas OPEC can give producers and exporters a strategic position" [28]. Another Committee member, Hamid Reza Haji-Babai, contends that a Gas OPEC would serve as a "center of power to resist Western influence, especially America's economic and political pressure" [28].

On the other side of the Persian Gulf, Qatar has taken a very different position on the formation of a Gas OPEC than the one adopted by Iran. The emirate holds the world's third largest gas reserves behind Russia and Iran. Most of its natural gas is located in the massive offshore North Field, a geological extension of Iran's South Pars. In 2006 Qatar surpassed Indonesia to become the largest exporter of LNG in the world.

Qatar is well placed to play a major role in natural gas policy and markets. Though its reserves are substantially below those of Russia, Doha's reserves are very high relative to its population and the need to maintain output to generate adequate revenue is not as urgent as in the case of Russia. Furthermore, Qatar's heavy reliance on LNG cargoes to export its natural gas gives it more flexibility than Russia, which exclusively relies on pipelines.

Despite these geological and commercial advantages, Qatar has no reason to wield its increased energy power against the EU and the United States. Doha hosts US military bases and troops in the Persian Gulf and enjoys close ties with Washington, Brussels, and several international oil companies. Accordingly, Qatari leaders have taken a cautious stance on the evolution of the GECF into a Gas OPEC. The country's ruler, Emir Hamad bin Khalifa al-Thani, expressed his doubts about "creating a cartel designed to influence natural gas markets in a similar way to that in which OPEC controls crude oil markets" [29]. Minister of Energy

Abdullah bin Hamad al-Attiyah expressed similar sentiments: "forming a gas cartel would be difficult, but it cannot be ruled out" [22].

Algeria holds less gas reserves than Russia, Iran, and Qatar. But it was the first country in the world to produce and export LNG and has since maintained its status as a leading LNG exporter, along with Qatar, Indonesia, and Malaysia. Most of these exports go to Europe, particularly France, Spain, Italy, and Turkey. Indeed, the EU imports most of its gas from Russia, Norway, and Algeria. In short, Europe depends on Algerian gas supplies, while Algiers relies on gas revenues.

This heavy interdependence between Algeria and its European partners has been under some pressure in recent years. Since the early 2000s, Europe has sought to liberalize its energy sector, by introducing more competition in its gas market. More competition would lead to lower prices. In late 2007, Spanish gas distributors tried, and failed, to prevent Sonatrach, Algeria's state-owned oil and gas company, from competing in Spain's energy market. The dispute was resolved when Algeria threatened to cancel the construction of a pipeline called Medgas that connects Algeria to Southern Europe.

In order to counter this growing liberalization and increasing competition in Europe's gas sector, Algerian leaders have reasserted their control over the country's hydrocarbon deposits and announced tighter terms for foreign investors since the late 2000s [30]. They also reached out to Russia and Iran and expressed their support for the formation of a Gas OPEC. At the end of a visit to Tehran, Chakib Khelil, Minister of Energy and Mining, stated: "In the long-term we are moving toward a Gas OPEC" [31].

9.7.3 Consumers' Reaction

Since the founding of the GECF in 2001, gas consuming nations have been nervous that it might evolve into an OPEC-like organization that might threaten the accessibility and affordability of their gas supplies. Talks between major gas producers conjure up images of energy vulnerability and broad economic slowdown.

Hydrocarbon reservoirs were discovered in the North Sea in the 1960s, but the area did not become a key producing region until the 1980s, when production from large discoveries began to come on line. Development of the North Sea's energy resources faced considerable hurdles, including the inhospitable climate and the great drilling depths, requiring highly sophisticated offshore technology and imposing high production costs. On the other hand, the region enjoys political and financial stability and proximity to large European consuming markets – advantages that have added to its significance as an oil and gas producing region. The North Sea's crude oil production peaked in the late 1990s and its natural gas production has shown signs of slowing down and is projected to peak in the second decade of this century. EU members possess very limited indigenous natural gas reserves. Norway, the Netherlands, and the United Kingdom hold the bulk of Europe's gas reserves. This limited indigenous energy production capacity underscores Europe's heavy dependence on foreign supplies.

In recent years European countries have sought to consolidate their relationship with major gas producers in North Africa, the Caspian Sea, and the Persian Gulf. Despite Iran's massive resources and potential, the confrontation over Tehran's nuclear program and US pressure have restrained European–Iranian cooperation and limited European investment in Iran's energy sector. Despite its efforts to diversify the sources of its natural gas supplies, Europe is still heavily dependent on Russia, which provides about one-quarter of Europe's gas needs.

Brussels has a "mixed relationship" with Moscow. Russia is the only country with which the EU organizes two summits a year, while on the commercial level Russia is the EU's third largest partner after the United States and China [32]. Despite this close political and economic cooperation, European leaders are concerned about their overdependence on Russia's gas supplies. Russia's use of energy leverage to influence political developments in neighboring Belarus and Ukraine since the early 2000s has heightened Europe's fears. In addition, the Russian government and Russian companies are involved in constructing pipelines that would further deepen Europe's dependence on Moscow. Energy Commissioner Andris Piebalgs and other leaders of the EU expressed deep concern about Russia' energy policy and the potential creation of a Gas OPEC.

The United States is less vulnerable than Europe to interruption of gas supplies. At the time of writing (2010), the United States produces more than 80% of its gas needs and imports about 16% [33]. Most of these imports are delivered by pipelines from Canada, but a growing volume of natural gas is coming to the United States as LNG from overseas. Washington is importing, and negotiating to import, LNG from a variety of suppliers including Trinidad & Tobago, Egypt, Nigeria, Algeria, Equatorial Guinea, Norway, Qatar, and Yemen.

In recent years US energy companies have developed and applied new technological techniques to explore and develop natural gas deposits. The new technology has had a promising impact on the US natural gas industry. Proven reserves have been substantially expanded and indigenous production is rising at an impressive rate. Still, the United States maintains its policy of opposing any governmental intervention in the free market system. Within this context, Secretary of Energy Samuel W. Bodman highlighted Washington's opposition to the formation of a Gas OPEC: "Initiatives new or old, which seek to control the flow of energy supplies to the market and circumvent the role of the market to set prices, are contrary to the long-term interests of both producers and consumers" [30].

9.8   Conclusion

Since the founding of the GECF in 2001, parallels have been drawn between it and OPEC. Gas producers have asserted that they do not intend to create a cartel to fix prices, while consumers have voiced their suspicion. In recent years major gas producers have taken important steps to consolidate their cooperation. In 2008, Iran, Qatar, and Russia signed a deal setting what is called a "Big Gas Troika" to coordinate market policies. The three countries also created a Supreme Technical Committee comprised of specialists and experts for discussing the implementation of several joint projects spanning the entire production chain from geological exploration to transportation and joint marketing of gas [34].

Equally important, GECF members have taken concrete initiatives to transform their GECF into an institutionalized international organization. In their 2008 meeting, held in Moscow, energy ministers adopted a charter making provisions for an organization headquarters, secretariat, and chief executive. They chose Doha, Qatar, as home for the headquarters (despite a strong push by Russia in favor of St. Petersburg). GECF members stated that one of the goals will be consulting each other on future investment so they would not produce more than the market needs. They also decided to consider alternative mechanisms for determining gas prices to the current method of tying them loosely to the price of a basket of oil products [35].

A year later (2009) in their meeting in Doha, GECF members chose Leonid Bokhanovsky, a Russian gas expert, to serve as Secretary General. Bokhanovsky pledged to turn the GECF into a leading advisory agency for both gas producing and consuming countries, major gas companies, and other international organizations. The Secretary General stated that the GECF will develop its own research team; promote new technology for gas exploration, production, and transportation; monitor and forecast supply and demand; coordinate gas shipments through pipelines and by tankers; study the relation between gas and other energy resources; and promote the growth of global gas consumption [36].

These ambitious goals aside, the fundamental differences between gas and oil markets as well as the seemingly conflicting interests of gas producers suggest that the GECF is not likely to evolve into a Gas OPEC, at least in the near future. Nevertheless, it would not be impossible for GECF to be more proactive in regulating how natural gas is traded, collect data, coordinate policies, and consolidate cooperation between its members [37]. The changing dynamics of natural gas markets away from regionalism and closer to being a global market (similar to oil) are likely to add more leverage to any group of gas producers.

References

[1] Bahgat, G. (2003) *American Oil Diplomacy in the Persian Gulf and the Caspian Sea*, University Press of Florida, Gainesville, FL, p. 178.
[2] OPEC, *General Information*. Available at http://www.opec.org (accessed January 8, 2010).
[3] Blaydes, L. (2004) Rewarding impatience: a bargaining and enforcement model of OPEC. *International Organization*, **58** (2), 213–237, 221.
[4] Fattouh, B. (2006) The origins and evolution of the current international oil pricing system: a critical assessment, In *Oil in the 21st Century: Issues, Challenges and Opportunities* (ed. R. Mabro), Oxford University Press, Oxford, pp. 41–100, p. 52.
[5] Bahgat, G. (2008) Supplier-user teamwork key to stable oil prices. *Oil & Gas Journal*, **106** (32), 20–24, 20.
[6] Fattouh, B. (2010) *Oil Market Dynamics through the Lens of the 2002-2009 Price Cycle*, Oxford Institute for Energy Studies, Oxford, p. 21.
[7] International Energy Agency (2009) *World Energy Outlook*, International Energy Agency, Paris, p. 42.
[8] Mabro, R. (2006) Introduction, In *Oil in the 21st Century: Issues, Challenges and Opportunities* (ed. R. Mabro), Oxford University Press, Oxford, pp. 1–18, p. 12.
[9] Mabro, R. (2005) The international oil price regime: origins, rationale and assessment. *Journal of Energy Literature*, **11** (1), 3–20, 14.
[10] OPEC, *OPEC Statute*. Available at http://www.opec.org (accessed December 12, 2008).
[11] OPEC, *General Information*. Available at http://www.opec.org (accessed January 8, 2010).
[12] OPEC Fund for International Development, *OFID at a Glance: A Brief History*. Available at http://www.ofid.org/about/about.aspx (accessed January 11, 2010).
[13] OPEC Fund for International Development, *The Agreement Establishing the OPEC Fund for International Development*. Available at http://www.ofid.org/publications/PDF/AE-engl_fund.pdf (accessed February 9, 2010).
[14] OPEC, *OPEC Solemn Declarations*. Available at http://www.OPEC.org (accessed January 8, 2010).
[15] OPEC, *OPEC Long-Term Strategy*. Available at http://www.opec.org (accessed February 10, 2008).
[16] International Energy Agency (2007) *World Energy Outlook*, International Energy Agency, Paris, p. 42.
[17] Wood, D. (2007) Could a future gas OPEC shape LNG import plans? *Oil & Gas Journal*, **105** (20), 22–34, 26.
[18] Energy Business Review, *GECF Unlikely to Emerge as OPEC Equivalent*. Available at http://www.energy-business-review.com (accessed May 4, 2006).
[19] Fattouh, B. (2008) *The Dynamics of Crude Oil Price Differentials*, Oxford Institute for Energy Studies, Oxford, p. 16.

[20] Hartley, P.R., Medlock, K.B., III, and Nesbitt, J.E., *A Global Market for Natural Gas? Prospects to 2035*, James A. Baker III Institute for Public Policy, Rice University, Houston, TX. Available at http://www.bakerinstitute.org (accessed May 28, 2004).

[21] Bainbridge, K. (2004) World markets speeding change for LNG shipping. *Oil & Gas Journal*, **102** (17), 98–100, 98.

[22] Gold, R. and White, G.L. (2007) Russia and Iran Discuss a Cartel for Natural Gas. Wall Street Journal (February 2).

[23] Fattah, H.M. (2007) Putin Visits Qatar for Talks on Natural Gas and Trade. New York Times (February 13).

[24] Medetsky, A. (2007) Putin Warms to Idea of a Gas OPEC. Moscow Times (February 2).

[25] Radio Free Europe, *Russia Says Gas Forum Will Not Be OPEC-Like Cartel*. Available at http://www.rferl.org (accessed December 23, 2008).

[26] Medetsky, A. (2007) Qatar Summit Will Likely Avoid Cartel. Moscow Times (April 9, 2007).

[27] Kupchinsky, R., Payvand, *Gas OPEC Moves Closer to Becoming Reality*. Available at http://www.payvand.com (accessed February 7, 2007).

[28] Sepehri, V., Payvand, *Iranian Politicians Support Establishment of Natural Gas Cartel*. Available at http://www.payvand.com (accessed February 7, 2007).

[29] Zwaniecki, A., Payvand, *US Energy Secretary Bodman Calls Natural Gas Cartel a Bad Idea*. Available at http://www.payvand.com (accessed February 16, 2007).

[30] Saleh, H. (2008) Algeria Tightens Foreign Investor Rules. Financial Times (August 11).

[31] Fars News, *Iran, Algeria Seeking to Establish Gas OPEC*. Available at http://english.farsnews.net (accessed August 12, 2008).

[32] Gomart, T., Center for Strategic and International Studies, *EU-Russia Relations: Toward a Way out of Depression*. Available at http://www.csisi.org (accessed July 12, 2008).

[33] Energy Information Administration, *What Is Liquefied Natural Gas and How Is It Becoming an Energy Source for the United States?* Available at http://www.eia.doe.gov (accessed February 10, 2010).

[34] Mehr News, *Iran, Russia, Qatar to Resume Gas Talks in Doha*. Available at http://www.mehrnews.com (accessed November 9, 2008).

[35] Moscow Times (2008) Gas Exporters Downplay Cartel Talk. Moscow Times (December 24).

[36] Medetsky, A. (2009) Gas Exporters Choose Russian Chief. Moscow Times (December 10).

[37] Hallouche, H. (2006) *The Gas Exporting Countries Forum: Is It Really a Gas OPEC in the Making?* Oxford Institute for Energy Studies, Oxford, p. 54.

10

International Energy Agency

Since its establishment in 1974 the International Energy Agency (IEA) has served as the main oil consuming countries' organization. The vulnerability of the industrialized countries to serious oil supply disruptions and to price shocks occurring largely outside of their control was demonstrated shortly in the aftermath of the 1973 Arab–Israeli War. The roots of this first oil price shock go back several decades earlier with the shift from coal to oil as the favored source of energy in Europe, the United States, and eventually most of the world. Prior to the 1930s, coal was the dominant source of energy, accounting for most of the industrial world's fuel consumption. Given the comparative advantages of oil, most industrialized countries replaced coal with petroleum products. Unlike coal, most industrialized countries lack large domestic proven oil reserves.

Thus the large and growing gap between consumption and production has been increasingly filled by imported oil. This heavy dependence on imported oil had made consuming countries vulnerable to disruption of supplies. The rapid economic growth and economic recovery in Europe and Japan were fueled mainly by cheap imported oil largely from OPEC producers. This heavy dependence had further deepened the industrialized world's energy vulnerability.

On the other hand, the industrialized world's energy vulnerability left oil producing countries with substantial leverage. Thus, political and strategic disagreements between the Arab oil producing countries on one side and the United States and a few other countries in the aftermath of the 1973 Arab–Israeli War paved the way for an intense showdown in which oil was used as a political weapon. This confrontation between oil producers and consumers and the apparent lack of coordination between consuming nations had led to the creation of the IEA.

In the following sections a detailed discussion of the geo-economic and geopolitical circumstances under which the IEA was born is provided. This will be followed by an analysis of the organization's membership and structure and an assessment of the IEA's performance, particularly during the second oil shock (1979–1980) and the third one (1990–1991).

10.1 The Founding of the IEA

For a long time, the idea of using oil as a political weapon by Arab producers against Western powers that support Israel had been considered. Thus, in the Suez crisis (1956) and the Six-Day

Energy Security: An Interdisciplinary Approach, First Edition. Gawdat Bahgat.
© 2011 John Wiley & Sons, Ltd. Published 2011 by John Wiley & Sons, Ltd.

War (1967), limited embargoes were implemented but with little effect. In 1956, the United States was able to compensate for the embargo against the United Kingdom and France from its own domestic production. In 1967, the Saudi government ordered Aramco to stop oil supplies to the United States and the United Kingdom during the Arab–Israeli War. The ability of the international oil companies to reroute supplies, however, made the embargo ineffective. Furthermore, Saudi leaders were not fully convinced of the validity of mixing oil with policy. Rather, they preferred to use oil revenues as a "positive weapon" to build up the military and economic strength of the Arab world [1]. Accordingly, Saudi Arabia and other Arab oil producers gave substantial financial aid to Egypt and Jordan from 1967 to 1973.

Debate about the potential of the oil weapon continued nonetheless. Shortly after the outbreak of the Yom Kippur War in 1973, President Richard M. Nixon asked the Congress to provide $2.2 billion in emergency security assistance to replace Israel's losses in the war [2]. On the following day, Saudi Arabia, which had not gone beyond cutting production and issuing warnings during the fighting up to that point, announced a total embargo on oil shipments to the United States. Denmark, the Netherlands, Portugal, Rhodesia, and South Africa were also subjected to a similar embargo. Other Arab states on the Persian Gulf followed suit. In the following months oil production from Arab producers dropped sharply creating circumstances that led to a skyrocketing in the posted price of oil. It was this cutback in production, not the embargo, that produced the shortage felt for several months in most industrialized countries.

The Arab producers' use of oil as a political weapon in the aftermath of the 1973 Yom Kippur War proved more successful than previous attempts in 1956 and 1967. Several factors contributed to this relative success. The prolonged period of secure oil supplies at low prices prompted industrialized countries to permit excessive and even wasteful and inefficient use of energy and of oil in particular. Energy conservation measures either did not exist or were rarely implemented. Secure and cheap imported oil left little incentives to invest in exploring and developing indigenous oil deposits in the industrialized world. In other words, it was largely cheaper to import oil from overseas (mainly from OPEC producers) than produce it at home. Finally, the availability of secure and cheap oil supplies left few incentives to diversify the energy mix away from oil. Little effort was made to develop alternative energy sources such as nuclear, solar, and wind power.

Equally important, the unusual high degree of cooperation and discipline between Arab oil producers in initiating and implementing the oil embargo and repeated production cuts was met by a lack of coordination and disarray among oil consuming nations. The immediate supply disruption had to be managed by both Western oil companies and governments. The oil companies scrambled to adjust available supplies when possible, but they were hampered by insufficient market information, organizational weaknesses, and lack of political and legal authority to take allocation decisions. On the other hand, governments were caught unprepared to cope with the mix of economic and political challenges. Both companies and governments suffered from a lack of necessary information and organizational weaknesses. Furthermore, conflicting national perceptions and interests prevented Western companies and governments from articulating a unified response and speaking with one voice.

Thus, it became apparent to consumer governments that the 1973–1974 oil crisis could not be addressed by private oil companies or individual consuming governments. Rather, international cooperation through permanent institutions would be necessary to meet the new challenges. Thus, major consuming countries agreed in November 1974 to establish the IEA with a broad mandate on energy security. The IEA was the vehicle to ensure cooperation in vital energy issues such as security of supply, long-term policy, information transparency, protecting the environment, promoting research and development, and consolidating international energy relations among consuming countries and between them and the producing regions.

Given the disarray in global energy markets and the serious threat to international economic prosperity, and indeed to the way of life in several consuming societies, time was of the essence. The industrialized countries reached a consensus on the agreement of an International Energy Program (IEP) which was utilized to launch the IEA. The Organization for Economic Cooperation and Development (OECD) provided rapid logistical support and was ready to accept the IEA as an autonomous agency. In short, the international system gave the IEA the benefit of a rolling start within months after the crisis.

The OECD evolved in 1961 from the Organization for European Economic Cooperation (OEEC), which initially had been established by Western European countries to coordinate Marshall Plan aid to Europe following World War II [3]. The OECD's membership expanded to include the United States and Canada and evolved from a regional to a broader functional organization of the free market democratic states. At the time of the 1973–1974 oil crisis, the OECD was the principal economic organization of the industrialized world (including Europe, North America, Japan, Australia, and New Zealand). As such, the OECD's responsibilities covered a variety of issues related to the economic prosperity of its Member States, including energy [4]. However, the OECD was not founded with an exclusive mandate to focus on energy and accordingly lacked the necessarily expertise. Little wonder that the OECD was not in a position to adequately address the 1973–1974 oil crisis. In short, major consuming countries concluded that creating a new organization – the IEA – with a mandate to focus exclusively on energy challenges was the most appropriate course of action.

The US Secretary of State Henry A. Kissinger proposed the establishment of the new energy agency in the middle of the 1973–1974 oil crisis. In presenting his proposal, Kissinger downplayed the association between the oil crisis and the Arab–Israeli War and instead underscored the underlying causes. He argued that global demand for oil had massively outstripped global supply leading to a fundamental imbalance in the energy markets. In order to address this imbalance he urged producers to increase their production and consumers to reduce their consumption and to develop alternative energy sources. In order to realize these objectives and to coordinate a long-term energy policy among industrialized countries, Kissinger also proposed that an "Energy Cooperation Group" (ECG) be established comprising Europe, North America, and Japan. This proposal was the first step toward the founding of the IEA. Other Western leaders endorsed Kissinger's plan.

In order to implement this plan the United States sponsored the Washington Energy Conference (February 11–13, 1974) which brought together representatives from the major oil consuming countries (Belgium, Canada, Denmark, France, the Federal Republic of Germany, Ireland, Italy, Japan, Luxembourg, the Netherlands, Norway, the United Kingdom, and the United States). The participants agreed on a number of themes including the need for a concerted international action program to deal with all facets of the world energy situation by cooperative measures, the need to develop oil stocks, control energy demand, and establish an effective organization [5].

Austria, Spain, Sweden, Switzerland, and Turkey joined the ECG and spent most of 1974 discussing different proposals to establish an energy organization. In November the group drafted an agreement, signed in Paris (November 18, 1974) and entitled Agreement on an International Energy Program (IEP).

The drafted agreement set the stage for the creation of the IEA as an autonomous unit of the OECD. The OECD Council adopted a decision to officially establish a legal link between the OECD and the IEA. The Council decided that the IEA should cooperate with other component of OECD bodies in areas of common interest and for those bodies and the IEA to consult with one another regarding their respective activities. The Council also decided that the IEA budget should be part of the OECD budget.

10.2   The International Energy Program

The agreement on the IEP reached between the major oil consuming nations in 1974, and since modified a number of times, outlined the main objectives and structure of the IEA. This agreement was driven mainly by two fundamental propositions: (a) sustained energy security can only be reached through continued cooperative efforts within effective organs; and (b) governments have a special responsibility to ensure energy security both in their home countries and in the global market. Based on these two propositions, the signatories to the IEP agreed that the IEA should pursue several objectives including:

- To promote secure oil supplies on reasonable and equitable terms.
- To take common effective measures to meet oil supply emergencies by developing an emergency self-sufficiency in oil supplies, restraining demand, and allocating available oil among their countries on an equitable basis.
- To promote cooperative relations with oil producing countries and with other oil con-suming countries through a purposeful dialogue, as well as through other forms of cooperation to further the opportunities for a better understanding between consuming and producing countries.
- To play a more active role in relations to the oil industry by establishing a comprehen-sive international information system and a permanent framework for consultation with oil companies.
- To reduce consuming countries' dependence on imported oil by undertaking long-term coop-erative efforts on conservation of energy, on accelerated development of alternative sources of energy, on research and development in the energy field, and on uranium enrichment.

In order to achieve these goals, the IEP signatories agreed on a set of specific commitments including the establishment of a common emergency self-sufficiency in oil supplies. To this end, each Member State agreed to maintain emergency reserves sufficient to sustain consump-tion for at least 90 days with no net oil imports. On the demand side, each country agreed to prepare a program of contingent oil demand restraint measures. The participants agreed to activate the emergency measures if the group as a whole or an individual member sustains a reduction in its oil supplies equal to 7% of the average daily rate of consumption.

Equally important, the IEP signatories decided to establish an information system to be operated on a permanent basis, both under normal conditions and during emergencies, and in a manner which ensures the confidentiality of the information made available. They also established a permanent framework for consultation with oil companies, under which they can request and share information on all important aspects of the oil industry.

In order to reduce their dependence on imported oil, the IEP signatories pledged to exchange information on energy conservation and promote cooperation on the development of alternative sources of energy such as domestic oil, coal, natural gas, nuclear energy, and hydroelectric power. They also agreed to share data on reserves, prices, and taxation. This sharing of information is not limited to major energy consumers, but extends to producers as well. The participants in the IEP decided that promoting industrialization and a broad socio-economic development in producing nations would contribute to economic prosperity and political stability with a significant positive impact on the global energy markets.

10.3   Structure of the IEA

The IEA's programs and mission are carried out by a number of organs, mainly the Governing Board, the IEA's supreme institutional organ, the Management Committee, and the four Standing Groups on Emergency Questions, Oil Market, Long-term Cooperation, and Relations with Producer and other Consumer Countries as well as the Secretariat. These are the sole organs established directly by the IEP agreement, and they may be abolished or modified only by amendment of the agreement.

10.3.1   The Governing Board

The Governing Board is composed of IEA Member States and Norway. New members immediately become members of the Governing Board as a legal result of their accession to the IEP agreement. The Governing Board enjoys the most power in the IEA and its decisions are legally binding on Member States. It has the last word on all matters. This power covers members' energy policies, oil supply shortfalls, internal organization and operations of the IEA, the admission of new members, and external relations. It provides the institutional mechanism to reach agreements on energy policy, to exchange views, to establish cooperative activities and projects, and to make formal international commitments. The Governing Board appoints the Executive Director, who provides the highest IEA leadership and direction. It also elects the officers of each of the Standing Groups and Committees as well as its own Chair and Vice-chair.

The Governing Board deals with a host of financial questions, including the adoption of the IEA budget, audit authorizations, and other budgetary questions. It fixes the scale of members' contributions each year and decides on other financial questions including the acceptance of voluntary contributions. It authorizes IEA conferences, seminars, workshops, experts meetings, and similar events which have an official IEA connection.

Member States are the principal participants in the Governing Board's meetings. In addition, observers are occasionally invited to attend with some limitations. Over the years, the EU and Norway have been the most notable observers. The EU enjoys a special status in the IEA. The EU has regularly attended Governing Board meetings and enjoyed the right to receive documents and to speak (upon the Chair's invitation). However, the EU has not received the right to vote and has no obligation to make a financial contribution to the costs of operating the IEA. On the other hand, Norway's participation is fixed under a separate agreement between the IEA and the government of Norway. Under this arrangement Norway is allowed to participate in plenary and restricted organs of the IEA, including the right to participate in discussions and to make proposals. Norway also has the right to receive agendas and other documents for IEA meetings but is not allowed to vote.

Other than for the EU and Norway, the admission of observers has been arranged on a case-by-case basis when there have been particularly cogent reasons for doing so.

The Governing Board has endeavored to reach a consensus on virtually every issue. Within the IEA, consensus refers to the absence of a serious objection or reservation which would lead a delegate to request a formal vote. Indeed, each member remains entitled to invoke the applicable voting rule at any moment it might find convenient. Instead of using the voting rules, the Governing Board has acted on the basis of consensus. Thus, consensus has reduced the possibilities of polarization and isolation of the minority, made workable compromises possible, and enhanced the atmosphere of cooperation in the general interest. The successful application of the consensus procedure also provided a means for strengthening institutional development.

Abstentions are explicitly recognized in the IEP agreement and are employed as a normal procedure in most deliberative bodies. The logic behind abstention is to permit a member to avoid direct participation in a decision, yet be bound by it. Textually, this means that absent or abstaining members cannot prevent a unanimous decision being taken by the rest of the members. Hence, there is no veto effect. To carry out a veto, the member has to appear and cast a negative vote. The implication is that the abstaining member is nevertheless bound by the decision in accordance with its terms. In order for the decision not to be binding on the absent or abstaining members, the decision has to contain language to that effect. This reflects the intention of the founders to avoid selective participation of members in difficult decisions to be taken by the Governing Board. In practice this type of "contracting out" has been used sparingly and reluctantly.

The IEP empowers the Governing Board to establish other organs and to delegate functions to them. The Governing Board has availed itself of this power to establish a number of committees and advisory boards and has delegated functions to them. The list includes the Committee on Budget and Expenditure, Committee on Energy Research and Technology, Committee on Non-Member Countries, Oil Industry Advisory Board, and Coal Industry Advisory Board.

10.3.2   Standing Groups

The Standing Groups were created by the IEP agreement and not by the Governing Board, therefore they enjoy "treaty status." The function of the Standing Groups has been essentially to carry out the mandates given to them respectively in the IEP agreement and by Governing Board decisions. The functions of the Standing Groups are to prepare reports and make proposals to the Governing Board. Decisions, recommendations, and other actions have been almost always taken by the Governing Board rather than by Standing Groups, although some delegation of power has been made to them from time to time. Since the Governing Board has been composed principally of high officials with political or broad technical responsibilities, there was need for expert-level preparation in the Standing Groups. This preparation has usually consisted of initiating proposals, researching, debating, refining, and drafting them into Governing Board formulations with preliminary political as well as technical expertise.

Specifically, the Standing Groups carry out specialized functions within their broad mandate and expertise. The Standing Group on Emergency Questions (SEQ) is responsible for all aspects of IEA oil emergency preparedness. In cooperation with the IEA Oil Markets and Statistics Divisions and other IEA Standing Groups and Committees, the SEQ examines oil security issues including global supply and demand prospects, production capacity, and refinery flexibility, and holds seminars, conferences, and workshops on oil security issues. The SEQ periodically tests and updates the IEA Emergency Response Mechanisms, which were set up under the 1974 IEP Agreement. The planning and execution of the work of the SEQ is carried out by the staff of the Emergency Planning and Preparations Division of the IEA Secretariat.

Training and testing involves personnel from Member States, the oil industry, and the Secretariat. The SEQ works closely with the international oil industry, notably through an Industry Advisory Board composed of senior supply, refining, and transport experts from oil companies operating in IEA countries. To ensure the potential of the IEA countries for rapid and effective response to oil emergencies in changing oil market conditions, the SEQ conducts a regular cycle of emergency response reviews of IEA Member States [6].

The Standing Group on the Oil Market (SOM) monitors and analyzes short- and medium-term developments in the international oil market to help IEA Member States react promptly and effectively to changes in market conditions. The SOM works closely with the SEQ in helping to develop plans for responding to oil supply disruptions. The SOM prepares current oil market assessments from information submitted by member governments, international oil companies, and others. Issues covered include: exploration and production developments, demand, price and refining trends, and international trade in petroleum products. The SOM collects and analyzes information on world oil supply and demand, stock levels and changes, oil imports and exports, refinery operations, and prices. It holds conferences on a biannual basis and is responsible for a number of the IEA publications, particularly the influential IEA monthly *Oil Market Report* [7].

The Standing Group on Long-Term Cooperation (SLT) encourages cooperation among IEA Member States to ensure their collective energy security, improve the economic efficiency of their energy sector, and promote environmental protection in the provision of energy services. In order to achieve these goals, IEA governments agreed in 1976 to a program of long-term cooperation in formulating and implementing national energy goals. The SLT makes efforts to increase the awareness of the Member States and to ensure that they will share the same understanding and main policy directions through several contributions including: forward-looking economic analysis of the energy sector through medium- to long-term world energy outlook; topical studies on energy prices; diversification of energy supply through increased shares of gas, coal, renewable, hydro, and nuclear, and better management of the gas, coal, and electricity markets; and protection of the environment including reductions of greenhouse gas emissions and improved energy efficiency.

Another major driver of cooperation is the process of in-depth reviews of the energy policies of Member States. The policies of six individual countries a year are reviewed by their peers. A team including peer countries' experts and the IEA Secretariat visits the country and prepares a report including recommendations to move toward the IEA shared goals as approved by the IEA ministers in 1993. The SLT meets four times a year to provide guidance to the Secretariat, to undertake broad policy analysis, to share the results of the in-depth reviews, and to provide a forum for the development of policies related to economic analysis of the energy sector, energy diversification, and regulatory reforms, climate-friendly energy use, and energy efficiency [8].

The Standing Group for Global Energy Dialogue (SGD) is responsible for work with countries and regions outside of the IEA membership. Many SGD projects draw upon both regional and sectoral expertise and are carried out jointly with other IEA divisions. The IEA has entered memoranda of policy understanding to strengthen cooperation beyond its membership, most actively with China, India, and Russia.

The SGD is focusing on areas such as security of supply, where its work is highlighted in the ongoing oil Producer–Consumer Dialogue, which holds a biennial ministerial summit organized by the IEA. Furthermore, the SGC studies oil developments in major emerging non-OPEC regions, such as the Caspian Sea, Russia, and West Africa. The SGC is also involved in energy efficiency and technology. This area is crucial for many non-OECD countries to meet their growing energy needs in an economically and environmentally sustainable manner. The IEA is drawing on its policy expertise for the promotion of energy efficiency and technology in many of its projects with non-Member States. These states are also encouraged to participate in the IEA Program of International Collaboration on Energy Technology [9].

The Committee on Energy Research and Technology (CERT) coordinates the development, demonstration, and deployment of technologies to meet challenges in the energy sector. The CERT program enables experts from different countries to work collectively and share results, which are usually published. The CERT has established four expert bodies: the Working Party on Fossil Fuels; the Working Party on Renewable Energy Technologies; the Working Party on Energy End-Use Technologies; and the Fusion Power Coordinating Committee. In addition, expert groups have been established to advise on research and development priority setting and evaluation, basic science research for energy, and on oil and gas [10].

Over the years the Standing Groups have provided the vehicle for the technical background work necessary for the IEA. Indeed, the heart of the IEA's work has taken place often in the Standing Groups which gave indispensable support to the Governing Board and thereby contributed generally to the stature of the IEA. The Secretariat has regularly provided essential assistance to the Standing Groups.

10.3.3 The Secretariat

The IEA Secretariat is equal to the bureaucracy in any government and is in charge of the IEA's day-to-day official business. Members of the Secretariat are chosen from highly qualified personnel from IEA Member States. Their job is not to represent their countries in the IEA. Rather, they perform their jobs in an impartial way under the authority of the Executive Director, without seeking or accepting instructions from their governments or from any other external source. The Executive Director and Deputy also form part of the Secretariat structure and, therefore, are subject to the same rule of impartiality and with responsibility solely to the IEA.

The role of the Secretariat is to carry out the tasks assigned to it in the IEP agreement and in Governing Board actions. This role extends broadly across all sectors of the IEA's responsibilities. The Secretariat's responsibilities include making the "finding" which could trigger the oil Emergency Sharing System, the initiation of policy directions, the preparation and presentation of action proposals to the different organs of the IEA, the provision of factual and policy research, preparation of reports and other documents, representation of the IEA in external relations, general logistical support, and execution of the instructions of the Governing Board.

In addition to the permanent Secretariat members, the IEA requires specialized support for particular projects and functions, which is provided in the form of "project staff" and consultants. The appointments in these categories of staff are tailored to meet special and often non-recurring needs. The appointment of project staff is authorized each year by the Governing Board. Project staff and consultants have been appointed in most sectors of the IEA's responsibilities.

The administrative integration of the IEA Secretariat into the OECD was one of the principal reasons for the decision of the founders to lodge the IEA in the OECD as an autonomous agency. That was one of the arrangements which enabled the IEA to commence operations immediately without the relatively slow startup that would have ensued from the need to create all the Secretariat support services from point zero, as would have been the case if the alternative of a wholly separate and independent agency had been adopted. Accordingly, the IEA staff became members of the OECD and were immediately set in place with all the necessary administrative elements already arranged.

The principal function of the Executive Director is to take part in the work of the Secretariat and to direct that work, whether it is specified in the IEP agreement or assigned to the Secretariat by the Governing Board. Overall, the Executive Director's function is to provide leadership and direction on energy policy, on IEA program activities, and on institutional development and operations. Within the general mandate of the IEA, the Executive Director is directly and personally engaged, with support from others in the Secretariat, in developing energy policy initiatives on a broad range of energy problems of interest to Member States, in advising and, if necessary, in mediating to help Member States reach agreement on the complicated issues facing them. Such functions carry political as well as operational and administrative responsibilities. The Executive Director serves the IEA indefinitely until resignation, replacement, or other decision of the Governing Board.

Geographical balance within the IEA overall and within the major parts of the Secretariat has always been an important and regular consideration. Balance reflecting the scope and size of the member's energy economy and its budgetary contributions has always been desired and, within the limits imposed by other recruiting norms, an effort is usually made to have at least one staff member appointed from each Member State. But this is not possible at all times. An overriding consideration is always the need to have the highest level of competence regardless of whatever the geographical balance of the moment might be. Furthermore, there are no posts assigned either formally or informally to any country. Indeed, all vacant posts are open to nationals of all members at all times. Thus, a certain degree of flexibility has always been maintained.

The Governing Board's preference for the appointment of government officials has been followed throughout the history of the IEA, but the appointment of others has been found necessary at times. Expertise found only in the oil companies has been required for the emergency preparedness and oil market sectors of the Secretariat; persons with public utility expertise have been recruited from that industry for different offices within the IEA. Other non-governmental recruitment sources have included universities, research institutions, consulting firms, and individual consultants, law firms, news organizations, and other intergovernmental organizations such as the OECD, the EU, NATO, and the IAEA.

10.3.4   Membership

Membership of the IEA has been established through two quite separate procedures: by signature of the IEP agreement and the procedures applicable to signatories; or by accession to the agreement and the procedures appropriate to accession. In either case, IEA members

Table 10.1 IEA Member States.

Country	Status	Country	Status
Australia	Joined 1979	Republic of Korea	Joined 2002
Austria	Founding 1974	Luxembourg	Founding 1974
Belgium	Founding 1974	The Netherlands	Founding 1974
Canada	Founding 1974	New Zealand	Joined 1977
Czech Republic	Joined 2001	Norway	Participated since 1974
Denmark	Founding 1974	Poland	Joined 2008
Finland	Joined 1992	Portugal	Joined 1981
France	Joined 1992	Slovak Republic	Joined 2007
Germany	Founding 1974	Spain	Founding 1974
Greece	Founding 1974	Sweden	Founding 1974
Hungary	Joined 1997	Switzerland	Founding 1974
Ireland	Founding 1974	Turkey	Founding 1974
Italy	Founding 1974	United Kingdom	Founding 1974
Japan	Founding 1974	United States	Founding 1974

These countries are also members of the OECD, as the IEA is an autonomous agency linked with the OECD. The European Commission also participates in the work of the IEA.

Source: International Energy Agency, *IEA Member Countries*. Available at http://www.iea.org/about/membercountries.asp (accessed February 16, 2010).

were initial parties to the OECD or they were required to accede to the decision to establish the IEA as a condition of membership.

The 16 original signatories were Austria, Belgium, Canada, Denmark, Germany, Ireland, Italy, Japan, Luxembourg, the Netherlands, Spain, Sweden, Switzerland, Turkey, the United Kingdom, and the United States. These countries participated in the Washington Energy Conference in February 1974, or in the Brussels Energy Coordinating Group, or in both during the preparation of the IEP agreement. Provision was also made for qualified countries which were not signatories to the IEP agreement to become members of the IEA by accession to the agreement. The Governing Board, acting by majority, is in charge of deciding on any request for accession. The condition that prospective members be able and willing to meet the requirements of the IEP has required in most cases a lengthy process of consultation and negotiation before a prospective member's request for admission could be acted upon by the Governing Board. At the time of writing (2010), more than 10 countries have joined the IEA as Table 10.1 indicates.

In all cases of new membership, the IEA is required to review the ability of the prospective member to carry out IEA commitments and, on the other side, the prospective member must familiarize itself with these commitments and IEA's practices and expectations. Over the years a procedural pattern has developed to simplify and expedite this process.

From the outset, Norway has had a special relationship with the IEA. Though technically Norway is not a member, with only a few exceptions it participates like a member in the work of the IEA. In 1974 Norway was actively involved with the diplomatic process that led to the establishment of the IEA, beginning with its representation at the Washington Energy Conference and continuing in the Brussels negotiation of the legal instruments which established the IEA. Nevertheless, it soon became apparent that despite the broad scope of

Norwegian interests which corresponded to those of the other industrial, market-economy countries in the OECD, the political environment in late 1974 made it impossible for it to join the IEA as a full member. Norway could participate in the IEA's energy work generally so long as appropriate provision was made to reflect its particular situation, especially its substantial proven oil reserves and its status as a major oil producer. Especially difficult were the issues of Norway's participation in the Emergency Sharing System and of its taking new commitments under IEA auspices. The resulting differences in the commitments of Norway and IEA members were so far reaching that an exception for Norway could not be written into the agreement or be permitted by way of reservation to it. Therefore, Norway could not be a full member. Yet, there was an interest in reaching a compromise that would provide for Oslo's fullest possible participation in most aspects of the IEP. In the end Norway and the IEA signed a comprehensive separate agreement that outlined this special relationship.

Finally, the relationship between the IEA and the EU provides the European Commission (EC) with the right to have access to IEA meetings, to receive regularly agendas and other documents distributed to IEA bodies, and to speak and make proposals, but not with the right to vote in those bodies. The EC does not enjoy the power to prevent the IEA from taking actions wished by its members, nor does it have any obligation to contribute financially to its budget. Theoretically, the EC could accede formally to selected Governing Board decisions. Indeed, the EC has been regularly and actively engaged in the work of the Governing Board, the Standing Groups, Committees and sub-groups of the IEA. Accordingly, the EC's relationship with the IEA can be described as an "active observer" [11].

10.4   Energy Security

The search for energy security was the main objective of the IEA's founders in establishing it and this remains the fundamental drive for all its activities several decades later. Generally, the IEA approaches energy security from the consumers' perspective as the availability of diversified energy supplies at affordable prices. Since its founding in 1974, the Governing Board has focused upon the specific policies, obligations, and mechanisms designed to realize the energy security objectives. In both the IEP agreement and Governing Board actions, energy security concerns extend not only to responses to short-run emergencies, but also to long-term solutions to the problems of reducing oil import dependence.

For short-term oil emergencies, the IEA maintains a treaty-based system for the physical sharing of oil – the Emergency Sharing System (to be discussed in the following section). This system requires the members to build and maintain oil stocks, to plan for and carry out a short-term reduction of demand for oil, and to gather and transmit emergency oil data. At the center of the system there is an institutional mechanism designed to enable these elements to work together smoothly, objectively, and reliably. When supply disruptions occur, all of these oil security measures require members to demonstrate a spirit of cooperation, a willingness to share sacrifices, and a resolve to avoid unilateral solutions, favoring instead multilateral and collective action. Moreover, the IEA has regularly refined, tested, and improved its array of oil disruption response and emergency data systems and has conducted country reviews to ensure the systems' completeness, readiness, and credibility.

The IEA's long-term energy security objectives focus on reducing its Member States' dependence on imported oil. In order to achieve such an ambitious goal, the IEA has pursued a number of strategies. Its conservation policy is concerned with reducing the rate of growth of energy and particularly of oil consumption, eliminating waste, promoting more efficient energy utilization, and applying energy price levels to reduce demand. The policy has also emphasized the environmental objectives and the need to cooperate with industry and accelerate the deployment of new technologies.

In order to diversify its Member States' energy mix, the IEA has pursued a strategy aimed at encouraging the development of alternative energy sources including coal, natural gas, nuclear power, hydroelectricity, and other renewable sources. Such a diversified energy mix not only would contribute to improved energy security, but also is likely to have a significant impact on global efforts to slow down climate change and reduce pollution.

10.4.1   Emergency Response Mechanisms

Emergency response to oil supply disruptions has remained a core mission of the IEA since its founding in 1974. The IEA Emergency Response Mechanisms were set up under the IEP agreement, which requires IEA Member States to hold oil stocks equivalent to at least 90 days of net oil imports and – in the event of a major oil supply disruption – to release stocks, restrain demand, switch to other fuels, increase domestic production, or share available oil, if necessary.

IEA collective response actions are designed to mitigate the negative impacts of sudden oil supply shortages by making additional oil available to the global market through a combination of emergency response measures, which include both increasing supply and reducing demand. Although supply shortages may bring about rising prices, prices are not a trigger for collective response action, as these can be caused by other factors and the goal of the action is to offset an actual physical shortage, not react to price movements.

Under a scenario of oil supply disruption, a close dialogue and cooperation with consuming countries that are not members of the IEA are maintained and collective action is taken in coordination with major producing countries.

In the event of an actual or potentially severe oil supply disruption, the IEA Directorate of Energy Markets and Security assesses the market impact and the potential need for a coordinated response. This market assessment includes an estimate of the additional production that oil producers can bring to the market quickly, based on consultation with producing governments. From this assessment, the IEA Executive Director consults with and advises the Governing Board. Once coordinated action has been agreed upon, each Member State participates by making oil available to the market, according to national circumstances. An individual Member State's share of the total response is generally proportionate to its share of the IEA Member States' total consumption. Throughout this decision-making process and the implementation stage of a decision, industry experts, through the IEA Industry Advisory Board, provide advice and consultation on oil supply/demand and emergency response issues.

Member States have different options for how best to meet their share of additional oil to be made available to the market by implementing a combination of emergency response measures which increase supply and/or reduce demand. The release of stocks is a major aspect of an IEA action. Member States are required to maintain total oil stock levels equivalent to at least 90 days of the previous year's net imports. They are also required to have demand restraint programs in place which can be implemented in a crisis to free up supply through reduced consumption. Furthermore, surge production and fuel switching are additional measures available to Member States to bring relief to markets during a supply disruption.

As mentioned above, each IEA Member State is required to maintain total oil stock levels equivalent to at least 90 days of net imports, but there is flexibility in meeting this requirement using both crude and refined products. Countries may guarantee this minimum obligation by holding stocks as government emergency reserves, through specialized stockholding agencies, or by placing minimum stockholding obligations on industry. Stocks held by agencies or owned directly by governments are referred to as public stocks. On the other hand, industry stocks include both stocks held to meet government stockholding obligations and stocks held for commercial purposes.

The use of stocks in an IEA coordinated action may involve public stocks, industry stocks, or a combination of both, depending on the stockholding system of the given Member State. Public stocks may be released through processes such as tenders or loan offers. Industry stocks held to meet minimum stockholding requirements are made available by temporarily reducing stockholding obligations. These industry stocks have the advantage of already being in the supply chain and therefore very rapidly available to the market in an emergency stock release. In recent years public stocks of IEA Member States have been growing both in terms of volume and as a share of the total stocks. In 2010, over 37% of the total IEA stocks are held as public stocks, compared to 24% in 1984. These public stocks provide a very visible response during a supply disruption, putting additional volumes of oil into the supply chain.

The implementation of demand restraint measures is another tool available to Member States in an IEA coordinated action. Demand restraint programs include temporary measures implemented in a crisis to free up supply by reducing consumption in the short term. The transportation sector accounts for more than half of total oil consumption in IEA Member States and offers the most potential for rapid reductions in demand through restraint measures. The initial emphasis is often on light-handed, voluntary measures, instigated through public persuasion campaigns. These can be particularly effective in a crisis, when consumers are more receptive to the need for saving oil. Compulsory measures can range from medium-handed restrictions such as speed reductions, to more heavy-handed policies such as fuel rationing.

The most significant oil supply disruptions in recent decades have occurred in the Middle East, the largest of which was associated with the 1979 Iranian Revolution. More recently, in early 2003, the market suffered disruptions from overlapping events: the effects of a strike at the national oil company in Venezuela and the outbreak of war in Iraq were exacerbated by strikes in Nigeria. In assessing the necessity to initiate a coordinated action, the IEA considers multiple factors beyond the gross peak supply loss caused by the event. The decision depends on the expected duration and severity of the oil supply disruption, and also takes into account any additional oil which may be put on the market by producing countries. Since its creation, the IEA has acted on two occasions to bring additional oil to the market through coordinated action: in response to the 1991 Gulf War and the hurricanes in the Gulf of Mexico in 2005.

As oil consumption outside IEA Member States is increasing rapidly, the IEA is promoting dialogue and information sharing on oil security policies and measures with key transition and emerging economies, particularly China, India, and other Asian states. The IEA shares information and experience about creating national strategic oil stocks and emergency response policies. It also offers training in statistics and emergency preparedness for China, India, Central Asian countries, and European countries who are not members of the IEA.

Geopolitical tensions and terrorism create uncertainty as to the continuous availability of supply. This "risk premium" adds to the volatility of an already tense market, where available oil supplies are increasingly concentrated in fewer countries. Although the oil delivery system has changed dramatically since the oil shocks of the 1970s, there is still a high risk of a supply disruption which could have great economic consequences for IEA members. Capacity constraints, in both production and refining, have increased the potential of supply falling short of demand. Given this delicate balance of supply and demand, even a disruption of relatively small volume can have a significant impact on the market. Global demand growth exacerbates market tightness, further reinforcing the need for investment in capacity expansion. Uncertain investment climates in some producing countries, often described as an aspect or "resource nationalism," may also hamper the development of future supply streams.

Finally, without continuous monitoring and regular updates, stocks, procedures, and other response measures would not be sufficient to deal with a supply disruption. To ensure that they are effective, the IEA uses several instruments including:

- *Monitoring the market*: The IEA constantly monitors the oil market. The IEA's Statistics Division collects and provides monthly oil data on supply, demand, balances, and stocks for OECD and non-OECD Member States for use by IEA oil market analysts. The analytical capabilities of the IEA enable it to make the necessary assessments of supply disruptions quickly and to provide Member States with advice on the most adequate response measures.
- *Emergency response reviews*: The IEA Secretariat and Member States' representatives participate in peer reviews of emergency preparedness of IEA Member States every few years. Procedures and institutional arrangements are checked. A report and recommendations are discussed with all member states and key recommendations are made public.
- *Emergency response exercises*: Every two years, the IEA carries out a series of workshops and exercises to train and test policies, procedures, and personnel. Major consuming non-Member States are invited to participate. The objective is to practice the decision-making process, review policies and procedures, and ensure readiness to act quickly and effectively.
- *Emergency stock levels*: Whether IEA Member States do in fact have oil reserves up to a minimum of 90 days of net imports is checked and disclosed on the IEA's public web site on a monthly basis [12].

At the time the IEA was established, the founders considered the oil Emergency Sharing System to be the industrial countries' first line of defense against serious oil supply disruptions. The establishment, development, and, in case of need, the operation of Emergency Sharing System were indeed the main objectives of the IEA. The consensus was that the Emergency Sharing System type of response could be effective only through an established cooperative institution, as the lessons of the 1973–1974 crisis taught so well. The operation of an effective oil sharing system required treaty obligations covering the essential rules and creating the necessary infrastructure, decision process, information services, trained personnel, and procedures. All of these had to be available on a permanent basis for urgent call to action. Industrial world leaders concluded that only a set of institutional arrangements like those created in the IEA could meet these requirements.

10.5 How Did the System Work?

The IEA was created in the midst of a global oil crisis triggered by geopolitical tension in the Middle East. The main mission of the then newly founded organization was to prepare its Member States to overcome the disruption of supplies and the surge of prices. In short, the goal was, and still is, to overcome the disruptive impact of future energy crises. In the long term, the IEA remains committed to enhancing its Member States' energy security. Since the IEA's inception in 1974, there have been two oil supply crises triggered by tension in the Middle East. They raised the question about whether the loss of supplies was sufficient to trigger the Emergency Sharing System and, if not, what other responses were appropriate.

10.5.1 The 1979–1981 Crisis

The 1979–1981 crisis caused severe economic damage to IEA countries and brought about a far-ranging reform movement in the IEA, leading to greater awareness of the possibilities of employing oil stocks and demand restraint and to the adoption of procedural changes designed to enable the IEA to respond more rapidly and flexibly to future oil supply disruptions. Although significant measures had been taken to reduce the members' dependence on imported oil, they still remained vulnerable to oil supply disruptions, as the 1979–1981 crisis would show. In the late 1970s industrialized countries (mainly the United States, Europe, and Japan) were almost as heavily dependent on oil supplies from OPEC as they were in the early 1970s.

Following months of political and labor unrest in Iran, the country's oil exports virtually stopped in December 1978. Though the IEA did not immediately find the need to take responsive measures, that was soon to change. Supply was falling and prices were rising, so largely independent actions were taken by a few countries to permit the drawdown of stocks below the 90-day IEA level, to cease the purchase of strategic stocks, and to persuade buyers to avoid paying the rising spot market prices.

By March 1979, the IEA Secretariat's assessment showed that the disruption was serious, though not sufficient to trigger the Emergency Sharing System for the group, but severe enough for some members to experience individual reductions above the 7% trigger level. Accordingly, the Governing Board called on the members to reduce their demand for oil on the world market. The reduction would be in the order of 2 million b/d, which would correspond to about 5% of IEA consumption. Each participating country was to regard this as guidance in the policies it would pursue to achieve its contribution to this reduction. The IEA action did not assign to members firm commitments as to the particular measures to be taken. Without specific commitments the responsibility for performance was diffuse and accountability was difficult. This poorly defined response to the 1979–1981 crisis provided important lessons to the IEA's more coherent action during the 1990–1991 crisis.

The second phase of the 1979–1981 crisis began in the fall of 1980 with the outbreak of the Iran–Iraq War, which resulted in the blocking of all exports from Iraq and caused the loss of a large proportion of the two countries' exports. Concerns about more price rises led the Governing Board to meet in late 1980 to take measures designed to reduce pressures on the oil market. IEA members agreed to urge private and public market participants to refrain from any abnormal purchases on the spot market and to undertake consultations with oil companies. The persistence of high prices prompted the IEA to call on its members to draw on stocks as necessary to maintain a balance between oil supply and demand in the world market. The IEA also sought to correct severe imbalances between countries or companies as a result of the Iran/Iraq supply disruption. Finally, the IEA supported high levels of indigenous oil and gas production.

While precise decisions as to the timing, rate, and duration of stock drawdown cannot realistically be taken in advance of a supply disruption, the IEA sought to establish clear and firm procedures for prompt decisions on stock drawdown and other measures. This was the background which led the Secretariat to propose the "Decision on Stocks and Supply Disruptions," adopted by the Governing Board in July 1984. This decision established the IEA's Coordinated Emergency Response Measures system, commonly known as the CERM, which not only highlights stock drawdown, but also retains the important IEA emphasis on demand restraint measures. The CERM also refers to the severe economic damage caused by exaggerated oil price increases. Although the CERM acknowledges that responses to the disruption will vary from country to country, it emphasizes that an aggregate of national responses is likely to achieve a coherent overall result if they are coordinated and are as complementary as the circumstances and individual national policies permit.

After the CERM was tested in early 1988, the Governing Board adopted the CERM Operations Manual in September the same year. The Operations Manual contains detailed information on the context of CERM consultations, the decision-making process, information requirements, response development methodology, timetable, and monitoring systems.

10.5.2   The 1990–1991 Crisis

The 1990 Iraqi occupation of Kuwait led to the UN embargo of all exports of oil from Kuwait and Iraq. When the international coalition forces began the military campaign for the liberation of Kuwait in January 1991, the IEA was well prepared to respond to the resulting threat to oil supplies. The Iraqi invasion of Kuwait and the UN embargo on Iraqi and Kuwaiti oil had removed the two countries' exports from the market. The amount and duration of any further loss of oil supply in the course of the military action was potentially extensive. The vulnerability of Saudi Arabia and the difficulty of forecasting the responses of other oil producers added to the uncertainty. Hence, when the war began, the IEA activated the Coordinated Energy Emergency Response Contingency Plan. All IEA countries joined by Finland, France, and Iceland had adopted the IEA Contingency Plan.

When the invasion occurred, the IEA had in place two principal oil emergency response systems. The first was the oil Emergency Sharing System, designed to respond massively to oil disruptions exceeding 7% of expected supply. The other was the CERM, intended to facilitate rapid agreement on stock drawdown and demand restraint in response to an oil supply disruption below the 7% level, or in conjunction with the Emergency Sharing System. Both systems were ready for implementation before the Gulf crisis began. Additionally, in support of these systems, the IEA had in place a number of oil supply information systems.

The initial review of the supply/demand situation in the face of the embargo of oil imports from Iraq and Kuwait led to the conclusion that compensatory oil supply would be available from other OPEC producers and from high stock levels, if necessary, to supply the market adequately, without need for recourse to either of the IEA emergency response systems. Throughout the crisis, the assessments continued to show that, due principally to increased oil production by OPEC and others, the market was generally adequately supplied.

Despite the continuing availability of ample oil supplies to the market, the IEA's Governing Board agreed in early January 1990 that the outbreak of hostilities in the Gulf could lead to heightened uncertainty and volatility in the market as a result of the possible temporary shortfall of some Gulf supplies. This assessment convinced the Governing Board to complete preparations for a coordinated response in the event of hostilities. The IEA viewed its proper role as requiring responses not to undesirable price movements, but to disruptions of physical supply that would lead to loss of oil volumes in the market.

The time for the IEA to act came when the air campaign against Iraq commenced in January 1991. On January 17, the IEA activated the Contingency Plan. The intention was not to influence price, rather, it was to avoid a possible far-reaching panic in the market after the outbreak of hostilities. The US, German, and Japanese authorities made available millions of tons of crude oil. This reassured market operators that oil supplies were sufficient to meet current needs. With the end of hostilities and the fall of oil prices closer to normal levels, the risk of supply reduction was reduced. As a result, the IEA terminated the Contingency Plan in March 1991 [13].

The IEA's successful response presented a model of how an international institution and the industrialized countries should respond to a supply crisis. The contrast between the foregoing levels of preparedness for the 1990–1991 Gulf crisis and the disarray of the industrialized countries in the period leading up to and during the 1973–1974 crisis is quite remarkable. The depth of the preparations as well as the aggregate cooperation and political readiness of the industrialized countries to act are credited to the IEA, which transformed the degree of cooperation and preparedness of the industrialized countries over the period 1974–1990 and made possible the vigorous response to the 1990–1991 Gulf crisis.

10.6   Conclusion

The initial role of the IEA was to help its Member States reduce their exposure to damage from any further oil supply shocks. This was to be achieved by equipping them with a collective response mechanism for the short term through the establishment of emergency oil stocks as well as the development of demand restraint mechanisms [14]. Since its inception in 1974, the IEA and global energy markets have changed. A list of these fundamental changes is as follows:

- The IEA has evolved from a reactive and defensive organization into a more proactive policy adviser. Protecting its Member States from interruption of oil supplies is no more the main focus of the IEA. Rather, it is involved in promoting energy security in the broadest sense, including conservation, climate change, research and development, and investment in technology, among others.
- In addition to the close relations with the OECD, the IEA has developed close partnerships with other oil consuming and producing nations as well as with OPEC. The rise of China and India on the global energy scene means that the IEA will not be able to maintain its status as a major international energy organization without close relations with these two major energy players who are not members of either the OECD or the IEA.
- The initial focus of the IEA was oil; however, in recent decades the IEA has widened its interest to include almost all sources of energy, most notably renewable ones. This growing interest in "green energy" is in line with the policy of most IEA Member States.

- The IEA's original approach to protect its Member States against oil shocks was based on collective action by consuming governments. This approach, government intervention, reflected the dominant economic policy in the 1970s. Gradually, more IEA Member States have lost faith in such government-intervention solutions and endorsed market-oriented strategies [15].

In 1974 the IEA was created as the oil consumers' club, but it has since evolved into a major international energy organization playing a vital role in multi-energy arenas and involving both consumers and producers. This evolution has not only enhanced the energy security of the IEA Member States, but equally importantly contributed to the stability of global energy markets. In recent years, the IEA has taken the lead in promoting a consumer–producer dialogue. This newly established partnership is the topic of the following chapter.

References

[1] Stork, J. (1975) *Middle East Oil and the Energy Crisis*, Monthly Review Press, New York, p. 44.

[2] Tillman, S.P. (1982) *The United States in the Middle East*, Indiana University Press, Bloomington, p. 168.

[3] OECD, *Organization for European Economic Cooperation*. Available at http://www.oecd.org/document/48/0,3343,en_2649_201185_1876912_1_1_1_1,00.html (accessed June 26, 2010).

[4] Organization for Economic Cooperation and Development, *About OECD*. Available at http://www.oecd.org/pages/0,3417,en_36734052_36734103_1_1_1_1_1,00.html (accessed July 5, 2010).

[5] European Navigator, *Final Communique of the Washington Conference*. Available at http://www.ena.lu/final-communique-Washington-conference-13-February-1974-020002734.html (accessed June 27, 2010).

[6] International Energy Agency, *Standing Group on Emergency Questions (SEQ)*. Available at http://www.iea.org/about/stanseq.htm (accessed February 16, 2010).

[7] International Energy Agency, *Standing Group on the Oil Market (SOM)*. Available at http://www.iea.org/about/stansom.htm (accessed February 16, 2010).

[8] International Energy Agency, *Standing Group on Long-Term Cooperation (SLT)*. Available at http://www.iea.org/about/stanslt.htm (accessed February 16, 2010).

[9] International Energy Agency, *Standing Group for Global Energy Dialogue (SGD)*. Available at http://www.iea.org/about/stansgd.htm (accessed February 16, 2010).

[10] International Energy Agency, *Committee on Energy Research and Technology (CERT)*. Available at http://www.iea.org/about/stancert.htm (accessed February 16, 2010).

[11] Scott, R. (1994) *Origins and Structure*, International Energy Agency, Paris, p. 153.

[12] International Energy Agency, *IEA Response System for Oil Supply Emergencies 2010*. Available at http://www.iea.org (accessed July 4, 2010).

[13] Scott, R. (1994) *Major Policies and Actions*, International Energy Agency, Paris, p. 140.

[14] International Energy Agency (2009) *Lessons Learned from the Energy Policies*, International Energy Agency, Paris p. 35.

[15] Van de Graaf, T. and Lesage, D. (2009) The International Energy Agency after 35 years: reform needs and institutional adaptability. *Review of International Organizations*, **4** (3), 293–317, 311.

11

Conclusion

Since the early 1970s, energy security has occupied the minds of policy-makers to a degree rarely seen regarding any other commodity. Almost every country in the world imports or exports a significant part (mainly oil) of its energy consumption or production. Indeed, energy and energy products are among the most traded commodities in the world. This high level of trade reflects the geographical mismatch between the availability of resources and the demand for them. This characteristic is particularly relevant to oil deposits. Presently, international trade in natural gas remains more regional than oil, with the European Union, North America, and North-East Asia as relatively separate markets. This high volume of trade means that changes in oil and gas prices have a big impact on the balance of payments of many countries where fuel is a high proportion of their imports or exports.

Some analysts view energy security in terms of price stability, vulnerability to disruptions, reliability of the grid, or environmental sustainability. The bottom line is that without abundant supplies of clean, reliable, and affordable energy and the infrastructure to distribute and deliver it, life in a modern society would be greatly disrupted. Thus, energy security is recognized as among the top challenges to the world's quality of life, economic prosperity, and political stability.

A detailed discussion of the evolution of the concept of "energy security" is provided in the next section. The analysis suggests that unlike the 1970s and 1980s, both consuming and producing countries have concluded that too high and too low prices are bad for all parties. Rather, cooperation between both sides and partnerships between national and international companies would contribute to the stability of global energy markets. In other words, energy security should not be seen in zero-sum terms. Instead it is a win–win situation. This realization is the drive for the ongoing consumer–producer dialogue that started in the early 1990s. The International Energy Forum is the child of this dialogue, seeking to broaden and deepen the cooperation between all players in the global energy market.

11.1 Energy Security

Energy security has many dimensions that encompass a range of concerns – long term and short term, domestic and foreign, economic and political. At least four areas of risk to energy security can be identified – geopolitical, economic, reliability, and environmental:

Energy Security: An Interdisciplinary Approach, First Edition. Gawdat Bahgat.
© 2011 John Wiley & Sons, Ltd. Published 2011 by John Wiley & Sons, Ltd.

- *Geopolitical*: Fossil fuels (oil, natural gas, and coal) vary in terms of trade and concentration. Most oil proven reserves are concentrated in a few countries, notably in the Persian Gulf. The volume of trade in oil and petroleum products is huge. The picture is similar when it comes to natural gas deposits, with the Middle East, Russia, and the former Soviet republics in Central Asia taking the lead. As a result of technological advances and lower prices, the share of LNG in international trade and the market for natural gas is likely to continue to change from regional to global. Coal is more evenly distributed. Still, as China, India, and other large economies expand, coal has increasingly become an international trade commodity. These huge volumes of internationally traded petroleum, natural gas, and coal raise concerns about political stability in producing regions.
- *Economic*: Money spent/received on importing/exporting energy and energy products represents a major proportion of a national budget. Most global recessions have been associated with volatility in oil prices. Thus, predictability and stability of prices at "reasonable" levels are crucial for national and global energy security.
- *Reliability*: It is hard to overestimate the significance of reliability of energy supplies. However, the fact that energy comes from different sources and is transmitted via a variety of means indicates that the threats to and potential for disruptions are serious. Disruptions to energy supplies – whether natural or human made – entail high costs both economically and politically.
- *Environmental*: Combusting fossil fuels releases carbon dioxide and other gases that contribute to global pollution and environmental risks. These uncertainties are appropriately included as part of an assessment of energy security. In recent years both consuming and producing countries have acknowledged the leading role that "green energy" is likely to play in the near future and have allocated substantial financial resources to reduce pollution and slow down climate change.

In addition to these four areas of risk, other dimensions should be taken into consideration in defining the concept of "energy security". First, the notion of energy security is not an "either–or" proposition. Rather, it should be understood as a "less–more" proposition in which the risks to energy security span a spectrum of possibilities ranging from very good to very bad [1]. Second, talks about "energy independence" are unrealistic, particularly in the twenty-first-century global economy. Energy insecurity in any country impacts negatively not only other consuming countries, but producing ones as well as the global economy. In short, insecurity anywhere leads to insecurity everywhere. Third, traditionally, the concept of energy security referred to disruption in oil supplies that led to surges in oil prices. True, oil is, and will continue to be, the dominant source of energy, but focusing exclusively on oil does not tell the whole story of energy security. In recent years the term has come to refer to the entire energy mix, including natural gas, coal, wind, solar, biofuel, and nuclear power. Conservation and efficiency are other significant components. The safety of pipelines and tankers, technological advances, and the availability of adequate investments are also important parts of a broad energy security strategy. Fourth, security of supply and security of demand cannot be decoupled. Security of demand is as important to producers as security of supply is to consumers [2]. Fifth, "export dependence" should be taken into consideration in any definition of energy security. It can be argued that international petroleum trade is more important to crude exporters than to crude importers. Because petroleum exports account for such a high proportion of total exports, and their value fluctuates, a good measure of exporters' "balance-of-payments" dependence is the share of their imports which is paid for by petroleum exports. In other words, most oil producers, particularly OPEC members, have made very modest progress in diversifying their economies away from oil revenues [3]. The International Energy Forum embodies most of these aspects of energy security.

11.2   The International Energy Forum (IEF)

One of the earliest attempts to launch a producers–consumers dialogue took place in Paris in 1975–1976 when the two sides held a meeting to discuss the rise in oil prices. Positions were polarized and consequently no concrete results came out of this meeting [4]. In the ensuing years, oil consuming countries sought to utilize the IEA to lower prices, restrain demand, promote alternative energy, and ensure non-interruption of supply. On the other side, producing countries sought to strengthen their newly found voice by collectively improving their collective bargaining position led by OPEC. They wanted to maintain high prices and resist any threat to demand security. In short, for more than a decade consumers' and producers' interests were perceived as mutually exclusive with little room, if any, for accommodation and compromise.

Despite this bleak assessment, it became increasingly clear that sharply fluctuating oil prices were detrimental to both producers and consumers and that there could be no long-term winners in a volatile environment. A growing realization by both consumers and producers was that stable prices at a reasonable level would serve their common interests. This realization of mutual interest, coupled with the geopolitical turmoil of the 1990–1991 Gulf War, furnished the ground for renewed efforts to establish a producers–consumers dialogue. The Iraqi invasion and occupation of Kuwait (both are OPEC members) highlighted the threat to global oil markets and the broader world's economic prosperity. A more cooperative framework between producers and consumers was born out of this conflict.

At the initiative of Presidents Francois Mitterrand of France and Carlos Perez of Venezuela, a "Ministerial Seminar" of producers and consumers was held in Paris in 1991. This initiative helped to clarify the atmosphere of mistrust that characterized the relations between producers and consumers and underscored the areas of mutual interest and the potential for cooperation to address common challenges. A follow-up meeting was held in Norway in 1992 where IEA and OPEC representatives were joined by delegates from other major consuming and producing regions. The IEF's roots go back to these informal meetings. Since then, the IEF has held a meeting alternately in an exporting and an importing country. The third meeting was held in Spain (1994), the fourth in Venezuela (1995), the fifth in India (1996), the sixth in South Africa (1998), the seventh in Saudi Arabia (2000), the eighth in Japan (2002), the ninth in the Netherlands (2004), the tenth in Qatar (2006), the eleventh in Italy (2008), and the twelfth in Mexico (2010). In these meetings delegates focused on topics of mutual interest such as global resource development, demand and supply outlook, market transparency, investment, energy security, the environment, taxation, poverty alleviation, and technology.

The IEF is unique not only in its global perspective and scope, but also in approach. It is not a decision-making organization or a forum for the negotiation of legally binding settlements and collective action. Nor is it a body for multilateral fixing of prices and production levels. The informality of this framework has encouraged a degree of frank exchanges, which cannot be replicated in traditional and more formal international settings [5]. Above all, the dialogue under the IEF umbrella is a global confidence-building process among the ministers of energy of producing and consuming countries, industrialized and developing countries, across traditional political, economic, and energy policy dividing lines [6].

In the IEF's twelfth meeting in Mexico (2010), participants from 66 IEF countries agreed to pursue several objectives:

- To identify principles and guidelines to enhance energy market stability and sustainability.
- To narrow the differences among producing and consuming countries, both developed and developing.
- To work together to promote transparency of data, stability of markets, and predictability of energy policy.
- To facilitate high-quality analysis and wider collection, compilation, and dissemination of data in order to focus debate more effectively [7].

The last two objectives highlight the IEF's long-term efforts to promote and disseminate data, known as the Joint Oil Data Initiative (JODI).

11.3   Joint Oil Data Initiative

The end of the 1990s was characterized by a high volatility of oil prices. The lack of transparent and reliable oil statistics was identified as an aggravating factor for this volatility. Both producers and consumers recognized the need for more data transparency in the oil market. At the seventh meeting of the IEF in Riyadh, Saudi Arabia (2000), then Crown Prince Aбdullah proposed the creation of a permanent secretariat to facilitate the work of the JODI. Following discussions and expert analysis over the next 48 months, the proposed secretariat was officially endorsed by the eighth meeting of the IEF in Japan (2002), and started working in Riyadh from December 2003 [8].

The JODI is a concrete outcome of the producer–consumer dialogue. In addition to the IEF, six other organizations have sponsored the JODI: the Asia-Pacific Economic Cooperation (APEC), Eurostat, the IEA, the Latin American Organization for Energy Cooperation (OLADE), OPEC, and the UN. Through the use of nationally sanctioned data, the JODI offers global coverage of oil consumption and production on a monthly basis. As a database, the JODI is instrumental to the pursuit of enhanced data transparency. By mitigating some of the uncertainties that can be detrimental to market functionality, the JODI aims to moderate price volatility, thereby increasing investor confidence and contributing to greater stability in energy markets worldwide.

When the JODI was first launched in 2001 (then known as the Joint Oil Data Exercise) its goal was not to build a database, but to raise awareness among oil market participants about the need for greater transparency in oil market data. In November 2005, the JODI partner organizations unveiled the JODI World Database to the public, marking a key milestone on the path to improved transparency.

The success and utility of the JODI are defined by the quality of data received and processed. To further improve the data submitted, and to build capacity among its participants, the IEF works with the JODI partner organizations to conduct regional training workshops which offer statisticians and experts from participating countries an opportunity to improve their knowledge of definitions, data quality assessment, and oil data issues. The workshops also offer a platform for JODI users to share their experiences and communicate best practices for oil data management. IEF ministers have called for an extension of the JODI to cover natural gas as well as annual data on capacity and expansion plans. The extension was initially advocated during the Ad-Hoc Energy Meeting in Jeddah in June 2008 and then later endorsed by Heads of State at the G8 Summit in L'Aquila (July 2009) and the G20 Summit in Pittsburgh (September 2009) [9].

Many challenges remain. The database is still a work in progress, but already for many countries, especially for the top 30 producers and consumers, timeliness, coverage, and reliability are at reasonable levels. The challenge for the organizations now is to increase the coverage to other countries, to reduce the delay in data submissions, and to rather enhance data quality. Due to initial differences in methodology and a lack of comparable sources of information in some countries, there is still much to be desired when it comes to date quality. The database, however, belongs not only to the organizations, but to all countries, oil companies, and analysts that can contribute to improving the quality of the database [10].

11.4   Conclusion: The Way Forward

One of the main themes of this volume is that energy security is a major challenge facing both consumers and producers. In order to maintain economic prosperity and political stability, the two sides need to further improve their cooperation and expand the joint institutions and programs already in place. In the late 2000s global demand for energy was reduced substantially and, as a result, fuel prices fell. The main reason behind this availability of supplies at low prices was the global recession. This temporary development should be seen in the right context. As the world economy recovers, the demand for energy will grow and prices will rise. For the long-term efforts to enhance global energy security a number of dynamics should be taken into consideration.

First, in the last few decades consuming and producing countries have been going in opposite directions. Most governments in consuming nations have reduced their role in the broad economic system, including the energy sector. Their role emphasizes providing incentives and regulations rather than active participation. Meanwhile, governments in most producing countries have achieved a little success in privatizing their public enterprises and still play a significant role in their economies, particularly the energy sector. Lessons from the global recession in the late 2000s are yet to be drawn. Nevertheless, efforts to enhance energy security, at both national and international levels, should be inclusive. Governments, the private sector, international and national companies need to work together.

Second, for domestic political consumption, some policy-makers in consuming countries, particularly in the United States, keep talking about "energy independence." These calls for energy independence were made in the early 1970s and have been proven unrealistic. In the twenty-first century all world economies are interconnected. Within the context of a global economy there is no room for "independence" in any commodity, energy or otherwise.

Third, for most of the decades following the first oil shock (1973–1974) energy security was largely the outcome of rivalry between consuming countries represented by the IEA and producing nations represented by OPEC. Asia's fast-growing economies (i.e., China and India) are not members of either the IEA or OPEC. Beijing and New Delhi have reshaped the global energy landscape since the early 2000s. Their role and leverage are projected to grow. Both OPEC and the IEA have approached China and India and sought to "engage" them and deepen cooperation with the two Asian giants. More is needed. Both nations have become major energy consumers and polluters. They need to be incorporated more politically and institutionally into the global energy system.

Finally, fossil fuels, particularly oil, have continued to dominate the global energy mix. Little wonder that political negotiations and the energy security literature have focused on oil. In other words, for most of the last few decades energy security and oil security meant largely the same thing. In recent years, oil security, though important, has increasingly become only a part of the broader energy security. Other sources of energy such as nuclear power, coal, natural gas, and particularly renewable sources, as well as efficiency and conservation, constitute important parts of the energy security story.

In the last few decades energy security has evolved from a zero-sum game where the producers' gains were seen as the consumers' losses to a win–win opportunity where cooperation and accommodation have replaced the old notions. Global energy security is likely to further improve by taking a more inclusive approach, under which all energy players (public and private, producers and consumers) work together in a more cooperative fashion.

References

[1] Institute for 21st Century Energy, *Index of U.S. Energy Security Risk.* Available at http://www.energyxxi.org (accessed June 14, 2010).
[2] El-Badri, A.S., *Energy Security and Supply.* Available at http://www.opec.org/opecna/speeches/2008/ energysecuritysupply.htm (accessed February 14, 2008).
[3] Mitchell, J. (2005) *Producer-Consumer Dialogue*, Chatham House, London, p. 3.
[4] Khadduri, W. (2005) Information and oil markets. *Middle East Economic Survey*, **48** (48). Available at http://www.mees.com (accessed November 28, 2005).
[5] International Energy Forum Secretariat, *The Case for Dialogue.* Available at http://www.iefs.org.sa/ pages/history.htm (accessed January 5, 2007).
[6] Walther, A. (2007) Dialogue for global energy security: the role of the IEF. *Middle East Economic Survey*, **50** (47). Available at http://www.mees.com (accessed November 19, 2007).
[7] International Energy Forum, *Cancun Declaration.* Available at http://www.ief.org/Events/Documents/ IEFPress-ReleaseCancunDeclaration_31_March.pdf (accessed March 31, 2010).
[8] International Energy Forum Secretariat, *IEFS Host Minister's Welcome.* Available at http://www.iefs.org.sa (accessed December 30, 2004).
[9] Joint Oil Data Initiative, *A Concrete Outcome of the Producer-Consumer Dialogue.* Available at http://www. jodidata.org (accessed July 8, 2010).
[10] Joint Oil Data Initiative, *About JODI.* Available at http://www.jodidata.org/aboutjodi.shtm (accessed July 8, 2010).

Index

Energy Security: An Interdisciplinary Approach, First Edition. Gawdat Bahgat.
© 2011 John Wiley & Sons, Ltd. Published 2011 by John Wiley & Sons, Ltd.